Transactions
of the
American Philosophical Society
Held at Philadelphia
For Promoting Useful Knowledge
Volume 100   Part 3, Sections 1 & 2

ALHACEN ON REFRACTION

A Critical Edition, with English Translation
and Commentary, of Book 7
of Alhacen's *De Aspectibus*

VOLUME ONE
Introduction and Latin Text

VOLUME TWO
English Translation

A. Mark Smith

American Philosophical Society
Independence Square • Philadelphia
2010

ISBN: 978-1-60618-003-7
US ISSN: 0065-9746

Library of Congress Cataloging-in-Publication Data

Alhazen, 965-1039.
  [Manazir Book 7. English & Latin]
  Alhacen on refraction: a critical edition, with English translation and commentary, of book 7 of Alhacen's De aspectibus./[edited by] A. Mark Smith.
    p. cm. — (Transactions of the American Philosophical Society, v. 100, pt. 3)
  Includes bibliographical references (p.) and indexes.
Contents: v. 1. Introduction and Latin text — v. 2. English translation and commentary.
ISBN v. 1 978-1-60618-003-7 (pbk.) ISBN v. 2 978-1-60618-006-8 (pbk.)
1. Optics—Early works to 1800. I. Smith, A. Mark, II. Title. III. Transactions of the American Philosophical Society; v. 100 pt. 3.
QC353
[.A32313 2001]
535'.09'021—dc21 2001041227

TRANSACTIONS

of the

American Philosophical Society

Held at Philadelphia for Promoting Useful Knowledge

VOLUME 100, Part 3, Section 1

Alhacen on Refraction

A Critical Edition, with English Translation
and Commentary, of Book 7
of Alhacen's *De aspectibus*

VOLUME ONE

In memory of Vincent Ilardi–
A good friend

CONTENTS

VOLUME I

VOLUME II

A. Mark Smith is Professor of History at the University of Missouri—Columbia. He teaches courses in Medieval History as well as the History of Science from antiquity to the late Enlightenment. Previous publications with the American Philosophical Society include all previous volumes of the Alhacen series (*Alhacen's Theory of Visual Perception*, 2001; *Alhacen on the Principles of Reflection*, 2006; *Alhacen on Image-Formation and Distortion in Mirrors, 2008*), *Descartes's Theory of Light and Refraction* (1987), *Ptolemy's Theory of Visual Perception* (1996), and *Ptolemy and the Foundations of Ancient Mathematical Optics* (1999).

PREFACE

Alhacen's treatment of light and sight in the *De aspectibus* culminates with the analysis of refraction in book 7. Far briefer than his account of reflection, which occupies three books, Alhacen's study of refraction follows essentially the same pattern of analysis, starting with various experiments designed to establish or confirm the governing principles of refraction empirically and proceeding to the application of those principles to specific cases according to the shape of the refracting surface. In many respects, in fact, reflection and refraction are sides of the same coin according to Alhacen. The main difference between the two phenomena lies in the relationship between the angle at which a ray of light strikes the reflective or refractive surface and the angle at which it leaves that surface. In reflection, the two angles are equal; in refraction they are not. Moreover, the inequality of those angles in refraction is not constant, so the law governing their relationship is considerably more complex than the governing law for reflection. Finding that law experimentally, as Alhacen attempts to do, proves to be no less complex, the result being that Alhacen eventually adduces seven laws rather than a single, comprehensive one.

That Alhacen was reduced to formulating seven governing laws instead of one is problematic enough. Even more problematic is that, although the laws are supposedly derived from experimental measures based on refraction from air to water, air to glass, glass to air, and glass to water, Alhacen provides no tabulations of those measures, so we have no way of knowing how accurate his experiments were or even whether he actually conducted them. Worse yet, of the seven laws Alhacen eventually comes up with, three turn out to be false because they are not universally applicable. Nor is this the only problematic feature of Alhacen's refraction analysis in book 7. The physical model upon which he bases his causal analysis of refraction is beset with interpretive and analytic difficulties, and his attempt to explain peripheral vision on the basis of rays refracted in the front portion of the eye complicates the original model of visual perception detailed in book 2 to the point of incoherence.

Not only does Alhacen's analysis of refraction raise logical issues, but it also raises two historiographical issues of particular significance. One has to do with his empirical and experimental approach to the phenomenon. The main issue here is twofold: whether that approach is truly innovative and whether it reflects the norms of modern experimental methodology. This issue has already arisen in the discussion of Alhacen's experiments to confirm the law of equal angles for reflection in book 4. The second issue centers on Alhacen's analysis of refraction through spherical interfaces in chapters 5 and 7. The issue here is also twofold: whether and how that analysis relates to Alhacen's study of spherical lenses in a later work entitled "On the Burning Sphere," and what impact that analysis might have had on the development of modern lens theory at the hands of Johannes Kepler. Because it is packed with so many issues in such a brief compass, book 7 is unquestionably the most interesting (perhaps in the Chinese proverbial sense) of the entire *De aspectibus*. It is certainly the most demanding in terms of commentary and explication.

As usual, I have gotten a good deal of help, both financially and intellectually, in producing this edition. On the financial side I want first to acknowledge the UM Research Board for supporting a research trip to Europe in the summer of 2006 so that I could consult the relevant manuscripts for the final transcription of the Latin text. Next I wish to thank the John Simon Guggenheim Foundation for generous support of a research leave for academic year 2007-2008 that allowed me to concentrate all my effort on this edition. Thanks, too, to the NSF (award number SES 819448) for its support of research assistance during calendar year 2009 to help me put this edition into publish-ready form. In supporting this project from the beginning, the NSF has eased the production and publication of the entire edition of the *De aspectibus* immensely. My gratitude is nearly boundless.

Warm thanks are also due the librarians and archivists in charge of the manuscript collections I consulted at the following libraries: Biblothèque Nationale, Paris; Biblioteca Nazionale Centrale, Florence; Bibliothèque de l'agglomération, St-Omer; Royal College of Physicians, London; Corpus Christi College, Oxford; Trinity College, Cambridge; and the Crawford Library of the Royal Observatory, Edinburgh. In one way or another each of these librarians has gone beyond the call of duty not only to facilitate my research but also to make me feel welcome at his or her institution.

On a more personal note, I want to acknowledge several colleagues, both intramural and extramural, for advising me on technical issues, offering editorial suggestions, or just giving me encouragement. In no particular order they are: Noel Swerdlow, Robert Hatch, Wilbur "Willy" Applebaum, David Lindberg, Abdullahi Ibrahim, Eileen Reeves, Rihab Sawah, Vincent Ilardi, and Richard Newhauser. If I've missed anyone, then I apologize profusely. I cannot, of course, ignore my wife, Lois, for bearing with me and my rather single-minded pursuit of this project. Finally, I cannot omit acknowledging my editorial assistants Autumn Dolan and Mark Singer for their patient work. Special thanks to M. Leo Shaw, of the Anthropology Department here at the University of Missouri. He has been involved with the publication of this edition from the very beginning and has served as an editorial jack-of-all-trades, ranging from double-checking manuscript readings to scoffing at my use of words like "nugatory" and "otiose." I haven't always followed his advice, but it has generally been advice well worth considering. He has been so much a part of this project that I feel almost compelled to list him as co-author.

INTRODUCTION

1. *Alhacen's Analysis of Refraction: An Overview*

Alhacen's account of refraction in book 7 follows essentially the same pattern as his account of reflection in books 4-6. After setting out the agenda for the book in chapter 1, Alhacen goes on in chapter 2 to establish the fundamental principles governing refraction. This he does on the basis of an experimental apparatus analogous to the one devised in book 4 for demonstrating the fundamental principles of reflection. Using the same apparatus in chapter 3, he describes an experimental method for determining the angles of refraction for particular angles of incidence when light passes obliquely from air into water, air into glass, glass into air, and glass into water.[1]

Alhacen next turns in chapter 4 to a set of empirical tests to prove that an object point seen by means of refraction is perceived along the refracted ray, its image thereby being displaced. He ends the chapter by showing that such displacement affects the observation of stars and the moon at horizon, atmospheric refraction causing them to appear elevated. Having made these points, he devotes chapter 5 to a discussion of image location in order to demonstrate empirically that the image of an object point seen at a slant through a refractive interface is located where the refracted ray intersects the normal dropped from the object point to the refractive interface. On the basis of this empirical demonstration, he concludes chapter 5 with a series of ten theorems proving that any object point seen through any refractive interface, whether plane, convex spherical, or concave spherical, will yield only one image.

In chapter 6 Alhacen explains how such images are actually seen and how their appearance is affected by refraction, closing the chapter with an empirical demonstration that everything is seen by means of rays refracted through the transparent tunics of the eye. He then brings book 7 to a close in chapter 7 with a brief account of misperceptions due to refraction, his focus in this chapter being the apparent magnification or diminution of images seen through refractive interfaces. The *pièce de résistance* of this concluding analysis is his explanation of the Moon Illusion on the basis of both psychological and optical factors.

Now that the bare bones of Alhacen's refraction analysis in book 7 have been exposed, it is time to flesh them out in chapter-by-chapter order.

Chapter 2: Alhacen's purpose in this chapter is to establish six fundamental principles of refraction, namely: 1) that refraction occurs when a ray of light passes at a slant from one transparent medium to another, the two media being of different densities or transparencies;[2] 2) that when it strikes a refractive interface orthogonally, light passes straight through without refraction; 3) that refraction occurs in a single plane perpendicular to the interface between the two media; 4) that when they pass obliquely into a denser medium, light rays incline toward the normal dropped to the refractive interface at the point of refraction, whereas when they pass obliquely into a rarer medium, they incline away from that normal; 5) that the greater the density differential between the two media, the more sharply the light will be refracted; and 6) that refraction occurs at the point of refraction only, so light rays pass along straight lines to that point and continue along straight lines after that point.

Alhacen opens chapter 2 in paragraphs 2.2-2.15, pp. 220-224, with a detailed description of the experimental apparatus to be used in establishing these principles. As depicted to scale in the top diagram of figure 1, p. 401, the central component of the apparatus consists of a shallow, circular bronze pan no less than one cubit in diameter (i.e., around 50 cm) with a rim two digits high (i.e., somewhat less than 4 cm). The flat, bottom face of the pan constitutes the register plate, across which diameter FG is drawn. At endpoints F and G of that diameter lines FK and GL are drawn on the inner wall of the rim perpendicular to the face of the register plane. On that same inner wall a circle is inscribed parallel to the face of the register plate at a height of one grain of barley above it (i.e., somewhat less than 1 cm).[3] At point M, where it intersects KF, the circle inscribed on the rim is divided into degrees, starting at M = 0° and continuing to the right. Each degree can be further subdivided, perhaps down to intervals of 10′. At point A, one digit to the left of F (i.e., somewhat less than 2 cm), a diameter is drawn through centerpoint C of the register plate. The quadrant on the rim between A and B is then excised. This open quadrant will allow an observer to see straight into the pan toward the rim on the other side.

At point M a hole one grain of barley in diameter is drilled so that the distance between its centerpoint M and the register plate is one grain of barley. The axial line passing through midpoint M of that hole will thus be parallel to diameter FG on the face of the register plate and will end at point M′, where line GL intersects the circle on the inner wall of the rim a full grain of barley above the face of the register plate. A small bronze panel two digits square and thick enough to stand perfectly upright is then formed, and a midline is drawn through one of its faces. On that midline a point one grain of barley from the bottom is marked, and through it a hole one grain of barley in diameter is drilled so that its centerpoint lies a full grain of barley above the panel's bottom edge on the midline drawn on its face. Radius CF on the register plate is bisected at R, a line is drawn perpendicular to CF at that point, and the panel with its hole is placed flush with that line so that the midline on the panel's face is perpendicular to CF. The panel is then glued to the face of the register plate in that position, as depicted in the top diagram of figure 1. Accordingly, the hole in the panel is perfectly aligned with the hole in the rim, which means that line MM' parallel to diameter FG at a height of one grain of barley above the register plate passes through the centers of both holes to form an axial line through them.

On the back of the register plate at its center an axle three digits long is firmly attached. A bronze bar two digits square in cross-section and well over a cubit long, is formed, and at its center a hole barely large enough to accommodate the axle on the back of the register plate is drilled. The bronze bar is slipped over the axle through that hole to fit quite snugly, as depicted in the middle diagram of figure 1. With the attached bar in that position, its overhanging portions are cut off so that the remaining ends of the bar are flush with the outer rim of the pan. The two excised pieces are affixed at the top of each end of the attached bar so as to protrude by at least a digit.

Then a copper ruler at least half a cubit in length, two grains of barley wide, and one grain of barley high is formed, as illustrated in the bottom right-hand diagram of figure 1. A midline is drawn along its upper face, and its end CD is cut so as to form an acute angle at D, the longer edge BD being left at least half a cubit long, preferably a bit longer. Finally, a vessel with a high upright rim is formed such that its diameter is slightly longer than a cubit, or at least slightly longer than the diameter of the shallow pan, its rim higher than half a cubit. The top edge of the rim is carefully planed so that, when the vessel is stood on the ground, that edge will be perfectly horizontal. Thus, when the pan is inserted upright into the vessel, it will hang from the vessel's rim by the pieces protruding at the end of the attachment bar at its back, as depicted in the top diagram of figure 2, p. 402. The face of the register plate will then be perpendicular to the plane of the vessel's brim, and thus to the plane of the horizon, and the pan will fit loosely enough in the vessel that it can be rotated about its axle without touching the rim or the bottom. Once rotated into a given position, the pan will remain firmly in place because of the snugness of the fit between its axle and the attachment bar.

With the pan hanging in it, the vessel is aimed in the direction of the sun, and the pan is rotated on its axle until the hole at F in the rim faces the sun, allowing a beam of sunlight to pass straight along radius FC through the hole in the panel and thence to centerpoint C of the register plate. The axial ray of that beam will therefore be parallel to radius FC at a height of one grain of barley above it. If the vessel is empty, the sunbeam will continue to the opposite side of the rim in a straight line parallel to diameter FG, where it will form a circle of light on its inner wall. The center of that circle of light will lie where the circle on the inner wall of the rim intersects perpendicular GL. In short, the axial ray of the sunbeam will follow line MM' in the top diagram of figure 1. All of this can be seen clearly by an observer looking straight down through the excised portion of the rim between A and B because the segment of the rim between A and the vessel's brim shades the inner wall of the rim from a point above G on down toward K and beyond.

Rather than describe every experimental test Alhacen suggests on the basis of this apparatus, we will limit ourselves to a few exemplary ones, starting with the test of refraction from air to water in paragraphs 2.16-34, pp. 224-230. For that test we need to fill the vessel with water up to centerpoint C of the register plate, as illustrated in the top diagram of figure 2. Then, rotating the pan on its axle until the sunlight passes through the holes in the rim and the panel, we notice at the outset that the beam of sunlight forms a circle on the water's surface, as well as on the inner wall of the rim below the water.[4] We also notice that the circle of light on the rim's inner wall under water does not lie in line with diameter FG but falls somewhat below it at point D, between G and endpoint K of normal CK. The light has thus inclined toward that normal after refraction. We also notice that the center of the circle of light at D lies on the circle inscribed on the inner wall of the rim with its center at C. This tells us that both the incident and refracted axial rays lie in the plane of that circle, which is perpendicular to the water's surface and which therefore defines the plane of refraction.

To prove not only that the light refracts at the water's surface and nowhere else but also that the incident and refracted axial rays are perfectly straight, Alhacen suggests a test aptly called "tagging" by Saleh Omar.[5] In order, first, to determine that the incident axial ray is straight, we are to take a needle and pose its point at the center of the hole in the rim. We will note that the shadow of its point is projected to the center of the circle (or ellipse) of light on the water's surface. We are then to move the needle's point inward to the center of the hole in the panel, and we will again find the shadow of its point projected to the center of the circle of light on the water's surface. This indicates that the axial ray has passed straight through the centers of both holes to the water without any deviation on the way. Next we are to take the copper ruler and apply its wider face to the register plate with its sharp point at the center of the circle of light at D, as represented in the top diagram of figure 2. In that position, the ruler's shadow will cut off a quarter of the circle of light on the wall of the rim under water, as illustrated in the lower right-hand diagram of figure 2, so the vertex of the ruler's leading edge will lie directly at the point of refraction on the water's surface. With the ruler in place, we are to put the needle's point right on its leading edge, where it cuts the water's surface, as illustrated in the lower left-hand diagram of figure 2. Looking at the wall of the rim under water while holding the needle in position, we will see the shadow of its point touching the shadow of the vertex of the ruler's leading edge at the center of the circle of light under water, as in the lower right-hand diagram of figure 2. We are then to place the needle's point at random spots on the ruler's edge between C and D below water, and we will see that the shadow of the needle's point continues to touch the shadow of the ruler's vertex throughout. From this tagging experiment, then, we have ocular proof that the refracted axial ray has followed the ruler's straight edge throughout its passage in the water. We also have ocular proof that the incident axial ray changes direction right at the water's surface and nowhere else.

So far we have demonstrated empirically that a ray of light passing at a slant from air to water is refracted at the water's surface (and only there), that it is refracted toward the normal in the denser of the two media, and that the incident and refracted axial rays are perfectly straight and lie in a single plane perpendicular to the refracting surface. All that remains is to show that light passes straight through the water's surface without refraction when it strikes it orthogonally. Here we are brought up short by a limitation in the experimental apparatus. In order to be conducted, the experiment requires that the two holes be lined up along normal CK so as to channel the sunlight along the orthogonal. This can only be done when the sun is at zenith, and that only occurs at latitudes at or between the tropics, where it happens at most twice a year.[6] For anyone at a more northerly or southerly point on the Earth the experiment is unfeasible. The best one can do in that case is demonstrate the obverse by lining the holes up along the normal, placing a light source at K, and sighting through the holes to determine that the light source can be seen directly through them.[7]

This particular demonstration poses no problem for the passage of light from air to glass, as is obvious from the first of the experiments Alhacen describes in paragraphs 2.35-46, pp. 231-234. That experiment requires us to form several glass cubes, each of whose faces is two grains of barley on a side. We are to draw a line through centerpoint C on the face of the register plate perpendicular to diameter FG and apply one of the cubes to the face of the register plate so that its bottom edge is flush with that line and the midpoint of the cube's bottom edge lies at C. We are to affix the cube in that position with a temporary adhesive, then take a second cube, nest it up against the first one, and affix it temporarily to the face of the register plate according to the configuration in the top diagram of figure 3, p. 403. If we continue the process until the line of cubes reaches as close as possible to the panel at the middle of radius FC, we will get the configuration in the bottom diagram of figure 3.

With the cubes in place, we are to apply ruler EK on its thinner edge to the face of the register plate so that its wider edge with the midline drawn through it lies toward the face of the glass cube at C and is parallel with it, as depicted in both diagrams of figure 3. After affixing the ruler to the register plate with a temporary adhesive, we are to mark the centerpoint of the face of the glass cube directly opposite with a small spot of ink. Keeping everything in place as described, we are to insert the pan into the empty vessel and adjust the entire apparatus so that sunlight streams through the two holes to the facing surface of the line of glass cubes. That light will continue straight through the glass cubes to form a circle of light on the face of the ruler. At the very center of that circle of light, on the midline through the ruler's edge, we should see the shadow of the ink spot projected from the center of the face of the cube directly opposite the ruler. Once that is established, we begin removing the cubes in order, starting with the one closest to the panel. Each time we remove a cube and look at the circle of light on the ruler's face, we will see the shadow of the spot on the face of the first cube projected to the center of that circle. From this we can conclude that the axial ray of light continues unrefracted not only through the full succession of cubes but also through each and every individual cube and combination of cubes on its way to the ruler. Moreover, since the axial ray is orthogonal to the back face of every cube, we can conclude that light striking a glass surface along the normal passes through it unrefracted.

The remaining experiments are based on a glass quarter-sphere with a radius of somewhat less than one-quarter of a cubit. According to Alhacen's description in paragraphs 2.47-49, pp. 234-235, this quarter-sphere is to be formed from a glass hemisphere whose diameter is somewhat less than half a cubit. The base of the hemisphere is pictured in the top left-hand diagram of figure 4, p. 404. Perpendicular diameters AB and HK are drawn through centerpoint D of that base, a point C′ is marked off at a distance of a grain of barley below D, and line LC′M is drawn through it parallel to HK. A cut is then made through LC′M perpendicular to the plane of the base to produce slightly more than a quarter-sphere, as depicted in the top right-hand diagram of figure 4. For all practical purposes we can call the resulting form a quarter-sphere. This quarter-sphere is then stood up on diameter FCG of the register plate so that point C′ on leading edge LC′M of the quarter-sphere lies on centerpoint C with the leading edge itself perpendicular to diameter FCG. Centerpoint D of the sphere encompassing this glass quarter-sphere will therefore lie precisely one grain of barley above the face of the register plate, so it will lie in the plane of the circle inscribed on the inner wall of the rim encircling the register plate, that being the plane of refraction.

We are now ready to undertake the set of experiments described in paragraphs 2.50-58, pp. 235-238. We start with the quarter-sphere affixed to the register plate with a temporary adhesive, insert the pan into the vessel, and point the hole at F in the rim toward the sun to allow a beam of sunlight to pass into the glass quarter-sphere through the hole in the panel. If the vessel is empty, as in the middle diagram of figure 4, the beam will pass straight through the glass and the air below it to point G on the other side of F. If water is poured into the vessel to reach slightly above point C, as in the bottom diagram of figure 4, the beam of light will still pass straight through the glass and the water below it to point G. In both cases, then, the light will have passed along the normal from the glass into the medium below it, so it will be unrefracted throughout its passage. This can be confirmed by the tagging procedure described earlier. Accordingly, when we place the needle's point at the center of each hole as well as at the center of curvature of the glass quarter-sphere, which lies a grain of barley above the face of the register plate, we will invariably see its shadow projected to the center of the circle of light on the rim below.

To test refraction from air to glass, glass to air, or glass to water, as suggested in paragraphs 2.59-78, pp. 238-244, we need only reattach the glass quarter-sphere to the register plate with point C′ on its leading edge at centerpoint C of the register plate and pose the leading edge at a slant to diameter FCG. Thus, as represented in the top diagram of figure 5, p. 405, the flat face of the quarter-sphere lies slantwise to the beam of light coming through the two holes. As a result, when the light passes from air into glass at point C, it will be diverted along refracted ray CK, which shows that it refracts toward normal ACL in the denser medium. On the other hand, if the glass quarter-sphere is fixed in the position represented in the bottom diagram of figure 5, then when the light passes from the glass into air at point C, it will be diverted along refracted ray CK, which inclines away from normal ACL, just as expected for refraction from a denser into a rarer medium.

Moreover, if the glass quarter-sphere is left in place and water poured into the vessel to a level somewhat above point C, as in figure 6, p. 406, the light will be diverted away from normal ACL along refracted ray CK when it enters the water, which demonstrates that water is less dense or less refractive than glass. Yet, since the ray does not incline as far away from normal

ACL in this case as it did when the light passed from glass into air, we can conclude that the density differential between glass and water is less than that between glass and air. Alhacen devises several other experiments to show that, whether it passes from air into the glass quarter-sphere or from the glass quarter-sphere into air or water, light follows precisely the same pattern when it strikes the convex surface of the glass obliquely. Hence, the shape of the refracting surface makes no difference.

Aside from confirming the principles already established in the tests for refraction from air to water, these tests with glass show explicitly that light passing along the normal from a denser into a rarer medium continues unrefracted. They also show that light refracts away from the normal when it passes obliquely from a denser into a rarer medium and that the greater the density differential between the two media, the more pronounced the refraction. All of these principles, Alhacen adds in paragraph 2.78, pp. 243-244, can be shown to apply not only to the refraction of sunlight, which he characterizes as essential, but also to the refraction of daylight streaming through a window, which he characterizes as accidental.[8] In addition, from all of the tests with air, water, and glass we can conclude that the principles apply generally not only to all types of light, but also to all types of transparent media according to their relative transparency or density, regardless of their surface shape.

From a modern perspective, these tests of refraction according to types of light and refracting surface shapes are wholly unnecessary, adding clutter to an otherwise straightforward experimental analysis. Furthermore, Alhacen has imported unnecessary technical difficulty into his tests for glass by insisting on the use of a quarter-sphere. Far easier would have been to use a semi-cylindrical section, which would have had no perceptible effect on the experimental results. As we have seen throughout the *De aspectibus*, though, Alhacen insists on such punctilio in order to make his experimental tests as general and as mathematically precise as possible. Such mathematical precision raises questions about the feasibility of the experiments as described and the physical limitations of the apparatus, but we will defer discussion of these questions until later in the introduction (see pp. lii-lvii).

Having confirmed the basic principles of refraction as exhaustively as possible for air, water, and glass, Alhacen concludes chapter 2 with a causal explanation of refraction based on likening light to a tiny, swiftly moving sphere that loses motion/momentum when it enters a denser, more resistive medium or gains motion/momentum when it enters a rarer, less resistive one. The resulting explanation, which looks surprisingly modern, assumes that, after entering a denser medium, the light particle conserves its motion, insofar as possible, by being physically shunted toward the normal, which constitutes the path of least resistance and thus the most forceful and natural—or, as Alhacen puts it, the "easiest"—path for the light to follow. To support this notion Alhacen offers two thought experiments.

The first is based on imagining a window covered by a thin wooden plank. Experience teaches us that, if we stand directly in front of the plank and hurl an iron ball straight at it, the ball will break through if the plank is thin enough. When, however, we stand to the side and hurl the ball with the same force but along an oblique trajectory, it may glance off the plank because it cannot overcome its resistance to penetration. In the second example we are to imagine striking a log at various angles with a sword. Again, experience teaches us that, when we strike the log directly along the perpendicular, the sword will make a deep cut, whereas the more oblique the sword-stroke, the shallower the cut until, finally, the sword glances off. Up to that point, moreover, the more oblique the blow, the more the sword tends to be diverted toward the normal as it penetrates the log.

In light of these examples, Alhacen breaks the overall motion of the light particle down into horizontal and vertical components or vectors in order to show how the motion along each vector is affected by the horizontal and vertical resistance of the medium. According to this analysis, the loss of motion along the vertical is recouped along the horizontal by the light's being pushed toward the normal and thus toward an easier path than it would follow were it to continue along the original line of incidence. The refraction of light from a denser to a rarer medium is explained by the lifting of horizontal and vertical resistance and the resulting "push" of the light away from the normal and a concomitant gain in motion. In both cases, the resulting path is determined by the natural tendency of light to conserve its overall motion insofar as possible. As will become clear when we analyze it in detail later in the Introduction, this dynamic model of refraction presents serious problems at a variety of levels (see pp. lvii-lxii).

Chapter 3: Alhacen's main purpose in this chapter is to show how to determine the angle of refraction for any given angle of incidence when the light passes from a rarer into a denser medium or from a denser into a rarer medium. As before, the three representative media are air, water, and glass, the latter medium being represented throughout by the quarter-sphere used in the previous chapter. We are also to employ the same experimental apparatus as before, the determination being based upon the division of the circle inscribed on the inner wall of the pan's rim into degrees, which are further subdivided as feasible. One point merits emphasis before we turn to the actual procedure: For Alhacen the angle of refraction is measured according to the continuation of the incident ray rather than the normal, so his angle of refraction is formed by that line and the refracted ray. In modern parlance this is the angle of deviation, and throughout the rest of book 7 it is this angle to which Alhacen refers when he speaks of the angle of refraction. In order to distinguish Alhacen's from modern usage, we will use "angle of refraction/deviation" to denominate his version in future discussion.

No matter which media are tested, the basic experimental procedure is the same. The pan is inserted into the carrying vessel, the entire apparatus is directed toward the sun, and the pan is rotated about its axle until sunlight streams through the two holes in the rim and panel. More specifically, those holes will be directed toward the sun in a succession of angles at increments of no more than ten degrees, each angle being an angle of incidence. Hence, the determination of angles of refraction/deviation requires that they be compared to no fewer than eight angles of incidence, starting at 10° and ending at 80°, as measured from the normal. Alhacen's experimental determination throughout this chapter is based on these eight angles of incidence, although he assures us that we can refine the determination as much as we want not only according to smaller angular increments but also according to subdivisions within each angle.

Described in paragraphs 3.14-20, pp. 251-253, the first and easiest determination to make is that for refraction from air to water. Accordingly, we are to take the pan with its register plate and mark a point ten degrees to the right of point F, i.e., point A in figure 7, p. 406. From that point we are to draw diameter ACA" on the face of the register plate. Then we are to mark point A' 90° to the right of A and draw line A'C. That line will therefore be normal to diameter ACA", and arc A'G will be 80°. Now that the relevant lines have been drawn and the appropriate parameters established, we are to insert the pan into the vessel and turn it on its axle until line ACA" is perfectly horizontal, as represented in the top diagram of figure 8, p. 407.

After filling the vessel with water right up to the level of ACA", we are to direct the entire apparatus toward the sun until a beam of light passes to the water through the holes at F and R. Suffice to say, this must be done either at early morning, when the sun has risen 10° above the horizon, or late in the evening, when it has descended to within 10° of the horizon. Under these conditions the light will strike the water at C to form an angle of incidence of 80°, as measured according to the normal. After being refracted at that point, the light will follow line CD to the rim under water, where it will form a circle of light with its centerpoint on the circle inscribed on the rim's wall. We mark that point and then measure the arc between it and point G at the end of diameter FCG along which the light reaches the water. This arc will subtend the angle of refraction/deviation GCD, so GCD will be the angle of refraction/deviation for an angle of incidence of 80°, which is equal to angle A′CG.

Once we have made this determination, we are to remove the pan, erase lines ACA″ and A′C, and mark a point 20° to the right of F, i.e., point B in figure 7. From it we draw diameter BCB″, mark point B′ 90° to the right of B, and drop line B′C normal to BCB″, which leaves angle B′CG equal to 70°. Inserting the pan into the vessel as before, this time with line BCB″ flush with the water's surface, we aim the entire apparatus toward the sun and wait until it has risen 20° above the horizon or has set to within 20° of the horizon. At that point, if the apparatus is properly aimed, the sunlight will stream through the two holes to point C in the water and be refracted thence to the inner wall of the rim. This time, of course, the angle of incidence will be 70°, so the center of the circle of light refracted to the inner wall of the rim under water will lie closer to the normal than in the previous case. The arc between that point and G will thus measure the angle of refraction/deviation for an angle of incidence of 70°, and it will be smaller than its counterpart for an angle of incidence of 80°. All we need do, then, is repeat the process for the remaining angles of incidence between 60° and 10°, and we will have the full array of angles of refraction/deviation tied to each angle of incidence at ten-degree increments. However, the range of angles of incidence i that can be tested by this method depends upon the latitude at which the test is conducted. If, for instance, Alhacen actually conducted these tests at Cairo, which lies 2 minutes north of 30°, there may have been enough leeway in the system of holes to allow him to test for $i = 30°$, but not for $i = 10°$ or 20°. This same limitation will affect the remaining tests for air to glass, glass to air, and air to water.

The tests suggested in paragraphs 3.21-22, pp. 253-254, for refraction from air to glass, glass to air, and glass to water through the flat face of the glass quarter-sphere involve the same process of marking off diameters at ten-degree intervals to the right of point F and dropping a normal from the bottom of the pan to those diameters. We then affix the glass hemisphere to the register plate with its flat face flush with the diameter and the centerpoint of the sphere enclosing it directly over centerpoint C of the register. That done, we insert the pan in the vessel with the respective diameter horizontal and thus flush with the water's surface. The middle diagram of figure 8, for instance, illustrates the setup for determining the angle of refraction/deviation for an angle of incidence of 80° when the light refracts toward the normal from air to glass so as to form a circle of light at D on the inner wall of the rim. Arc DG therefore subtends the angle of refraction/deviation for the represented angle of incidence, which happens to be 80°. The lower diagram illustrates the equivalent setup for the refraction of light away from the normal when it refracts from glass to air. Here, too, arc DG subtends the angle of refraction/deviation for the represented angle of incidence. Using this same setup, we can also determine the angle of refraction/deviation when light passes from glass to water at an angle of incidence of 80°. For that determination we need only fill the vessel with water until its surface is flush with, or reaches slightly above the flat surface of the glass quarter-sphere. In all these cases, then, we simply redraw the relevant diameters and normals at ten-degree intervals, reposition the glass quarter-sphere accordingly, and reinsert the pan into the vessel with the diameter horizontal.

Alhacen's procedure in paragraphs 3.24-28, pp. 254-256, for testing the angles of refraction when the light passes into air through the convex face of the glass quarter-sphere need not detain us. We can also pass over the method he details in paragraphs 3.29-3.33, pp. 257-259, to test those angles when the light passes from glass into air through a concave surface. Suffice it to say, these procedures are close cousins—with some minor adjustments—to the ones outlined above. Suffice it to say, as well, that the results obtained for refraction through these interfaces should be precisely the same as those obtained for refraction through the flat face of the glass quarter-sphere. In short, the shape of the refracting surface has no bearing on the outcome of the experiment.

Among the conclusions to be drawn from all these experimental determinations, one in particular is critical for subsequent analysis. Refraction, Alhacen tells us in paragraph 3.23, p. 254, is symmetrical and thus subject to the principle of reciprocity. In other words, light follows the very same path through two media no matter the direction of radiation. Hence, as illustrated in figure 9, p. 408, if light radiates through a rarer medium along AC and is refracted in a denser medium along DC to form angle of refraction/deviation A"CD, then if the light were to radiate through the denser medium along DC, it would be refracted along CA in the rarer medium to form angle of refraction/deviation ACD' equal to angle A"CD.

Before turning to the actual results of the experiments described above, we should briefly revisit the case illustrated in the bottom diagram of figure 8. As shown there, the light arriving through the holes along FC is refracted away from normal A'C when it enters into the air below ACA", just as we would anticipate from the previous analysis. But although the refraction represented in the diagram may seem to conform with theory, it does not conform with fact. Rather than pass into the air through the flat face of the glass quarter-sphere, the light will actually be reflected at point C, and the reflection will be total. The reason is that the angle of incidence, which is 80°, far exceeds the critical angle for refraction from glass to air, which is around

42°. This angle represents the limit at which refraction will no longer occur, so any incidence at or above roughly 42° will result in total reflection. In fact, the critical angle is a factor in refraction from any denser to any rarer medium, and its size depends on the density differential between the two. The critical angle for refraction from water to air, for instance, is slightly under 49° and for glass to water around 61°. Yet, oddly enough, Alhacen appears to have been unaware of this fact, or so it would seem from his failure to forwarn us that carrying out the experimental determination of angles of refraction/deviation for glass to air at the four angles of incidence between 50° and 80° would be a waste of effort. Equally pointless would be conducting the experiment from glass to water at angles of incidence of 70° and 80°, and equally silent is Alhacen about that point. The implications of this silence will be discussed in detail later in the introduction (see pp. lv-lvi). As to Alhacen's experimental results in chapter 3, the most salient thing about them is their total absence. Nowhere does he provide tabulations for any of the experimental determinations he describes. The implications of this silence will also be discussed later in the introduction (see pp. lii-lv).

At the very end of chapter 3, in paragraph 3.34, pp. 259-260, Alhacen lists seven general rules governing the relationship between angles of incidence and refraction, all of them inferred from the experiments supposedly conducted to that point. Rule one (p. 259) states that for any two given media the same angle of incidence i will always yield the same angle of refraction/deviation r/d, that a larger angle of incidence i will yield a larger angle of refraction/deviation r/d, and, finally, that for two different angles of incidence i and i', the difference between the larger and smaller angle of refraction/deviation r/d and r/d' will be less than the difference between the respective angles of incidence. In short, if $i > i'$, and if i yields r/d while i' yields r/d', then $r/d - r/d' < i - i'$. Although generally true for the media Alhacen tests, this particular claim turns out not to be universally valid for refraction from glass to air (see note 68 to paragraph 3.34, p. 361).

Rule 2 (p. 259) asserts that, given the same two refractive media, if $i > i'$, and if i yields r/d while i' yields r/d', then $r/d:i > r/d':i'$, which is tantamount to saying that the smaller i becomes the closer r comes to equality with it. According to the third rule (p. 259), if $i > i'$, and if i yields r/d while i' yields r/d', then $i - r/d > i' - r/d'$. In essence, the fourth rule (p. 259) claims that, when light passes obliquely from a rarer to a denser medium, r/d is always less than i. The fifth rule (p. 260) is that, when light passes obliquely from a denser to a rarer medium, r/d is equal to one-half $i + r/d$. This rule is patently erroneous, and, as it turns out, the source of the error lies in a misapprehension or mistranslation of the Arabic text. This point will be explained when we take a close look at the problems arising from rules 1, 4, and 5 (see pp. lxii-lxvi). The last two rules (p. 260) state in a rather round-about way that the greater the density differential between the two media, the more pronounced the refraction through them will be.

Chapter 4: Alhacen has three main, interrelated goals in this chapter. The first is to demonstrate empirically that the image of any object seen through a refractive interface is seen along the continuation of the refracted ray. The second, which follows from the first, is to show that such an image is displaced from its object. And the third is to prove that such displacement affects astronomical observation at the horizon because of atmospheric re-fraction. After a fairly lengthy analysis in paragraphs 4.1-15, pp. 260-265, to show that the forms of object points and object surfaces propagate through transparent media according to cones of radiation, Alhacen addresses the first two goals with a series of tests based on the experimental apparatus used in the previous chapters. The method for conducting those tests is es-sentially the reverse of the method used in chapter 2 to analyze the refraction of light through water and glass.

The first and simplest test, which occupies paragraphs 4.15-18, pp. 265-266, is for refraction from water to air. We begin with the apparatus disposed at it was for the test of refraction from air to water in chapter 2. The appropriate setup is illustrated in the top diagram of figure 2, where the pan is inserted into the vessel and water is poured in until its surface reaches centerpoint C of the register plate. In this case, however, the ruler will not be applied to the register plate between C and D. When things are so arranged, we rotate the pan about its axle until diameter FG is normal to the water's surface. We then take a needle and pose its point directly above G, where the axial line through the two holes intersects the middle circle inscribed on the inner wall of the rim, that axial line being parallel to diameter FG and one grain of barley above it. In order to restrict the line of sight as much as possible, Alhacen suggests slipping a reed through the two holes and boring a narrow channel through it along their axis. As expected, when we look through the hollow reed along that axial line, we will see the needle's point directly below and in line with the normal, which is where it actually is. This test of course demonstrates that an object point seen through a refractive interface along the normal is seen at its true location, so its image is not displaced. It also demonstrates that the object point is seen along the axial ray extending from the refractive interface to the eye.

Once that has been established, we rotate the pan clockwise in what-ever position we please between the normal and the brim of the vessel and conduct the same test. For instance, after choosing the position illustrated in figure 10, p. 408, we put the needle's point at G', where axial line EC'G' passing through the channel in the reed intersects the circle inscribed on the inner wall of the rim. Then, when we sight through the reed along EC'G', we will not see the needle's point. If, however we gently move the needle's point to and fro along the circle on the rim, we will eventually see it when it is placed somewhere between G and K. Leaving the needle's point in place at that juncture and looking at where it actually lies, we find it at D, which is at a significant remove from G' toward normal C'K'. Hence, although we saw the needle's point along line of sight EC'G' parallel to diameter FG on the face of the register, it clearly does not lie on that line. That line, more-over, coincides with the refracted axial ray C'E that the luminous color from the needle's point follows after reaching refraction point C' along DC', so we have determined that the image of the needle's point is seen along the continuation of refracted ray C'E and is therefore displaced from its object at D. No matter how we pose line FC toward the water, the results will be the same: the needle's point will appear to lie directly in line with refracted axial ray C'E, and where it appears will not be where it actually lies, except along the normal. It bears noting, by the way, that this test is limited by the critical angle of around 49° for refraction from water to air.

Alhacen's subsequent tests in paragraphs 4.19-26, pp. 266-270, for images seen from glass to air and air to glass are illustrated in figure 11, p. 409. In the top diagram, the flat face of the glass quarter-sphere is posed aslant to line of sight EC'G' passing through the channel in the reed, so if the needle's point is placed at G', it will not be seen, but it will eventually come into view when it is moved toward normal AC'K to point D. The luminous color from the needle's point thus radiates straight through the convex face of the glass along DC' to refraction point C' and is then refracted in the air above along axial ray C'E, which inclines away from normal AC'K. Hence, we see the needle's point along line of sight EC'G' although it in fact lies elsewhere. The case illustrated in the bottom diagram is opposite, insofar as the luminous color from the needle's point passes through the air along DC' to point C' on the flat face of the glass quarter-sphere, where it is deflected toward normal AC'K to follow refracted axial ray C'E straight through the convex face of the glass quarter-sphere to the eye. Whether its luminous color is refracted from glass into air or vice versa, then, the needle's point always appears to lie along line EC'G' when it actually lies elsewhere, except, of course, when it is viewed along the normal passing through both faces of the glass.

Aside from confirming that the image is seen along the refracted ray and is therefore displaced from its object, these experiments could be used to check the values for the angles of refraction/deviation obtained in chapter 3 by translating those angles into angles of refraction according to the normal. Take, for example, the value of around 39° for the angle of refraction/ deviation that Alhacen should have obtained when testing the refraction from air to glass at an angle of incidence of 80°. This translates to 41° for the angle of refraction with respect to the normal. Accordingly, we can situate the glass quarter-sphere as represented in the bottom diagram of figure 11, and then rotate it clockwise around C' until normal AC'K passing through its flat face makes an angle of 41° with sighting-line EC'G' (i.e., angle AC'E = angle G'C'K = 41°). Affixing the glass quarter-sphere temporarily in place according to that disposition, we can then put the needle's point on the circle on the inner wall of the rim at a point precisely 80° away from K in a clockwise direction. Then, when we look through the sighting-hole in the reed along EC'G', we should see the needle's point. Alhacen in fact never suggests such a test—and for good reason, as will become evident when we discuss the limitations of his experimental apparatus later in the introduction (see pp. lii-lvii).

Alhacen brings chapter 4 to a close in paragraphs 4.28-35, pp. 270-274, by demonstrating that the stars and the moon appear to lie higher at the horizon than they actually do and that this displacement is due to atmospheric refraction. In the first case, which is analyzed in paragraphs 4.28-29, pp. 270-271, we are to select a star that passes directly overhead at zenith during its passage through the night sky. We are then to position an armillary sphere so that we can measure the star's arcal distance from the north pole at horizon and zenith, and we will find that the distance from horizon to pole is slightly less than that from zenith to pole. In the second case, described in paragraph 4.30, pp. 271-272, we are to determine the precise time that the moon peeps over the horizon and, knowing its orbital period, predict the precise time at which it will pass through our meridian. Placing the armillary sphere with its sighting-holes directed at that point, we are then to wait until the predicted time and look. Contrary to expectation, the moon will not be seen there but will in fact lag somewhat behind that point. In both instances, Alhacen concludes, the apparent upward displacement is due to atmospheric refraction, which is most severe at horizon, considerably less severe in mid-sky, and non-existent at zenith when the star or the moon is seen along the normal. In order to make this point crystal-clear, Alhacen offers a descriptive demonstration in proposition 1, paragraphs 4.31-34, pp. 272-273. On the basis of both experience and reason, therefore, we will have determined that the light from celestial bodies is refracted at the interface between the heavens and the atmospheric shell surrounding the earth. We will also have determined that the aerial medium within this shell is denser and thus more refractive than the aetherial medium beyond it.

Chapter 5: Early in this chapter, in paragraphs 5.4-19, pp. 275-280, Alhacen suggests two experiments to demonstrate that the image of an object point seen through a refractive interface appears to lie at the intersection of the refracted ray and the normal dropped from the object point to the refractive interface. Both experiments require a wooden disk roughly a cubit in diameter. In the first experiment we are to draw two intersecting diameters on the disk's face and demarcate them clearly with bright white paint. We are also to mark the point of intersection at its center with a clear, constrasting color. We are then to stand the disk, with all its markings, upright in a vessel that is deep enough to allow water to be poured in to a level above the center of the disk, and we are to set one of the diameters normal to the water's surface. When everything is properly arranged and an appropriate line of sight is established, we will see what is depicted in the top diagram of figure 12, p. 410. Accordingly, the perpendicular diameter MX will appear unbroken at the water line, the other diameter LX will be bent upward at that point, and intersection-point X will lie in line with both the perpendicular diameter and the broken segment of the slanted diameter. From this we can infer that the luminous color from point X has been refracted away from normal MX in passing from the water into the air and, therefore, that we are seeing it along the refracted ray. In addition, we see that the image of X lies on normal MX, which confirms that the image lies at the intersection of the normal and the refracted ray. By extrapolation we can conclude that the image of every other other point on the broken segment of the slanted diameter under water lies at the intersection of the refracted ray and the normal dropped from it to the refractive interface—hence the perfect straightness of its composite image.

The second experiment is depicted in the bottom diagram of figure 12. We are to form glass brick ABCDEFGH pictured on the right-hand side of that diagram, and on the back face of the wooden disk we are to mark diameter MX in white and intersecting diameter LX in another color, also marking intersection-point X clearly. We are then to attach the glass brick to the face of the disk as illustrated in the left-hand diagram so that one of its edges protrudes beyond diameter MX and the other extends beyond the disk to point D. If we place our right eye as close as possible to the glass brick in line with endpoint L of the slanted diameter and look down through the bottom face of the brick along edge BC, we will see normal diameter MX unbent and slanted diameter LX bent downward below the glass. In this case, then, the apparent position of X is lower than its actual position, whereas in the previous case it was higher. Thus, the image of X is seen along a ray refracted toward normal MX when it enters the glass from the air below it, but that image is also seen on normal MX, so we can conclude that it lies at the intersection of the refracted ray and the normal. The same can be inferred by extrapolation for every other point on the broken segment of diameter LX seen below the glass.

After briefly explaining in paragraphs 5.20-24, pp. 280-282, why images seen at a slant through refractive interfaces appear to lie at the intersection of the refracted ray and the normal, Alhacen devotes the remainder of the chapter to a series of ten rather mundane theorems, one of them a lemma, that are meant to demonstrate that an object seen through a refractive interface either along the normal or at a slant will yield one, and only one image, no matter if the refractive interface is plane, convex spherical, or concave spherical. The first three theorems, propositions 2-4, pp. 282-285, prove the point for plane interfaces both when the eye lies in a rarer medium and when it lies in a denser medium, the object point occupying the other medium. These propositions are so simple and straightforward that they need no explanation here.

Proposition 5, pp. 285-287, is a lemma proving that, if two chords intersect inside a circle, the angle of intersection will be equal to an angle with its vertex on the circle's circumference that is subtended by an arc equal to the sum of the arcs cut off on the circle by the intersecting diameters. For example, in the top diagram of figure 13, p. 411, if AG and BD intersect at point E in circle ABD, angle of intersection DEG is equal to angle DBZ subtended by arc DZ, which = arc DG + arc AB. A corollary to this is that, if the chords intersect outside the circle, as in the bottom diagram of figure 13, angle of intersection DEG is equal to angle DAZ on the circle, which is subtended by arc DZ equal to the difference between arcs DG and AB cut off on the circle by the intersecting lines, i.e., DZ = DG – AB. This lemma is actually a generalized version of Euclid, III.20, which proves that, if two diameters intersect in a circle, the angle of intersection is twice as large as an angle with its vertex on the circumference of the circle that is subtended by an arc equal to the arc subtending the angle of intersection. This lemma comes into play in propositions 6 and 11.

Proposition 6, pp. 287-289, addresses the situation in which the eye lies in a rarer medium and faces a convex spherical surface beyond which lies a denser medium and proves that, whether the object point and eye do or do not lie on a normal, only one image will be seen. Proposition 7, pp. 289-290, follows up by showing that, depending on the relative position of eye and object point in the previous case, the image location will vary according to whether and where the normal from the object and the refracted ray intersect. If they do not intersect, there is no true image, whereas if they do intersect, the image can lie at, behind, or in front of the center of sight itself. Only an image directly in front of the center of sight is seen clearly and accurately. Proposition 8, pp. 290-291, is the converse of proposition 7 in that it addresses the case in which the eye lies in the sphere containing the denser medium and the object point lies outside the sphere in the rarer medium.

Proposition 9, pp. 291-292, is a departure from all the other theorems in the chapter because, unlike them, it deals with object lines rather than object points. It also has some ramifications that will be explored later on in this overview, when we deal with a particular theorem in chapter 7 (see the discussion of proposition 17, pp. 318-319). The ostensible point of proposition 9 is to demonstrate that, if A in figure 14, p. 412, is a center of sight facing a sphere consisting of a denser medium, if an axis is dropped from A through the sphere's center Z and extended indefinitely beyond it, and if the medium surrounding the axis beyond the sphere is of the same density as the medium filling the sphere, then there is a line segment on that axis beyond D with B as an endpoint that will be seen by A as a ring on the convex surface of the sphere.

Alhacen begins by assuming that point E is the limit at which the luminous color from B will refract from the glass into the air to reach the center of sight A. In this situation, then, angle of incidence BEZ yields the largest possible angle of refraction/deviation AEL, which is to say that, if the luminous color from B strikes the refractive interface at some point X below E to form angle of incidence BXZ > BEZ, it will not pass through the interface.[9] From this it follows that there is some other point B on the axis and some other point E on arc GE such that a ray of light from B will be refracted to A from new point E at a new angle AEL smaller than maximum angle AEL. Since there is an infinitude of points on arc GE that fulfill this condition, there is an infinitude of corresponding points on the axis beyond D from which the light reaching the appropriate point on arc GE will be refracted to A. All the points on the line segment between original point B and new point B on the axis will therefore be refracted to A from the arc between original point E and new point E.

Hence, if we rotate the entire figure about axis AB, the circles described by limiting point E of refraction and the new point E will form a ring on the convex face of the sphere, and that ring will form the locus of refraction for all the points between original point B and new point B. The image location for those points will be where the refracted rays and normal AB passing through those points intersect, and that happens to be center of sight A. But according to Alhacen, when the image location is at the center of sight itself, the image will actually appear on the refractive interface, so the composite image of all the relevant points on the line segment will be the ring that defines the locus of refraction.[10] From this analysis it would appear to be the case that, since there is an infinitude of possible points on arc GE that can serve as points of refraction for an infinitude of corresponding points on the axis, the entire arc GE may serve as the image for all those points. Hence, the largest possible image would seem to be the full circular segment on the sphere defined by limit-point E of refraction in its rotation about the axis. This is in fact not the case because the line segment producing the image is limited.[11]

Propositions 10 and 11, pp. 292-295, round out chapter 5 by addressing the cases in which the eye faces a concave refracting surface behind which lies a denser medium containing the object point. As in all the other cases, so in this one, the object point can be proven to yield one, and only one image.

Chapter 6: Alhacen opens this chapter in paragraphs 6.1-7, pp. 296-299, by explaining how the forms of object surfaces radiate through refractive interfaces according to cones converging at the center of sight. The portions of these cones that extend from the refractive interface to the vertex at the center of sight Alhacen defines as "cones of refraction." On this basis he extends the analysis of image location in book 5 from object points and object lines to object surfaces by showing that the image of any object surface lies at the intersection of the cone of refraction and all the normals dropped to the refractive interface from individual points on the object surface. He then goes on in paragraphs 6.8-9, p. 299, to explain how this model of image location works for object surfaces seen through plane, convex spherical, and concave spherical refractive interfaces. Paragraphs 6.10-16, pp. 300-301, cap the analysis to this point with an explanation of why the image of an individual object seen through a refractive interface is almost always perceived as single in binocular vision.

The remainder of the chapter is taken up with Alhacen's defense and explanation of the startling claim in paragraph 6.17, pp. 301-302, that everything is seen by means of rays refracted through the transparent tunics of the eye, no matter whether it is seen directly, or by reflected rays, or by refracted rays. What makes this claim startling is that it seems to contradict the account of visual selection Alhacen so carefully constructed in chapter 7 of the first book on the basis of the cone of radiation. This cone, he argued there, has its base on the visible object and its vertex at the center of sight, and it consists of the rays from every point within the base that strike the anterior surface of the lens orthogonally. Such rays constitute what he calls "radial lines" (*linee radiales*). The lens, for its part, is designed according to both its transparency and sensitivity to accept only those rays because, after passing unrefracted to its anterior surface through the cornea and albugineous humor, they continue straight through toward the center of sight. Being oblique, the remaining rays are refracted through all the interfaces between the outer surface of the cornea and the lens so as to be shunted away from the center of sight. Moreover, being orthogonal, those particular rays make the strongest impression on the lens' surface, so it senses their impingement while ignoring or failing to sense the weaker oblique impingements. The lens is therefore uniquely suited to filter out oblique radiation and thereby select a coherent point-by-point representation of any facing object from all the luminous color reaching it from that object's surface.[12]

Alhacen attempts in paragraphs 6.17-24, pp. 301-304, to resolve the apparent contradiction between this model of visual selection and the supposition that everything within the field of view is seen by means of refracted rays. It is true, he concedes at the outset, that we see every point in this field along orthogonal lines converging at the center of sight. But it is also true that this field is not limited by the cone of radiation. After all, that cone is circumscribed by the pupil, which is quite narrow, and it is evident that we see things outside the visual field defined by it. Indeed, our peripheral vision extends to things quite far to the side. That being the case, we can only see those things by means of rays reaching the surface of the eye obliquely, and those rays will perforce be refracted not only at the cornea but also at the anterior surface of the lens.

A brief look at figure 15, p. 413, will make this point and its implications obvious. Represented there is the front of the eye and its components according to Alhacen's description of its anatomical and geometrical structure in the sixth chapter of book 1.[13] The lens and cornea are concentric with the eye as a whole, C being the shared center of curvature. This point also constitutes the center of sight, from which the visual faculty sees everything. Rays DC and FC are at the extreme edges of the cone of radiation circumscribed by the pupil, to which Alhacen refers as the "aperture in the uvea" (foramen uvee). EC is the axial ray within that cone. Points D, E, and F are thus seen directly by the visual faculty along orthogonals DC, EC, and FC, and the same holds for any point between D and F.

Now take some point O outside the cone. It will radiate its luminous color to every point on the facing cornea to form a cone with its base on the cornea's surface. Like every other ray within that cone, rays OA and OB at its extreme edges strike the corneal surface obliquely. But since the cornea extends past the pupil, only some of the radiation within that cone will pass through to the exposed surface of the lens. Being oblique, the rays at the extreme edges of this inner cone (represented in heavier black lines) will be refracted through the cornea, then through the albugineous humor behind the cornea, and thence to points R and R' at the lens' anterior surface. Refracted yet again along RG and R'H, the luminous color conveyed by those rays will bypass center of sight C. This is true for all the other rays within that cone because they, too, are oblique.

How, then, is point O seen? Alhacen's response depends upon three suppositions. First, all the luminous color radiated from O and reaching the lens' anterior surface obliquely makes an impression upon it, each impression varying in intensity according to the obliquity of impingement. Second, the lens actually senses those impingements, no matter how obliquely and weakly they are made. And finally, in the aggregate the resulting impressions cause the visual faculty to perceive O as if it its luminous color had passed straight to C along radial line OC, which extends through the tissues surrounding the eye as well as through the opaque sclera and uvea. This is physically impossible, of course, so OC is a virtual rather than an actual ray and, as such, constitutes what Alhacen calls a "radial line by analogy" (*linea radialis transumptiva*).[14] Everything we see in the field of view outside the cone of radiation is therefore perceived along such fictive orthogonal rays.

Furthermore, Alhacen continues, each point within the field of view defined by the cone of radiation itself sends its luminous color to every point on the facing cornea to form a cone with its base on its surface. Only one of the rays within that cone is orthogonal. The rest are refracted at the corneal surface to reach the lens at a slant, so the luminous color striking the lens along each of these slanted rays makes a weaker impression than that made along the orthogonal. Collectively, however, these oblique impressions serve to reinforce the more intense orthogonal one. For instance, the luminous color from E in figure 15 radiates in a cone with its base on the corneal surface directly in line with the pupil, and every ray within that cone, except EC, passes through the pupil to strike the lens obliquely. The impressions made along these oblique rays will reinforce the one along EC, so EC serves as both a real radial line for the orthogonal impression and a virtual radial line for the oblique ones. Likewise, F radiates its luminous color in a cone based on the corneal surface, so the oblique impressions made within that cone reinforce the orthogonal one made along FC. Taken overall, though, the rays within F's cone are more oblique than those within E's, so their overall impression is weaker. That is why F, being off to the side, appears less clear than E, because its orthogonal impression is more weakly reinforced.

In order to test this analysis empirically, Alhacen invites us in paragraphs 6.25-32, pp. 304-307, to conduct two experiments, the first of which is a simple test of peripheral vision. We are to place a small object, such as a needle point or the tip of our finger, to all sides of one or the other eye at the very fringe of the field of view. In every case that field will be limited either by the tissues surrounding the eye, including the nose, or by the plane tangent to the surface of the eye where the visual axis intersects it. Hence, our horizontal field of view for both eyes will extend through an arc of 180°. Since any object point within this arc that lies above the plane tangent to the surfaces of both eyes will be able to radiate its form to the corneal surface and thence to the anterior surface of the lens, it will be seen, although its image will appear increasingly ghostlike the farther the object point lies from the visual axis at the center of the eye because of the increasing obliquity and weakness of the impressions it makes on the lens.

The second experimental test is equally simple. We are to take a needle, pose it directly in front of the pupil of one eye, and bring it as close as possible to the eye's surface while staring at a distant white wall. In that case, we will see the portion of the wall occluded by the needle as if through an imperfectly transparent body. Since we cannot actually see the wall directly through the needle, we must be seeing it by means of oblique rays reaching those portions of the cornea's surface exposed to the wall on each side of the needle. Meanwhile, the luminous color from all the points on the needle's surface facing the eye radiates to the lens, making its own impression on the lens. The combination of impressions, one set from the wall, the other from the needle, causes the illusion of transparency in the needle. It also causes the portion of the wall occluded by the needle to appear somewhat blurred because of the weakness of the oblique impressions by means of which it is seen. That blurring ceases as soon as the needle is removed. Moreover, if we replace the needle with an object just thick enough to block the entire pupil, none of the wall will be seen. All these tests, Alhacen concludes, demonstrate that everything within the cone of radiation is seen by means of both orthogonal and oblique rays. If there is nothing to block the pupil, the impressions along the orthogonal will be reinforced by those along the oblique to yield the clearest possible perception of whatever lies on that line within the cone of radiation.

Aside from explaining peripheral vision and the needle illusion, Alhacen adds in the concluding paragraph of the chapter (6.33, p. 307), this revised model of visual selection also accounts for variation in visual acuity for either eye. Because the impression along the visual axis is most reinforced, that one yields the clearest possible perception. Less clearly perceived are the impressions made along radial lines to the side of the visual axis because they are less reinforced, and for that reason the farther the radial lines lie from the visual axis, the less clear the perception along them will be. Even less clearly perceived are the impressions made outside the cone of radiation along virtual radial lines because those impressions are all made along oblique rays and are therefore relatively weak. Least clearly perceived will be impressions along virtual radial lines extending to the very edge of the field of vision because the rays producing them are as oblique as possible. Combined with Alhacen's analysis of the variation of visual acuity for binocular vision on the basis of imperfect image fusion, this explanation makes clear why visual acuity drops off so sharply from the center to the periphery of the field of vision and, therefore, why visual scanning is so critical for an accurate perception of any visible object.[15] We will see later on in the introduction that, while this revised model of visual selection explains a great deal, it also raises serious issues about the psychology of visual perception (see pp. lxvii-lxx).

Chapter 7: The analysis in this, the final chapter of book 7, is essentially bipartite. The first section consists of a set of seven theorems intended to demonstrate both how and why objects appear magnified when seen through refractive interfaces. The initial three theorems in the set, propositions 12-14, pp. 309-314, show that any object line lying in a denser medium and facing a center of sight in a rarer medium appears magnified when seen through a plane refractive interface. The proofs themselves are fairly trivial, designed to confirm in a simple, straightforward way that, regardless of where the object line and the center of sight lie in relation to the refractive interface, the image of that object line will subtend a greater visual angle than the object line itself. Hence, the image will appear closer to the center of sight than its object.

   This point is illustrated in figure 16, p. 414, which accompanies proposition 12, pp. 309-311. A is the center of sight, ED the refractive interface, and GB the object line in the denser medium, which is bisected at Z by normal AZ dropped from the center of sight. The luminous color from endpoint G of that line radiates to H along GH and refracts to A along HA. Because B and G are equidistant from Z, the luminous color from B radiates to T and refracts to A along AH in symmetrical fashion. Since the images of G and B are seen along refracted rays HA and TA, the image of GB as a whole will be KL, and it will seen under visual angle KAL. Since angle KAL > angle GAB under which object line GB would be seen without refraction, the image will appear closer to the center of sight than its object.

   It would seem to follow automatically that, in appearing closer, the image will be perceived as larger than its object. But in fact Alhacen is prevented from drawing this conclusion because he subscribes to the size-distance invariance principle, which holds that identical objects seen from different distances will be perceived to be the same size even though they are viewed under different visual angles.[16] Accordingly, since image KL and object GB are the same size, they should be perceived to be the same size. For that reason, Alhacen is forced to attribute the apparent magnification of KL to an apparent lengthening of its distance from the center of sight caused by the natural weakening of light and color due to refraction. Such weakening makes the object appear dimmer than it actually is, the result being that the image appears farther away and therefore larger than it would appear if it were brighter and more distinct.[17]

Among the last group of four theorems in the set, the first two, i.e., propositions 15-16, pp. 314-316, prove the same point for any object line lying behind a convex spherical interface facing a center of sight when the medium behind the interface is denser and the object line lies between the interface's center of curvature and the center of sight. The simplest case is illustrated in figure 17, p. 414, which accompanies proposition 15, pp. 314-315. Again, the center of sight is A, which faces convex spherical interface NME, and object-line GB is posed between center of curvature D and center of sight A. The luminous color from endpoint G follows GT to the interface and is refracted to A along TA, and the luminous color from endpoint B follows BH to the interface and is refracted to A along TA. The image of GB as a whole is therefore KL, and it is seen under angle KAL, which is larger than angle GAB under which GB would be seen without refraction. In this case, unlike that for a plane-refractive interface, the image actually is larger than its object, so its magnification will be all the greater than that due to refraction through a plane interface.

Although related to the two preceding theorems, the third theorem of the set, proposition 17, pp. 318-319, is anomalous in several ways. For one thing, the object line at issue does not face the center of sight; rather, it coincides with a normal dropped to the sphere from the center of sight. In addition, the object line does not lie between the center of curvature and the center of sight but, instead, on the other side of the center of curvature from the center of sight. Most anomalous of all, the object line lies entirely outside the sphere in a medium of the same rarity as that in which the center of sight lies. What is at issue, then, is the image of an object line seen *through* rather than *in* a sphere consisting of a denser medium.

In both purpose and approach, proposition 17 is intimately related to proposition 9 in chapter 5 (see pp. 291-292). Harking back to the analysis of proposition 9 based on figure 14, we recall that point E on the refractive interface was taken to be the limit at which the luminous color from B on normal AB will refract to center of sight A, which means that angle BEZ is the largest possible angle of incidence at which the luminous color from B will penetrate the refractive interface. We then concluded that there will be some other point on arc EG from which the luminous color of some point other than B on normal BD will refract to A because the angle of incidence will be smaller than BEZ. By extension, therefore, we deduced that the luminous color from an infinitude of points on a line segment bounded at one end by B on the axis will refract from associated points on arc EG to reach A. In theory, then, arc EG will form a locus of refraction for an infinitude of points on BD. Since A is the image location of all those points, arc EG will also be their image, and if the entire figure is rotated about the normal to carry arc EG with it, the composite image of all those points as seen from A will form a circular segment on the face of the sphere.[18]

Whereas proposition 9 appears to demonstrate that all the points on normal BD from B on will yield a complete circular image on the spherical refractive interface, proposition 17 demonstrates that a line bounded by particular points on the normal will yield an annular image on that interface. Thus, as represented in figure 18, p. 415, which is copied from the actual figure accompanying the proposition (figure 7.7.66, p. 204), if A and extension DL of the normal lie in the same rare medium while the sphere consists of a denser medium, our object is to prove that there is a line segment on the normal beyond D whose image as seen from A will form an annular image on the convex spherical surface facing A.

We start by assuming that the medium surrounding segment DH of the normal is of the same density as the medium filling the sphere, which is the case for proposition 9. We then pick points L and H on the normal that define a line segment whose form will refract to A, the luminous color from endpoint L refracting at T and that from endpoint H at G. It follows, then, that angle LTE is either the largest possible angle of refraction or less than it. In fact, as represented in the figure, it is less. Now let us assume that A is a source of radiation. According to the principle of reciprocity, the luminous color passing from A along AT will strike the refractive interface at T, where it will be diverted toward normal ET in the sphere along refracted ray TZL. Likewise, the luminous color passing from A along AG will follow ray GMH after refraction toward normal EG at G. In both cases, then, the original incident rays will become refracted rays, and vice-versa, which of course follows from the principle of reciprocity.

With that in mind, let us assume that the medium surrounding segment DH on the normal is of the same rarity as the medium within which A lies, and let A continue to be a source of radiation. In that case, the luminous color from A will pass along AT and, as before, will follow ray TZL after refraction toward normal ET at T. Similarly, the luminous color from A will pass along AG and, as before, will follow GMH after refraction toward normal EG at G. When it reaches points Z and M on the inner surface of the sphere, however, the luminous color conveyed along TZ and GM will be refracted away from normals EZF and EMC, respectively, when it passes into the rarer medium beyond the sphere. By how much? According to the geometry of the situation, angle TZE of incidence for the luminous color reaching Z along TZ is equal to angle ETZ of refraction for the luminous color reaching T along AT because triangle EZT is isosceles. Hence, by the principle of reciprocity, the luminous color emerging into the air at point Z will follow a path OZ that is perfectly symmetrical with AT. In other words, angle ETA = angle EZO. So, too, when it emerges into the air at point M, the luminous color conveyed along GM will follow path MK, which is perfectly symmetrical with AG.

Consequently, if we reverse the direction of radiation yet again, so that K and O are sources of radiation, it follows that their luminous color will radiate to the sphere along KM and OZ, will then refract toward the normal along MG and ZT in the sphere, and will refract again away from the normal along GA and TA on the other side of the sphere. Within the plane of refraction represented, the image of KO will therefore lie on arc GT as seen from A, so if we rotate the entire figure about axis AH, that arc will describe a ring-like spherical segment on convex surface GT. Moreover, the cross-section GT at any point in the rotation will be larger than KO, so KO will appear magnified and somewhat convex according to the convexity of the surface on which its image is formed.

Taken at face value this proposition is entirely unremarkable and unexceptionable. That a certain line segment seen through a sphere forms an annular image on its convex face may be interesting, but it hardly seems significant enough to warrant the proof. As far as the remainder of book 7 is concerned, in fact, it is irrelevant; nowhere is the point demonstrated in proposition 17 used as evidential or theoretical support for any subsequent analysis. At a deeper level, though, proposition 17 is extraordinary for its implications in at least two respects. For one thing, it is not just about the formation of annular images. It is also about the passage of light from air through transparent spheres, and in that regard it has potential bearing on the analysis of lenses and their focal and image-projecting properties. For another thing, figure 7.7.66, which reproduces the diagram that accompanies proposition 17 in all the Latin versions of the text, including Risner's edition, is misleading in a crucially telling way. Rather than explore these points here, however, we will defer the discussion to a later section of the Introduction (see pp. lxx-lxxi).

Before turning to proposition 18, Alhacen suggests an empirical verification of the analysis in proposition 17. To that end, he suggests that the reader take a ball of black wax the size of a chick pea and stick it firmly on the end of a long needle. He is then to look with one eye through a clear glass sphere and, with the wax ball suspended on the needle, pose the ball behind the sphere in line with the axis running from the center of his eye through the sphere's center. When he has the wax ball appropriately situated on the axis, he will see its image as a black ring on the front surface of the sphere.

Completing the first main section of chapter 7, proposition 18, pp. 319-320, extends the analysis in proposition 17 to cylinders, in which case the image will be seen not as a ring but as two horizontal lines toward the edge of the cylinder and separated in the middle. The reason for this truncation of the image is that, when the entire figure in the diagram is rotated about axis AL, arc TG will only touch the cylinder in the plane that passes perpendicular to the cylinder's axis at centerpoint Z. At every other point in its rotation on circle BTD that arc will lie inside an ellipse formed on the cylinder's surface by the cutting plane, or it will lie in the plane of a line of longitude on the cylinder's surface. In none of those positions, therefore, can it actually touch that surface.

The second and concluding section of Alhacen's analysis in this chapter deals with the effect that atmospheric refraction has on the apparent size of celestial objects. Capping this analysis is an explanation of why such objects appear larger at horizon than at zenith and whether this apparent enlargement is due in whole or in part to atmospheric refraction. Before launching into the actual analysis of these issues, Alhacen offers a brief account in paragraphs 7.42-46, pp. 320-321, of the atmospheric model on which the analysis will be based. Following Aristotle, he assumes a sharp dichotomy between the celestial realm and the terrestrial realm below it, this latter defined by the outer surface of the atmospheric shell surrounding the Earth. It is at this surface, where celestial and terrestrial realms meet, that atmospheric refraction occurs according to the density differential between celestial and terrestrial atmospheric media. The atmosphere itself consists of two transparent bodies, air (including the vapors that pervade it) and the shell of fire above it, but Alhacen insists that there is no clear differentiation between them. On the contrary, the atmosphere becomes continuously rarer as it progresses from the crass air nearest the Earth to the pure fire abutting the heavens. Pure fire, not air, is thus the medium responsible for atmospheric refraction, and once it is refracted by that medium, the light from celestial objects continues straight through the remainder of the atmospheric shell without further bending.[19] How high above the Earth this shell extends Alhacen leaves unspecified, but he probably believed its upper limit to lie much closer to the Earth than to the moon.[20]

Following his account of the atmosphere and its structure, Alhacen presents three theorems, propositions 19-21, pp. 321-325, whose purpose is to show that the image of any celestial body seen anywhere in the sky is reduced in size by atmospheric refraction. For instance, proposition 19, pp. 321-323, demonstrates this point for celestial objects viewed at zenith. The proof is simplicity itself. Let the large circle in figure 19, p. 416, represent the outer boundary of the celestial sphere, the smaller shaded circle MEZ the atmospheric shell, and the small inner circle the Earth, all three circles centered on G. Let T be the position of an observer on Earth, B his zenith, and KL the cross-section (grossly exaggerated) of some star or an interval between stars viewed at zenith. The light from K will extend along KM to M through the celestial medium, and at M it will be refracted toward the normal to reach the observer at T. Likewise, the light from L will extend along LZ to Z and will be refracted toward T along ZT. Hence, the image of KL as seen along the continuations of refracted rays MT and ZT will be QR, and it will be seen under angle QTR, which is smaller than angle KTL under which KL would be seen without refraction. Essentially the same line of reasoning is followed in propositions 20 and 21, pp. 323-325, for proving the same point for horizontal or vertical cross-sections of any celestial object viewed between zenith and horizon. Having shown that any celestial body appears diminished according to both its horizontal and vertical diameters and that the conditions of diminution are the same in both cases, Alhacen concludes at the very beginning if paragraph 7.62, p. 325, that those bodies appear round wherever they may be in the sky.

And yet, Alhacen continues in paragraphs 7.62-71, pp. 325-330, despite such diminution in apparent size (albeit negligible), celestial objects look much larger at horizon than they do at zenith. The principal cause of this phenomenon, Alhacen explains, is psychological and has to do with the way we correlate distance and size as well as distance and surface shape. Normally, we judge the size of things according to the visual angle it subtends and the distance it lies from us. Relatively moderate distances we gauge according to known intervals, such as "a pace away," on the ground and a recognition of a limited succession of such intervals. That way we recognize that something is around 100 meters away. Longer distances we gauge by means of intervening landmarks, such as a succession of fig trees or houses of a relatively well-known size and well-known distance from each other. That way we recognize that something is two or three kilometers away. There comes a point, however, at which we can no longer gauge a distance with any certainty. At that point, we simply estimate. Some distances are so immoderately long that we cannot even venture a reasonable guess. Celestial distances are of this sort.[21]

As to size, the more familiar we become with various things, the more familiar we become with how large they generally are. Humans, for instance, are more or less six feet tall, so when we see such things, we recognize that they are roughly the same size, no matter how far away they lie. The farther away they lie, the smaller the visual angle subtended by them, so we recognize that the difference in visual angle is due to the difference in distance and, therefore, that the two objects are roughly the same size. Conversely, if the two subtend the same visual angle but one is perceived to lie farther than the other, the farther one will be perceived as larger. If the two are so remote that the difference between their distances can no longer be perceived, the object subtending the larger visual angle will be perceived as larger, even though it may actually be closer and smaller.[22]

As to surface shape, if a convex object lies close enough, we can tell it is convex by perceiving that its central portions lie closer to us than its outer edges and, therefore, that it bulges toward us. By the same token, we recognize concavity by perceiving the central portions to lie farther from us than the outer edges. There comes a point, however, at which such objects are so remote that we can no longer recognize the disparity in distance between their central and outer portions. At that point, we perceive them as flat disks. That, for instance, is why we perceive even the closest heavenly body, the moon, as flat, because it lies so far away that we cannot detect its globular shape.[23] The same holds for the celestial vault. It lies so far away that we cannot detect its concavity. Indeed, all we can perceive of it is its color, so we perceive it as flat and, morever, as extending overhead like a ceiling.

Hence, as celestial bodies, such as the sun, moon, and stars, pass overhead, we perceive them as moving not along an arc but along a line on a plane parallel to the plane of our horizon. Moreover, since we are unable even to estimate their distances because of their vast remoteness, such bodies appear closest to us when they are at zenith, just as they would if they were directly overhead on a ceiling, and farthest from us when they are at horizon, just as they would at the distant end of a long ceiling. Yet throughout its passage the same celestial body is seen under the same visual angle, so, in correlating the apparently longer distance at horizon to a visual angle equal to the one the object subtended at zenith, we perceive the object as larger at the horizon than at zenith.

Figure 20, p. 417, illustrates the model just articulated, as well as a fundamental problem with it. The white disks represent the celestial body moving along its proper arc according to five spots between the viewer's zenith (position 1) and his horizon (position 5). Since the viewer cannot detect the concavity of that arc, he perceives the body, now represented in light gray, as moving along a line parallel to the plane of his horizon, as if it were moving on a ceiling. When it lies directly overhead at position 1, therefore, it will appear to be closest to him, whereas at position 5 it will appear farthest from him. But throughout its passage the body will be seen under the same visual angle, so when the viewer correlates that angle to the perceived distance, he will see the body as larger as it progresses through its arc. Hence, when it reaches position 4 on its arc, the body will be perceived at position 4 on the apparent horizontal path, and the disparity between the real and apparent distances is so great that the body will be perceived as that much larger than it actually is.

The most obvious problem with this model is that, as the body approaches the horizon, the disparity between real and apparent distances becomes so great that the body should be perceived as impossibly large. One way of obviating this problem is to assume that we see the heavenly vault not as a ceiling but as a flattened dome within which the apparent path of the body is reduced to the elliptical arc pictured in the diagram, the apparent body represented in darker gray. The amplitude of this arc can then be adjusted so that, when the body reaches horizon at position 5, its apparent size accords with experience, which is in fact quite variable.[24]

Our inability to detect the concavity of the celestial vault, coupled with our inability to perceptually gauge the distances of celestial bodies, thus leads us to perceive, or rather misperceive, those bodies as larger than they really are at the horizon. Moreover, the illusion is inescapable. Wherever on Earth an observer views a celestial body, it will always appear enlarged at horizon. Nevertheless, although this is the principal cause for such enlargement according to Alhacen, it is not the only one. Refraction through thick vapors rising slightly above the horizon can also cause the images of celestial bodies to be magnified.

Alhacen's explanation of this magnification in paragraphs 7.72-73, pp. 330-331, is evidently based on the model illustrated in figure 21, p. 417, where the observer stands at point E on the Earth's surface. EA' is a line in the plane of his horizon within the plane of the page, Z is his zenith, and the gray arc is a band, or partial band, of vapor rising northward above his horizon and lying some indeterminate distance above the Earth's surface. Its thickness is also indeterminate. BR' and AR are rays extending from the endpoints of the cross-section of some heavenly body after the light passing along them has been refracted at the outer surface of the atmosphere. Consequently, when the two rays reach points R and R' at the outer surface of the vapor band, the light conveyed along them will be refracted toward the normal along RR1 and R'R1'. Upon entering the rarer air at points R1 and R1' on the inner surface of the vapor band, the light following those rays will be refracted away from the normal so as to reach the observer along R1E and R1'E. The endpoints of the cross-section will thus be seen along refracted rays ER1A' and ER1'B' under angle B'EA', which is larger than angle BEA, so the image of AB will appear larger through the vapor than it would appear were it to be seen in the vapor's absence.

The amount of magnification will of course depend on how thick the band of vapor is and how great the density differential is between the vapor and the air above and below it. Alhacen has nothing to say on either score, but surely he would have supposed the band to be fairly thin and the density differential to be fairly slight, in which case the magnification would be totally imperceptible. Thus, compared to the psychologically based enlargement due to misperception of the heavenly vault's concavity, this optically based enlargement is imperceptible. But that is not why Alhacen characterizes the refraction of celestial light through vapor as a secondary cause of the Moon Illusion. Rather, it is because the presence of such vapor at the horizon is intermittent; in some places on Earth it is always present and in others only occasionally present. Thus, as Alhacen sums it up in paragraph 7.74, p. 331, the primary cause of the Moon Illusion is psychological, and it is "fixed and invariant," whereas refraction through vapors is accidental insofar as it only obtains in some places and at some times. With this distinction, Alhacen brings book 7, as well as the entire treatise, to an abrupt end.

2. *The Ptolemaic Underpinnings of Alhacen's Refraction Analysis*

Had I followed the pattern established in the Introductions to the previous editions of books 1-3, 4-5, and 6 of the *De aspectibus*, I would have entitled this section something like "The Sources of Alhacen's Refraction Analysis." By now, however, the number of sources—or at least clearly recognizable ones—has dwindled to two: Ptolemy's *Optics* and Euclid's *Elements*.[25] Alhacen's use of the *Elements* has already been amply discussed in book 5, and there is nothing new to say on that score here.[26] Indeed, there is less to say because the proofs in book 7 are considerably less complex and sophisticated than those in book 5. Let us therefore turn directly to the relationship between Alhacen's refraction analysis in the *De aspectibus* and that of Ptolemy in book 5 of the *Optics*.

In terms of both topical focus and approach, Alhacen's account of refraction in book 7 is fairly closely linked to Ptolemy's account in book 5 of the *Optics*. Indeed, much, if not most of Alhacen's refraction analysis represents an elaboration on Ptolemy's. The experimental derivation of the angles of refraction in chapter 3 is a case in point. There is no question, of course, that the apparatus Alhacen uses for that derivation differs markedly from Ptolemy's, but those differences are due primarily to the fact that Alhacen was measuring the refraction of light rays, not visual rays. For that reason, Ptolemy's apparatus is considerably simpler and more flexible than Alhacen's.[27]

Ptolemy starts with the bronze disk used for the reflection experiments in book 3 of his *Optics*. Represented in the top diagram of figure 22, p. 418, this disk is divided into quadrants by diameters DB and AG. Opposite quadrants AD and BG are further subdivided into 90 degrees. At centerpoint E of the disk is a small colored peg or marker, whose extension on the back side of the disk may also serve as a pivot pin for a thin rod, at the end of which is attached a sighting-device Z, perhaps a small piece of metal with a slit through which a sighting line can be established. To that same pivot pin may be attached another thin rod with a small marker H at its outer end.[28]

Like Alhacen's, Ptolemy's determinations of the angles of refraction are based on angles of incidence at ten-degree increments, starting with $i = 10°$ and ending with $i = 80°$. To determine the angles of refraction for a visual ray passing from air to water, the experimenter is to stand the disk upright in a pan (Latin = *baptistir*) large enough to accommodate it and pour water into the vessel until its surface is flush with diameter DB. The sighting-slit Z is then lined up at one of the prescribed degree marks—50° in the case illustrated in figure 22—and, while the experimenter holds the appropriate line of sight along ZE at angle of incidence ZEA, he pivots marker H along arc GB until it disappears behind the marker at E. Marker H will therefore appear to lie at point T in a direct line with ZE, and the appropriate angle of refraction will be GEH. The same can be done for all the other angles of incidence, including 0°, in which case normal AE will be the line of sight. H will therefore lie at point G, the visual ray having passed straight to that point through the interface and thus having undergone no refraction.

By Alhacen's reckoning, of course, what has been measured is the refraction of a light ray passing from marker H through the water into the air above it, GEH being the angle of incidence and ZEA the angle of refraction according to Ptolemy's definition, or ZET' according to Alhacen's definition. By the principle of reciprocity, a beam of light passing along ZE will refract to H, yielding an angle of refraction/deviation HET equal to ZET', so for all practical purposes Ptolemy's tabulations for refraction from air to water will also serve to measure refraction from air to water according to Alhacen's radiative model. It should be noted, however, that if we reverse the direction of radiation in Ptolemy's experiment so that the angles of incidence and refraction are interchanged, the final value of $r = 50°$ Ptolemy gives for $i = 80°$ exceeds the critical angle of slightly under 49° for the passage of light from water into air (see Smith, *Ptolemy's Theory*, p. 233).

In order to determine the angles of refraction for the passage of light from air to glass, the experimenter is to affix to the bronze disk a glass semicylinder of radius KE < radius DE of the disk so that its flat side TL is flush with diameter DB, as represented in the bottom diagram of figure 22. Instead of placing an actual peg at centerpoint E of the disk, the experimenter is to make a small but visible mark on the glass at that point. Then, posing sighting-slit Z at the required angle and maintaining the appropriate line of sight along ZE, he is to pivot marker H along the edge of the disk until it

xlviii ALHACEN'S *DE ASPECTIBUS*

disappears behind the mark on the glass at E so as to seem to lie on straight line ZET. GEH will be the resulting angle of refraction for angle of incidence ZEA, and the same procedure will give the angle of refraction for any angle of incidence between 10° and 80°, including 0°, in which case the ray will strike the refractive interface orthogonally and pass straight through without refraction. Here, too, what is actually being measured by Alhacen's lights is refraction from glass to air according to angle of incidence GEH and angle of refraction/deviation ZET'. But again, for all practical purposes Ptolemy's tabulations for refraction from air to glass can also be taken as tabulations for refraction from air to glass according to Alhacen's radiative model. In this case, moreover, Ptolemy's value of $r = 42°$ for $i = 80°$ is very nearly the actual value of the critical angle for the passage of light from glass into air.

Ptolemy's third and final determination is for refraction from water to glass, although practically speaking it is a test for refraction from glass to water. This determination calls for affixing the glass semicylinder as illustrated in the top diagram of figure 23, p. 419, inserting the bronze disk upright in the vessel used earlier for the test from air to water, and filling the vessel with water until its surface is flush with diameter DB and thus also with the flat surface of the glass semicylinder. Instead of posing sighting-slit Z at ten-degree intervals on arc AD, though, the experimenter is to pose marker H at ten-degree intervals on arc GB below the water and then rotate sighting-slit Z until H appears to line up behind mark E along line ZET. In essence, then, GEH will be the angle of incidence and ZED the angle of refraction, although according to Ptolemy's visual-ray model the two are reversed, since the ray passes along ZE and is refracted along EH.

Why did Ptolemy follow the anomalous procedure just described rather than continue testing angles of incidence at ten-degree intervals on arc AD? For a start, had he based his tabulations on arc AD, he would have been measuring the refraction of visual rays from glass to water, not water to glass. Also, according to our reconstruction, his apparatus did not allow for a direct view through water into glass without having the observer immerse his head in the pan and then line his eye up with the marker at H. Since this would have been unfeasible, Ptolemy was forced to measure the refraction indirectly according to the principle of reciprocity, which permitted him to assume that the respective angles of incidence and refraction in refraction from glass to water are perfectly interchangeable in refraction from water to glass.

Whether by design or accident, Ptolemy's experimental procedure was limited to situations that avoided the problem of critical angle. In the first two experiments, for instance, where he tested the refraction of visual rays from air to water and air to glass, he was actually testing the refraction of light from water to air and glass to air, i.e., from a denser to a rarer medium.

Accordingly, the refraction of light from marker H in the top and bottom diagrams of figure 22, p. 418, would have been limited by the critical angle, which is around 49° for water to air and 42° for glass to air. Still, for any angle of incidence i between 0° and those values along arc GB, the light will be refracted to some point on arc AD between A and D. Consequently, by basing his measurements of i in both cases on arc AD, Ptolemy was able to see the image of H throughout the range of angles from $i = 0°$ to $i = 80°$ on that arc. In the third experiment testing refraction from water to glass the situation is opposite in that the light passes from a rarer into a denser medium. Consequently, since the light actually refracts toward the normal, the maximum angle of refraction at $i < 90°$ on arc GB in the top diagram of figure 23, p. 419, is the critical angle of around 61° on arc AD. By basing his measurements of i on arc GB, then, Ptolemy was able to see the image of H throughout the sweep of angles from $i = 0°$ to $i = 80°$ on arc GB.

Given the experimental setup for the third experiment, Ptolemy had readily at hand the means to test refraction in the opposite direction for all three media. For instance, in order to make the test from glass to water, he could have used the experimental setup in the top diagram of figure 23 and reversed his procedure for the third experiment by sighting along ZE at intervals of ten degrees along arc AD. Had he done so, though, he would have discovered that the image of H disappears at the 70° and 80° marks because the light from H undergoes full internal reflection at the critical angle of around 61°. In the equivalent test for refraction from glass to air, with the water emptied from the vessel, he would have noticed the loss of image from $i = 50°$ to $i = 80°$. In the test for refraction from water to air, finally, he would have removed the glass semicylinder and measured the angles of incidence along arc BG underwater while sighting along ZE, in which case the image would have disappeared by the 60° mark on BG, as illustrated in the bottom diagram of figure 23. In all three instances the disappearance of the image is due either directly or indirectly to the critical angle. Did Ptolemy fail to include these tests in his analysis because he thought they were unnecessary or because he realized they would not work according to his protocol? There is no way to be certain, but it is difficult to believe that he did not try at least one of the tests, only to come a cropper because of the critical angle. That being the case, it is also difficult to believe that Ptolemy did not actually conduct the experimental tests for refraction from air to water, air to glass, and water to glass.[29]

The similarities between Ptolemy's experimental derivation of the angles of refraction and Alhacen's are obvious, and so are the fundamental similarities between the experimental apparatus they used. Ptolemy's bronze disk has become Alhacen's bronze pan; Ptolemy's sighting slit has become Alhacen's system of two holes; Ptolemy's marker on the opposite side of the sighting slit has become the point of a needle or a mark on the rim of the pan; Ptolemy's glass semicylinder has become Alhacen's glass quarter-sphere; and Ptolemy's movable sighting slit and marker have been translated into the axial rotation of Alhacen's pan. Both, moreover, deploy their respective apparatus in essentially the same way, inserting the disk or pan into a vessel and filling the vessel with water until the water's surface reaches the horizontal diameter.

Granted, Alhacen's description of the apparatus and its construction is considerably richer in detail than Ptolemy's, but it is clear that it takes the form it does because, unlike Ptolemy, Alhacen designed his to allow the refraction of light to be measured directly, on the basis of the incoming light itself, rather than indirectly, on the basis of the light's visual effect. In short, Alhacen's apparatus is an elaboration on, not a real departure from, Ptolemy's. Moreover, as mentioned above, Ptolemy's experimental apparatus is considerably more flexible than Alhacen's precisely because it uses indirect rather than direct measures. Accordingly, the marker at H, which constitutes the light source in Ptolemy's experiments, is infinitely easier to manipulate than the sun (through rotation of the pan on its axle), which is the light source in Alhacen's. It is somewhat surprising in fact that Alhacen did not check Ptolemy's results using the procedure in chapter 4, where the point of a needle, which can be moved anywhere on the rim of the pan, serves as the light source.[30]

The fundamental principles of refraction that Alhacen attempts to demonstrate experimentally are also Ptolemaic in origin. Some of these principles Ptolemy states explicitly, and some of them are implicit in the experiments detailed above for determining the angles of refraction for the passage of visual rays from air to water, air to glass, and water to glass. Early in book 5, for instance, Ptolemy establishes explicitly 1) that the severity of refraction depends on the density of the medium into which the visual ray is refracted; 2) that refraction occurs not only when the visual ray enters a denser medium but also when it enters a rarer medium; 3) that refraction is a surface phenomenon; 4) that the image in refraction is seen along the incident visual ray (Alhacen's refracted light ray), its specific location being where that ray intersects the orthogonal dropped from the object to the refractive interface; 5) that refraction occurs in a single plane perpendicular to the refractive interface; 6) that refraction is perfectly reciprocal.[31] Implicit in Ptolemy's experiments, in turn, is that light refracts toward the normal when it passes into a denser medium and away from the normal when it passes into a rarer one and that light entering a denser or rarer medium along the normal passes straight through without refraction.[32] Aside from his attempt to establish them empirically, then, Alhacen added nothing to the set of fundamental governing principles of refraction already adduced by Ptolemy.[33]

Alhacen's physical model of refraction is also linked to Ptolemy's, both being predicated on the notion that, when light or visual radiation strikes the surface of a given medium, the medium resists penetration according to its relative density or rarity. If the medium is dense to the point of complete opacity, and if it is adequately smooth, it will pose enough resistance to force rebound (Ptolemy's term, in Latin translation, is *reverberatio*) and thus cause the light or the visual radiation to reflect at equal angles. If it is less dense, it will allow the light or visual radiation to penetrate but will impede its progress, forcing it to follow a broken path determined at the point of entry. The denser the medium, the more broken the path, the most broken path of course being the reflected one. The same holds for light or visual radiation passing from a denser to a rarer medium; the greater the density differential, the more broken the path.

Thus, Alhacen and Ptolemy both view reflection as a special case of refraction, the two phenomena sharing essentially the same principles. Both phenomena, for instance, are assumed to occur in a single plane perpendicular to the reflective or refractive interface. In both, the image is assumed to appear on the line of reflection or refraction, and its specific location is governed by the cathetus rule. And in both, the effect is assumed to occur instantaneously at the refractive or reflective interface. Furthermore, Alhacen's vectorial analysis of refraction according to composite motion is at least hinted at by Ptolemy in his effort to explain the difference between the breaking due to reflection and that due to refraction.[34] Even Alhacen's principle of conservation of composite motion, which shunts refracted light toward an easier path, is presaged in Ptolemy's claim that the similarity of radial breaking in reflection and refraction is based on "the course of nature in conserving the exercise of power" (*cursus nature in conservandis actibus virtutibus*).[35]

There are a number of other links between Alhacen's refraction analysis in book 7 of the *De aspectibus* and Ptolemy's in book 5 of the *Optics*, but these need not detain us here.[36] One final set of links is worth discussing, though, because it extends beyond Ptolemy's *Optics* to the *Almagest*. The first link in this set involves Alhacen's account at the end of chapter 4, pp. 270-271, of how celestial bodies seen at the horizon appear displaced above their actual position because of refraction through the interface between the atmosphere and the aither surrounding it. For a start, his empirical demonstration of such displacement on the basis of observing the position of a star at horizon and zenith with an armillary sphere is clearly suggested by Ptolemy in paragraph V.24 of the *Optics* (Smith, *Ptolemy's Theory*, pp. 238-239). More to the point,

Alhacen's mathematical explanation of that displacement in proposition 1, pp. 272-273, is essentially the same as Ptolemy's in theorem V.2 of the *Optics* (Smith, *Ptolemy's Theory*, pp. 240-241). The second link is between Alhacen's psychological explanation of the Moon Illusion and Ptolemy's vague account in *Optics*, paragraph III.59 (Smith, *Ptolemy's Theory*, p. 151), of why celestial objects look smaller at zenith than at horizon. The reason, Ptolemy claims, is that at horizon they "are seen in [a] way that accords with custom," whereas "high up [they] are seen with difficulty." The third and final link takes us to *Almagest*, I, 3, where Ptolemy mentions in passing that

> ... the apparent increase in [the] diameters [of heavenly bodies] is
> ... caused ... by the exhalations of moisture surrounding the earth
> being interposed between the place from which we observe and the
> heavenly bodies, just as objects placed in water appear bigger than
> they are, and the lower they sink, the bigger they appear.[37]

It is to vindicate or at least clarify this claim that Alhacen devotes the concluding paragraphs of book 7 (pp. 330-331), where he explains how celestial bodies can appear magnified when their light is refracted through a wall of vapor lying between the object and the air surrounding the observer. The magnification of celestial bodies by such vapor was evidently a recurring issue for Alhacen.[38]

These examples should suffice to show how closely tied Alhacen's refraction analysis is to Ptolemy's and, therefore, why the former can for the most part be characterized as an elaboration on, or specification of, the latter. Not that Alhacen's refraction analysis is a mere elaboration or specification. On the contrary, much of his analysis is based on hints and vague suggestions in Ptolemy's *Optics*, an obvious example being his psychological explanation of the Moon Illusion, which betrays an originality and sophistication far beyond its source. The same holds for Alhacen's empirical demonstrations, two prime examples being those in chapter 5, pp. 275-280, designed to show that the image in refraction lies on the normal dropped from the object to the refractive interface. Wholly original, and thus wholly independent of Ptolemy, is Alhacen's explanation in chapter 6, pp. 302-307, of peripheral vision and the variation of visual acuity within as well as outside the cone of radiation. Indeed, Alhacen made no bones about his originality here, an originality that extends beyond the analysis itself to the recognition that peripheral vision needs explaining in the first place. Ptolemy certainly ignored it. In terms, therefore, of rigor and comprehensiveness, Alhacen's refraction analysis is far superior to its Ptolemaic source.

3. *Problematic Aspects of Alhacen's Refraction Analysis*

In the course of surveying Alhacen's refraction analysis and discussing its Ptolemaic underpinnings in the previous two sections, we noted several points at which that analysis raises methodological and logical issues, as well as questions about historical import. In this section we will deal with six of these issues, starting with a close look at Alhacen's experimentalism and its potential shortcomings. From there we will pass to a discussion of certain problematic aspects of his physical model of refraction, after which we will briefly evaluate the governing rules of refraction listed by him at the end of chapter 3. Next we will examine his account of peripheral vision and the difficulties it raises for his overall account of visual perception. After that we will determine how his analysis of refraction through spherical interfaces in chapters 5 and 7 might or might not have contributed to the development of modern lens theory at the hands of Johannes Kepler. We will conclude with a brief look at how Ibn Sahl's account of lenses in the "Treatise on Burning Instruments" might have influenced Alhacen's analysis of refraction through spherical interfaces.

The Limitations of Alhacen's Refraction Experiments: As we saw in the opening section of the Introduction, chapters 2-4 of the seventh book are devoted to an empirical verification of the principles of refraction on the basis of the experimental apparatus described in chapter 2. In typical fashion, Alhacen provides a detailed account of how the apparatus is to be constructed, starting with the shallow pan containing the register plate and concluding with its proper insertion into the carrying vessel. In typical fashion, as well, he specifies the materials from which the apparatus is to be made, the exact measurements according to which it must be formed, and the various ways in which it should be deployed. For the most part, however, instrumental precision is not an issue in his experiments because the majority of the ones he prescribes are qualitative. A slight deviation from perfect verticality when the instrument rests in the carrying vessel, for instance, will not affect the general observation that light bends toward the normal when it passes from air to water or to glass, or away from the normal in passing from glass to air or to water. The same holds for determining that the refracted image of a small object is seen directly along the line of refraction or that this image appears at the intersection of the line of refraction and the cathetus of incidence.[39]

There is one set of qualitative experiments, however, in which precision is crucial: the tests to show that light striking a refractive interface along the normal passes straight through without the slightest deviation. In this regard the test for the normal passage of light from air to water is problematic in two respects. First, it can only be conducted at or between the tropics, and only when the sun is at zenith, so it is feasible two times at most in any year and at a latitude of no more than 23.5° north or south.[40] Second, it requires that the shallow pan containing the register be perfectly vertical so that the sunlight passing through the two holes is perfectly perpendicular to the water's surface.

The equivalent tests for glass to air and air to glass are also problematic, primarily because the glass has to be of extraordinarily fine quality and shape. Take, for instance, the test showing that light travels straight through a line of small glass cubes when it strikes the leading edge of that line orthogonally. In order for this test to be valid, the edges of each cube must be absolutely perpendicular to one another so that, when the cubes are all applied successively to one another, they fall into a perfectly straight line. The slightest skew in the edges will cause the line of cubes to deviate ever more from rectilinearity as each cube is added, and that deviation in turn will cause the leading edge of the line of cubes facing the incoming light to be aslant rather than perpendicular to it. Furthermore, the glass itself must be virtually, if not literally, free of bubbles, waves, and other imperfections that could affect the light passing through.

Likewise, the glass quarter-sphere used in several of the tests must be free of internal flaws. Not only that, but its convex face must be perfectly spherical, and its flat faces must be perfectly plane, as well as perfectly perpendicular to one another. Otherwise the light may be slightly deviated when it passes into or out of those faces. These material limitations raise serious doubts about whether Alhacen could have ever conducted the experiment as described, particularly given the large size of the glass quarter-sphere. Indeed, as Gérard Simon observes, even if constructing the shallow pan and register plate within the necessary tolerances was technologically feasible in early-eleventh-century Cairo, forming and milling the glass cubes and quarter-sphere to the necessary level of fineness and precision was probably not.[41]

While instrumental precision may not have been of utmost concern in most of the qualitative experiments described by Alhacen in chapters 2 and 4, it would have been absolutely paramount for the experimental determination in chapter 3 of the angles of refraction when light passes from air to water, air to glass, glass to air, and glass to water. The accuracy of the results in all these determinations depends on a variety of interdependent factors. For a start, the shallow pan must be inserted into the carrying vessel so as to be perfectly vertical, and in order for that to happen the brim of the carrying vessel must be perfectly horizontal, as must the bottom edge of the hanging strip attached to the back of the register plate through the axle. In addition, the hole in the rim of the pan and the hole in the panel facing it must be aligned so precisely that the axial ray of the shaft of sunlight passing through both holes is parallel to the face of the register plate within the plane of the middle circle inscribed on the inner wall of the pan's rim.[42]

Furthermore, the division of that middle circle into degrees and minutes must be perfectly accurate if the measure of the angles of refraction on the circle is to be as precise as possible. Even then, Simon cautions, the diffraction of the light caused by its passing through the holes would be severe enough to make finding the exact center of the circle of light cast on the inner rim of the pan all but impossible.[43] At best, then, the resulting measures of refraction would be based on a subjective judgment of where the axial ray strikes the rim, and to make matters worse, that judgment would be based on observation that can be distorted by the refractive interface through which it is made.

The deployment of the entire apparatus, with the pan inserted into the carrying vessel, also poses problems because the arrangement is extraordinarily cumbersome, especially when the vessel is filled with water. First, the pan with its register plate must be rotated on its axle until it is set at whatever angle of incidence is to be tested. Then it must be turned about on the rim of the carrying vessel until the hole in its rim and the hole in the panel on the register plate line up more or less perfectly with the sun, which will have reached the required angle from the horizon in its ascent or descent through the sky.

In theory this is a simple enough procedure, but in practice it involves a number of fine adjustments, most of them in anticipation of where the sun will be rather than where it is at the time of the adjustment. This is made all the more difficult by the fact that, except at the equator, and then only twice a year, the sun never ascends straight overhead from the horizon but follows an offset arc. Furthermore, each such adjustment roils the water in the vessel, and the experiment cannot be conducted properly until the water is absolutely still. As for the glass quarter-sphere, not only must it be temporarily affixed to the register plate, then lifted and re-affixed at ten-degree intervals, but at each stage it must be lined up with absolute precision if the results are to be as accurate as possible.

Moreover, the range of angles of incidence that can be tested with the apparatus as described is limited by the latitude at which the determination is conducted. If, for instance, Alhacen conducted the experiments in Cairo, which lies roughly two minutes north of 30°, he might have been able to test for 30°, but not for 20° or 10°. This problem becomes even more acute if, as Alhacen suggests, we refine the determination by carrying it out at increments of 5° or less for the angle of incidence.

Were these physical, material, and technological limitations enough to prevent Alhacen from constructing the apparatus and its various appurtenances to the level of precision necessary for the experiments he describes in such detail, particularly those measuring the angles of refraction? Simon thinks so, basing his conclusion on Alhacen's failure to provide any actual tables of refraction. He reasons that, because of instrumental and observational limitations, the results Alhacen was able to obtain were no more precise than those listed by Ptolemy in book 5 of the *Optics*. Forced, therefore, to accept Ptolemy's tabulations *faute de mieux*, Alhacen presumably saw no point in repeating them in chapter 3. This conclusion is borne out at least to some extent by the fact that in his "Treatise on the Burning Sphere," which was written some time after the *De aspectibus*, Alhacen actually cites book 5 of Ptolemy's *Optics* and uses a couple of the tabulations for air to glass provided there rather than supply his own.[44]

From Simon's argument one can infer that Alhacen tried and failed to get adequately precise results from his experiments and therefore did not bother to record them, but there is reason to doubt that he even tried, at least not according to the protocol described in chapter 3. That protocol calls for measuring the angles of refraction for eight angles of incidence, starting at 10° and culminating with 80°, although Alhacen assures us that we can refine the measure by decreasing the intervals between successive angles of incidence from 10° to 5° or even less. We can even subdivide each degree into finer increments to increase the accuracy of the resulting observations. This protocol is to be followed for every test, including those for refraction from glass to air and glass to water in chapter 3, paragraph 3.22, p. 254.

But both these tests are limited by the critical angle, which is slightly less than 42° for glass to air and slightly more than 61° for glass to water. Consequently, in the test for glass to air there is no point in attempting the measurement for angles of incidence greater than 42° because the light will not pass through the refractive interface. Likewise, in the test for glass to water it is fruitless to try the measurement for angles greater than 61°. It is difficult to believe that, had he actually tried the measurements for glass to air and glass to water according to this protocol, Alhacen would have failed to notice these limitations. It is equally difficult to believe that, had he noticed them, he would have failed to alert the reader. Yet in fact he makes no mention of the critical angle anywhere in book 7 or, for that matter, anywhere else in the *De aspectibus*. This is puzzling because the critical angle plays a central role in the construction for proposition 9, pp. 291-292, where Alhacen acknowledges the existence of a limiting angle of incidence for refraction from a denser to a rarer medium.

Given all these limitations and lapses, can we conclude that Alhacen never conducted the experiments as described in chapters 2-4? Here we are faced with many of the same issues that arose when we evaluated Alhacen's experiments to confirm the equal-angles law of reflection in book 4.[45] There are several clear parallels between the two cases. In the experimental confirmation of the equal-angles law, for instance, the feasibility of the test involving the convex and concave spherical mirrors was cast in doubt because of the technical difficulty of forming and deploying such mirrors with the precision mandated by Alhacen. In the refraction experiments based on the glass quarter-sphere, the same doubts about technical feasibility arise in regard to both the quality of the glass and the precision of the quarter-sphere's formation.

Furthermore, the reflection experiments in book 4 are designed not to discover but to confirm a presupposed law, and the same holds for the qualitative experiments with refraction, which are designed to confirm the unrefracted passage of normal light and the direction of refraction in the passage of light into and out of a denser medium. Except for the unrefracted passage of normal light through water—which Alhacen could not have tested at the latitude of Cairo—and the same passage through the glass cubes and the glass quarter-sphere, these experiments do not require the instrumental precision that Alhacen specifies—except, of course, for confirming that refraction occurs in the plane of the middle circle inscribed on the inner wall of the rim. Where such precision is required, in the measurement of angles of refraction, Alhacen provides no results to indicate whether that precision was ever achieved and, if not, how closely it was approximated.

Finally, in both the reflection and refraction experiments the mandated exactitude is carried to a pointless extreme. In the reflection experiments that extreme is reached with the .36 mm adjustment in the position of the spherical mirrors. Not only does this adjustment involve boring a tiny hole in the concave mirror, but it makes no perceptible difference whatever in the final results.[46] In the refraction experiments the extreme is reached with the stipulation that the experiments with glass be conducted with a quarter-sphere, when for all practical purposes a semicylinder would do just as well.

As in the case of the reflection experiments in book 4, then, so in this one, there are solid grounds for supposing that, while Alhacen may have conducted experiments more or less similar to the ones described in chapters 2-4, he did not conduct, nor could he have conducted, the experiments as actually described because his apparatus could not have been constructed to the requisite tolerances. If, indeed, this is so, then it follows that the experiments as given are idealizations based on a mathematical exactitude unattainable in physical reality. They are, in short, elaborate and compellingly constructed thought-experiments.

Alhacen's Physical Model of Refraction: As explained earlier on pp. xxii-xxii, Alhacen's physical model of refraction is based on conceiving of light as composed of tiny, hard spheres moving in straight lines at great speed and interacting with various media according to their compactness or density. The more compact or dense the medium, the more it resists penetration by light. Full resistance results either in absorption and "scattering," when the surface of the medium is rough, or in total reflection, when it is perfectly smooth.[47] Any medium that allows light to penetrate its surface is therefore transparent, and its relative transparency is a function of its relative density. The resistance due to the medium's density affects the motion of the light at the point of entry, causing it to decrease or increase depending upon whether it passes from a rarer to a denser or from a denser to a rarer medium. This change is instantaneous and proportionate to the density differential between the two media. The medium's density also affects the motion of the light after refraction, so that it continues to move more slowly in a denser medium or more swiftly in a rarer one, presumably because it is more or less resisted in its passage.

In an absolute sense, the quantity, or intensity, of light's motion seems to be a function of its speed, irrespective of direction. Hence, light moves through any given medium at a constant speed determined by the medium's density. For lack of a better term, this can be called its "natural" speed. When light encounters a medium of different density, however, its intensity becomes a function of both speed and the angle at which it strikes the medium's surface. The closer to the perpendicular the impingement, the greater its intensity and, therefore, the more easily and forcefully the light traverses the refracting surface. The easiest and most forceful path is along the orthogonal itself because in that case the only resistance the medium poses to the incoming light is along the vertical. Alhacen uses the examples of an iron ball thrown straight at a thin wooden plank and a perpendicular sword-stroke to exemplify this point.

On the other hand, when the light strikes the medium's surface at a slant, the medium resists its entry along both the vertical and horizontal. For instance, let the medium above interface DRC in the top diagram of figure 24, p. 420, be rarer than the medium below it, and let the light from A radiate straight to R at whatever speed is appropriate to the medium through which it passes. When it encounters refractive interface DRC at R, its passage along the vertical will be resisted at that point, and the light will lose speed in proportion to that resistance. It will therefore continue along path RB less swiftly than it moved along path AR.

Kinematically speaking, the light will have traveled distance AR in the rarer medium during some given time period, struck the surface of the denser medium at R, been immediately slowed down by the vertical resistance posed at that point, and then continued straight through the denser medium at a commensurately slower speed along shorter distance RB in the same time period. Dynamically, the slowing effect of the medium's vertical resistance will occur instantaneously at R so that, when the light enters the denser medium, it will continue through it at the "natural" speed appropriate to it. Thus, if the slowing effect of the vertical resistance is measured by EC, the resulting speed along RB will be AR − EC.

Now let the light pass into the denser medium at a slant, as in the lower diagram of figure 24, and let it travel along AR at the same speed as before. According to Alhacen, its overall motion toward the refractive interface—Alhacen refers to this as its "composite" motion (*motus compositus*)—can be divided into vertical component AC and horizontal component AN. Its downward motion along AC will thus be impeded by the medium according to the vertical resistance along EC, and its horizontal motion will be impeded according to the resistance along FN. If it were to lose speed according to both resistances and were to continue straight through the denser medium, then its resulting path RG′ would be shorter than RB, so its speed would be less than the appropriate "natural" speed along RB. In order, therefore, that its appropriate "natural" speed be maintained in the denser medium after refraction, the light is shunted by the horizontal resistance along FN toward normal RB so as to follow easier, more forceful path RG. As a result, its absolute intensity along RG, which is a function of speed alone, will be the same as its absolute intensity along RB, but its relative intensity along RG will be less than its relative intensity along RB because of RG's obliquity. Thus, if the light along RG strikes the physical surface represented by the horizontal black line, it will have a less intense effect on it than RB would have on the same surface.

No doubt the above analysis is overdetermined in its quantitative precision, but it brings the intrinsic difficulties of the model into stark relief.[48] The most obvious difficulty involves the relationship between transparency and refractivity. Throughout his analysis in book 7, Alhacen imputes refraction to the passage of light at a slant from one transparent medium to another of different transparency, the difference lying in the relative degree of transparency. Air is more transparent than water, which is more transparent than glass, so light refracts toward the normal to a given extent when it passes from air to water and to a greater extent when it passes at the same angle from air to glass. The resulting change in path is commensurate to the difference in the light's "natural" intensity in the two media, that intensity being less in the less transparent medium. This correlation makes sense at a superficial level. Objects are seen more clearly through clear air than through water because their luminous color appears more intense in air than it does in water. But surely we can imagine water murky enough and glass clear enough that the water is less transparent than the glass. In that case Alhacen's model would have the light refract toward the normal when it passes from the glass into the water, but this seems counterintuitive.

Then there is the correlation between refractivity and density. Again, this correlation makes sense at a superficial level. A given volume of air clearly weighs less than a given volume of water, which weighs less than the same volume of glass. Thus, if density is a function of how concentrated or compact the material substrate is, glass is obviously denser than water and water denser than air. Moreover, the relative density of each is clearly related to physical permeability; it is far easier to throw a stone through air than through water and easier yet to throw it through water than through glass. But if permeability is a function of density, then why are water and glass also highly reflective when, according to Alhacen, reflective surfaces are such because they are too dense and resistive to allow light to penetrate? Problematic as well is the correlation between density and opacity because many opaque substances, such as wood, are less dense that transparent ones, such as water or glass.

Furthermore, the correlation between density and light's "natural" speed through a given transparent medium is problematic. If, as Alhacen claims, that speed is a function of the medium's density and the impedance due to it, then, being continuously impeded, the light should lose speed continuously as it traverses the medium. This problem comes to the fore in Alhacen's insistence that, once it enters the atmospheric shell, light continues straight through, despite the continuously increasing density of the fire and air it traverses on its way to the Earth's surface (see p. xlii). But if that is the case, then light passing obliquely toward the Earth's surface ought to lose speed and intensity as it approaches that surface and, in the process, be shunted toward the normal along a continuously easier and more forceful path in order to conserve its intensity insofar as possible.

The analogy between light and a swiftly moving particle complicates matters further. Earlier, in book 4, paragraph 3.97, p. 320, Alhacen adverts to a quantum of "least light" (*lux minima*), which is the smallest spot of effective light on any luminous surface. "Effective" is the operative word here because, as Alhacen observes, if this quantum "is divided, neither part of it will constitute [actual sensible] light; rather, both will be [effectively] extinguished and will [therefore] not be visible" (Smith, *Alhacen on the Principles*, p. 320). The size of the quantum thus seems to be a function of visibility; a quantum of least light for a brightly luminous body, such as the sun, will be smaller than that for the less brightly shining moon. Its size will also vary with distance; the farther the luminous source lies from the eye, the larger the minimally visible spot on it must be. Implicit in this notion of *lux minima* is that light loses intensity with distance because its effect is dissipated as it spreads out during propagation, so the intensity of light is a function of its concentration on whatever opaque or dense surface it encounters. Light emanating from any spot of *lux minima* can thus be imagined to form a cone with its vertex at that spot and its base defined by that surface. The closer the surface is to the source, the more light it will intercept and, consequently, the more it will be visibly or visually affected.

If, on the other hand, light is thought to consist of particles of *lux minima* shot off at unimaginable speed from luminous sources, then its loss of intensity with distance must be due to a commensurate loss of speed as it travels through the transparent medium. That loss, in turn, must be due to the constant resistance the medium poses to the light's passage. But what about the loss of intensity due to dispersion over space? If light is thought to consist of discrete particles shot off omnidirectionally from luminous surfaces in successive spherical fronts, then in the aggregate the particles comprising those fronts should have a diminishing visible or visual effect as they diverge while moving outward from the source. If so, then the decrease in light intensity with distance from the source will be doubly determined by loss of speed in combination with dispersion.

Then there is the issue of why some light sources are more intense than others. If, on the one hand, the particles constituting *lux minima* are the same size for all luminous sources, the intensity of those sources must be a function either of the speed at which the particles are shot off from a given point on those sources or of the number of particles shot off from that point in any given time. If, on the other, the speed of light and the rate of its emanation are constant, the intensity of the source must be function of the size of the particles shot off from any point on its surface. Whichever option is chosen, it follows that light must be infinitely variable according to its source-specific physical, kinetic, and dynamic properties, and that contravenes Alhacen's effort throughout the *De aspectibus* to show that light is and acts the same regardless of source or type.

Most of the problems discussed to this point arise because of Alhacen's failure to specify what he means by density in the first place. If, for instance, it is due to the "thickness" (*spissitudo*) of the material substrate, as Alhacen suggests in paragraph 6.17, then presumably it involves the consistency of that substrate; the more concentrated it is, the thicker (*spissior*) it is.[49] Conversely, the less concentrated it is, the thinner (*subtilior*) or rarer (*rarior*) it is. In line with this interpretation is Alhacen's equation of opacity with "solidity" (*soliditas*), which seems to trace opacity to a fundamental physical characteristic of the material substrate.[50] On the other hand, Alhacen contends that adulterants, such as haze and smoke, can render air thick (*spissus*) and less transparent.[51] Although this thickening causes the air to become more opaque, the change from more to less transparent has no bearing on the elemental composition or structure of the air itself.

In addition, if the density or thickness of the medium is taken as a function of the concentration or compactness of its material substrate, it is unclear how such concentration is to be understood. If, for instance, the substance of any transparent medium, whether air, water, or glass, is perfectly continuous, then it is difficult to see how one such substance can be more concentrated than the other because each is already as concentrated as possible by virtue of its continuity.[52] On the other hand, if these substances differ in concentration according to how tightly or loosely packed their material substrate, then they must be composed of discrete particles, which are subject to condensation or rarefaction. In that case, the refractivity of the transparent medium will be a function of its physical structure rather than its elemental composition, and its permeability will be a function of how easily the incoming light can push through its constituent particles. According to such a structural analysis, the combined reflectivity and refractivity of glass might be explained according to the intermittent porosity of its otherwise smooth, compact surface.[53] Although this terminological confusion about *densitas, grossities, spissitudo, soliditas, subtilitas, and raritas* in the Latin text is not necessarily reflected in the Arabic text, it nonetheless bespeaks an underlying conceptual confusion about the various connotations and implications of density and rarity that is common to both texts.[54]

Finally, the dynamics of Alhacen's refraction analysis raise difficulties. To start with, that analysis is based on supposing that refraction occurs right at the point of entry into a denser or rarer medium. That being so, the forces of resistance along the vertical and horizontal must be exerted instantaneously, and, as a result, the change in motion must also be instantaneous. Not only is such an instantaneous change of motion philosophically problematic, but it has no counterpart in the physical world, where all change in motion, no matter how abrupt, involves some transitional stage of deceleration or acceleration. In addition, the horizontal and vertical resistances seem to act in fundamentally different ways. The vertical resistance pushes directly against the light to slow it down; the horizontal resistance pushes obliquely against it to shunt it toward the easier, more forceful path that allows it to conserve its natural speed through the medium. In a sense, then, the vertical resistance slows the light while its horizontal counterpart speeds it up.[55]

The manifold shortcomings of Alhacen's physical model of refraction as an explanatory device are evident from this critique, but for all its flaws it should not be counted an abject failure. The very effort on Alhacen's part to construct such a model according to a relatively clear and simple, if ultimately incoherent, causal structure represents a significant step in creativity beyond the vague allusions offered in Ptolemy's *Optics*.[56] More important, though, in attempting to explain refraction physically, even mechanically, Alhacen set the analysis of refraction on a path that took important turnings during the seventeenth century, as various Western thinkers from Descartes to Newton attempted to refine and, at times, redefine the model.

Alhacen's Governing Rules of Refraction: At the end of chapter 3, Alhacen gives a set of general rules or laws that govern the relationship between angles of incidence and refraction. Numbering seven according to the arrangement on pp. 259-260, these rules are supposedly based on the results Alhacen obtained from the experiments designed to measure the angles of refraction when light passes from air to water, air to glass, glass to air, and glass to water at various angles of incidence. Whether he supposed that these rules apply universally to all possible transparent media, or whether he meant to restrict them to the three media actually tested is an open question, but all indications are that he took them in the unrestricted sense on the assumption that air, water, and glass are adequately representative of all transparent media.

For the most part, Alhacen's governing rules are unexceptionable as articulated in the Latin text. Three rules, however, turn out to be problematic, two of them because in Latin they do not properly reflect the Arabic versions from which they are derived. One of these is rule 4, which in its Latin form claims correctly (if tritely) that, when light is refracted from a rarer to a denser medium, the angle of refraction/deviation is invariably less than the angle of incidence—i.e., $r/d < i$. In the Arabic text, however, this rule states that, when light is refracted from a rarer to a denser medium, the angle of refraction/deviation is always less than *one-half* the angle of incidence—i.e., $r/d <$ .5i.[57] Likewise, the Latin version of rule 5 states literally that, when light is refracted from a denser to a rarer medium, the angle of refraction/deviation is equal to one-half the sum of the angle of incidence and the angle of refraction/deviation—i.e., $r/d = .5(i + r/d)$.[58] In Arabic the rule is that in refraction from a denser to a rarer medium the angle of refraction/deviation is *less than* one-half the sum of the two angles—i.e., $r/d < .5(i + r/d)$.[59]

Appropriately restated according to the Arabic text, though, neither rule is universally valid. Let us assume that the glass available to Alhacen had an index of refraction of at least 1.5 and that the index of refraction for air is virtually 1. That being the case, when light passes from air into glass at any angle of incidence i between 0° and around 82.5°, the angle of refraction/deviation r/d will be less than .5i, just as rule 4 claims. When i increases to slightly under 83°, however, r/d reaches equality with .5i, and from that point to the limit of $i = 90°$, r/d actually outstrips .5i. If, for example, $i = 83°$, then $r = 41.43°$, leaving $r/d = 41.57°$, which is greater than 41.5°—i.e., one-half 83°. Conversely, when light passes from glass into air according to the conditions for rule 5, then as i increases from 0°, r/d approaches equality with .5$(i + r/d)$ until i is just short of 41.4°. At that point i and .5$(i + r/d)$ reach equality, and from there to the limit posed by the critical angle of roughly 41.8°, r/d outstrips $.5(i + r/d)$. For instance, if $i = 41.5°$, then $r = 83.68°$, leaving $r/d = 42.18°$, which is greater than $.5(41.5° + 42.18°)$—i.e., 41.74°.

Bear in mind, however, that Alhacen's protocol calls for measuring r/d according to increments of 10° for i, so if he followed that protocol, $i = 80°$ would have been the last angle tested for refraction from air to glass and $i = 40°$ the last angle tested for refraction from glass to air. In the first case, $i = 80°$ yields $r/d = 39°$, which is less than one-half i, and in the second case, $i = 40°$ yields $r/d = 34.6°$, which is less than one-half $(i + r/d)$. In both cases, then, the results would have been consonant with the respective rules, so Alhacen would not have encountered the anomalies just discussed, which occur at the very extreme of the range for i, and he would have therefore had no reason to doubt the universality of the resulting rules.[60]

Such is not the case for the third problematic rule, i.e., rule 1, at the end of which Alhacen claims that in all instances of refraction, the difference between any two angles of refraction/deviation will always be less than the difference between their respective angles of incidence. In short, if $i > i'$, and if r/d is the angle of refraction/deviation for i and r/d' the angle of refraction deviation for i', then $i - i' > r/d - r/d'$. This rule is clearly contravened in refraction from glass to air when the angles of incidence lie toward the upper limit posed by the critical angle of 41.8°. For instance, when $i = 40°$, then $r = 74.6°$, leaving $r/d = 34.6°$, whereas when $i' = 30°$, then $r = 48.6°$, leaving $r/d' = 18.6°$. Accordingly, $i - i'$ (40° – 30°) = 10°, whereas $r/d - r/d'$ (34.6° – 18.6°) = 16°. Not only is $i - i' = 10°$ less than $r/d - r/d' = 16°$, but the discrepancy between the two values is so blatant that Alhacen could hardly have missed it had he actually conducted the experiment to measure refraction from glass to water. But if he did notice the discrepancy, then surely he would have dispensed with rule 1 or at least revised it to fit the anomaly. That he did not suggests strongly that he never conducted the experiment.

Furthermore, on the basis of the principle of reciprocity, Alhacen could easily have inferred that law 1 is not universally valid from Ptolemy's tabulations of refraction from air to glass in book 5 of the *Optics*. Take, for instance, the last two values of those tabulations, according to which $r = 42°$ when $i = 80°$, and $r' = 38.5°$ when $i' = 70°$. If we reverse the direction of radiation and interchange the angles, then it follows by the principle of reciprocity that $r = 80°$ when $i = 42°$, and $r' = 70°$ when $i' = 38.5°$. Translating the values of r according to Alhacen's measure of r/d, we get $r/d = 38°$ when $i = 42°$, and $r/d' = 31.5°$ when $i' = 38.5°$. Hence, $i - i' = 42° - 38.5° = 3.5°$, whereas $r/d - r/d' = 38° - 31.5° = 6.5°$, which clearly contravenes rule 1. Moreover, if we follow the same procedure with Ptolemy's values of refraction for $i = 60°$, $i = 50°$, and $i = 40°$, we find that $r/d - r/d'$ outstrips $i - i'$ until we reach $i = 40°$, when the two values reach equality.[61] Given Alhacen's familiarity with Ptolemy's tabulations for air to glass, and given his understanding of the principle of reciprocity, it is astonishing that he failed to see the potential in those tabulations for invalidating rule 1.

Even more astonishing is that Alhacen may well have had the means not only to avoid the erroneous conclusions drawn in rules 1, 4 and 5, but also to reduce the entire set of seven rules to one. These means, according to Roshdi Rashed, were available in a treatise entitled *On Burning Instruments* written between 983 and 985 by Abū Saʿd al-ʿAlāʾ ibn Sahl. As its title indicates, this work analyzes various optical devices designed to bring incoming light to such a concentration that it will ignite tinder. More to the point, the devices Ibn Sahl deals with are designed to bring incoming light to true focus at a single point without any aberration along the axis.[62] The first such device treated by Ibn Sahl is the paraboloidal mirror, whose study falls within a long tradition of analysis that includes Archimedes, Diocles, and Anthemius of Tralles.[63] Ibn Sahl took a significant step beyond the mainstream of that tradition by including an analysis of the focal property of ellipsoidal mirrors, but what sets him definitively apart from his Greek and Arabic predecessors is his analysis of plano-convex and bi-convex hyperboloidal lenses as focusing devices.[64]

The crucial case is the plano-convex lens. Let STBPO to the right in the upper diagram of figure 26, p. 422, be such a lens, whose plane face is SOP and upon whose convex face hyperbola SBP lies, and let it be suspended in air. Let XT parallel to axis ABO of the lens be a ray of light entering the plane face along the normal. Let it strike interface SBP at point T and be refracted along TA to point A on the axis, and let rays XT and TA, axis ABO, and refractive interface STBP all lie in the same plane. It is therefore to be proven that all incoming rays parallel to XT that strike interface SBP will be refracted to point A.[65]

Ibn Sahl lays the foundation for his demonstration as follows. Let FCG in the lower diagram of figure 26 be a plane interface between air below and glass above. Let DC be a ray of light striking the interface at angle of incidence DCQ, let CH be its continuation through the glass, and let CE be the ray along which it is refracted at angle RCE. Drop normal GEH to the refractive interface, and let it intersect CE and CH at E and H. Cut off segment CI on CH equal to CE, and bisect IH at J. Now, returning to the upper diagram of figure 26, take line AB, and divide it at K such that AK:AB = CI:CJ. Since CI = CE by construction, it follows that AK:AB = CE:CJ. Add segment BL to AB such that BL = BK, and cut off segment BM = AK. Consequently, since BK = BL by construction, and since and IJ = JH by construction, it follows that AK:AB+BL = CE:CJ+JH, so AK:AL = CE:CH. Drop BN perpendicular to AL, and let its length be such that rectangle BN,BM = 4 rectangle ML,LB. Then, taking BN as the latus rectum and AL as the axis, form the appropriate hyperbolic section SBP (by Apollonius, *Conics*, II.54), and produce its opposite branch S'MP'. Accordingly, L will be the focus for branch SBP and A the focus for branch S'MP' (by Apollonius, *Conics*, III.51).

Let ray XT strike interface SBP at point T, and draw line TA. Cut TA at U such that AU = AK, and draw line LU, extending it to intersect the continuation of XT at Y. Since triangles AUL and TUY are similar, TU:TY = AU: AL. But AU = AK by construction, so TU:TY = AK:AL. But we have already established that AK:AL = CE:CH, so TU:TY = CE:CH. Therefore, TU is the refracted ray for XT, and it is continuous with TA, so the light entering the lens along XT refracts to the focus A. The same holds for any other parallel ray, such as X'T' in figure 26a, p. 423; when its refracted ray T'A is cut at U' such that AU' = AK, and when line LU' is extended through that point to intersect the continuation of X'T' at Y', T'U':T'Y' = AK:AL = CE:CH. Thus, the refraction is determined throughout by the ratio CE:CH.[66]

Two crucial points follow from this analysis, according to Rashed. The first is that CE:CH is an expression of the relative densities of the two media, so it is constant. Taken as a quotient in reverse order, CH/CE is therefore the index of refraction for glass as we currently understand it. The second point, which is a concomitant of the first, is that CH:CE = sine *i*:sine *r* and, therefore, that sine *i*:sine *r* is also constant. This, of course, is Snell's law as it is commonly formulated today.[67] It bears noting that Ibn Sahl never articulates the law explicitly in the form of, say, "in all refractions CH:CE is constant, so the ratio of the sines of the angles of incidence and refraction is constant." Nevertheless, that law is clearly embedded in the demonstration.

That Alhacen was familiar with Ibn Sahl's *Treatise on Burning Instruments* can be inferred, Rashed assures us, by the similarity between Ibn Sahl's treatment of paraboloidal burning mirrors in that treatise and Alhacen's later treatment of the same mirrors in his *Treatise on the Parabolic Burning Mirror*.[68] Nor was this the only work of Ibn Sahl's known to Alhacen; there is no question that at some time in his career he had read Ibn Sahl's short treatise *A Proof that the Celestial Sphere is not Perfectly Transparent* because he cites it in his *Discourse on Light*.[69] But if Alhacen had in fact read the "Treatise on Burning Instruments," why did he fail to exploit the constancy of CH:CE in his analysis of refraction, particularly when formulating the governing rules at the end of chapter 3? Why not, in fact, condense all seven rules into that one governing relationship? All he needed to do, after all, was establish one trustworthy set of angle-pairs for refraction from air to water and air to glass, and he could extrapolate from that set to any angle-pair whatever for the given medium.[70]

Rashed suggests that Alhacen failed to exploit the constant relationship between CH and CE because he failed to understand it properly and thus failed to see its applicability to his own analysis. This failure, Rashed explains, stemmed from Alhacen's emphasis, perhaps overemphasis, on experiment and quantitative results in the analysis of refraction as a physical phenomenon. The resulting fixation on experiment and discrete physical instances of refraction, which harks back to Ptolemy, prevented Alhacen from seeing or appreciating the implications of Ibn Sahl's governing ratio.[71]

While this may be the case, other possibilities suggest themselves. For instance, Alhacen may not have come into possession of Ibn Sahl's *Treatise* until after he composed book 7. The one work in which Alhacen's use of Ibn Sahl is uncontested, the *Discourse on Light*, was definitely written after the *De aspectibus*, and the source used in that case was Ibn Sahl's piece on the transparency of the celestial sphere, not the *Treatise*. The other work in which Alhacen's use of Ibn Sahl can be inferred—this time with Ibn Sahl's *Treatise* as the source—is in Alhacen's study of parabolic burning mirrors, but whether this study was written before or after the *De aspectibus* has yet to be determined. It is also possible that Alhacen consulted Ibn Sahl's *Treatise* before writing book 7 but only read the section on burning mirrors without advancing to the analysis of burning lenses. Or perhaps he had read and fully understood Ibn Sahl's analysis but remained skeptical about the ratio itself as a true, physical principle. Of all these explanations, the ones appealing to Alhacen's not having properly understood or appreciated Ibn Sahl's analysis strike me as least likely because he was too gifted a mathematician to have missed the point of that analysis. Contrary to Rashed, then, I see no reason why Alhacen could not have incorporated Ibn Sahl's rule into his analysis.

The Problem of Peripheral Vision: Alhacen's explanation of visual perception in book 2 is based on accommodating the cone of radiation (*piramis radialis*) to the anatomical and physiological structure of the eye, particularly its anterior portion. Illustrated in figure 27, p. 425, that portion consists of the transparent cornea FF', the transparent lens enclosed by anterior surface GG' and posterior surface HH', the albugineous humor between the cornea and lens, the glacial humor filling the lens itself, and the vitreous humor extending from the lens's posterior surface to the optic nerve at the back of the eye. The eye as a whole forms a perfect globe centered on C, which is the center of curvature for both the outer and inner surfaces of the cornea and the anterior surface of the lens. These components form a succession of refractive interfaces according to their relative density or transparency. The cornea is denser than the air outside but rarer than the albugineous humor, the albugineous humor rarer than the glacial humor occupying the lens, and the glacial humor rarer than the vitreous humor beyond the lens.[72] Hence, light entering the eye encounters four refractive interfaces in the following order: the anterior surface of the cornea, the posterior surface of the cornea, the anterior surface of the lens, and the posterior surface of the lens.

Suppose that some luminous object with points A, B, and D on its surface faces the eye. The luminous color from each of those points will radiate to the anterior surface of the cornea in the form of a cone with its base on the cornea and its vertex at the point itself. Within each such cone there will be one, and only one, ray that strikes the cornea's surface orthogonally. This ray—designated by Alhacen as a "radial line" (*linea radialis*)—will pass straight toward center of curvature C through all the refractive interfaces concentric with it. AA'C, BB'C, and DD'C are such radial lines. When the luminous color passing along radial lines AC and DC meets refractive interface HH' between the glacial humor and the denser vitreous humor, it will strike it obliquely and therefore be refracted toward the normal along rays HA" and H'D", eventually to reach the opening in the hollow optic nerve at the back of the eye. The only ray not refracted at that interface is axial ray BB'C.

Furthermore, every point between A and D on the object's surface will radiate its luminous color in the form of a cone based on the cornea. For each such point of radiation there will be a unique radial line extending toward C and refracted toward the normal at interface HH' so as to reach the opening of the hollow optic nerve. The radial lines from all the points on object-surface AD will therefore form a cone of radiation with its base on the object and its vertex at C. Constituting the center of sight, this is the viewpoint from which everything in the field of vision is seen and visually judged. The continuation of the radial lines after refraction at HH' will also form a cone with one base on HH' but truncated by the circle at the opening in the optic nerve.

Now let the dotted lines originating at points A and D on the object surface be rays that strike the anterior surface of the cornea obliquely. The luminous color passing along them will be refracted toward the normal on entering the cornea's body and will reach its posterior surface at a slant. There it will be refracted toward the normal when it enters the denser albugineous humor and will eventually reach point B' on the anterior surface of the lens. Striking that point obliquely, it will be refracted yet again toward the normal along B'X and B'Z to bypass point C. The same holds for all the luminous color reaching the corneal surface at a slant; it will be refracted at all four interfaces so as to bypass point C and thus fall outside the cone of radiation. Given its anatomical structure and its optical properties, then, the eye is expressly designed to select out the radiation reaching it orthogonally because only that radiation is properly channeled by refraction into the hollow optic nerve.

The optical selectivity of the lens is complemented by its sensitive selectivity, which is due to the constant flow of visual spirit passing from the brain through the hollow optic nerve to its anterior surface. Appropriately animated by that spirit, this surface is able to feel the impinging luminous color at every point, but its exercise of this capacity is not indiscriminate. On the contrary, the lens only accepts impressions made along the orthogonal because those are the most forceful. The rest it ignores or fails to feel because they are less forceful on account of the obliquity of the lines along which they are made. This sensitive selection is crucial because it allows the lens to make visual sense of the chaos of luminous color reaching it at every point by selecting out only the orthogonals and thereby abstracting a pointillist representation of the original object that is in virtually perfect correspondence with it. That representation is then passed radially in proper point-by-point order through the lens to the opening in the optic nerve after refraction at the lens's posterior surface.

Furthermore, since everything within the field of vision defined by the cone of radiation is perceived and judged according to the viewpoint at center of sight C, no luminous color reaching the lens at a slant can be visually perceived because, after being internally refracted in the eye, it will follow a path that bypasses this viewpoint. Overall, then, the anatomical and physiological structure of the eye, as well as the optical properties of its component parts, is perfectly suited to the geometrical and dynamic characteristics of light and color impinging on the lens within the cone of radiation. The resulting model of visual perception is not only logical and mathematically simple, but it also supports the more or less intuitive notion that vision is based on faithful and faithfully replicated depictions of external objects.

Alhacen has nothing to say in book 2 about the cone of radiation's amplitude, leaving the reader to infer (falsely) that it is capacious enough to encompass the entire field of vision. Not until chapter 6 of book 7 does he put that inference to rest by pointing out that the cone's amplitude is restricted by the size of the pupil, which is quite small in relation to the radius of the eye. In short, the field of vision defined by the base of this cone falls far short of the normal field of view, which Alhacen eventually extends to 180° in front of the two eyes.[73] Hence, anything seen outside the cone of radiation must be seen by means of rays that reach the corneal surface obliquely, and such rays must strike the anterior surface of the lens at a slant after refraction at the cornea's posterior surface. In order to be seen, however, the luminous color arriving along such rays must make a sensible impression on the lens.

To concede this point, as Alhacen was forced to do in accounting for peripheral vision, is to undermine the original model of visual perception to its foundations. No longer is the optical selection of visual representations by the lens confined to orthogonal rays; nor, for that matter, is the sensitive selection so confined. But if the lens is to abstract a coherent, point-by-point impression of the represented object, then both kinds of selectivity must be limited to the orthogonals. Otherwise, the lens will suffer a chaos of infinitely overlapping impressions at each point on its surface. In order to account for the perception of oblique radiation, moreover, Alhacen had to fall back upon virtual lines of radiation (*linee radiales transumptive*) that exist only in the imagination, not in physical reality. If it is to be affected by each such line of radiation according to its relative intensity, though, the lens must be sensitive to every possible impingement from the very weakest and glancing to the very strongest and most direct.

The visual faculty must have an infinitely fine power of discrimination if it is to distinguish among the infinitude of such impingements at each point on its surface. It must also be able to determine the direction of impingement if it is to distinguish among impingements of equal intensity striking the same point on the lens's surface but coming from different points in the field of vision. Only then can it tell that the light reaching point B' from points A and D in figure 27, belongs to different points of radiation in the field of vision. Not only that, but it must be able to tie particular impingements to particular points in the field of vision. To do this it must somehow know the law of refraction and apply it inerrantly to each and every ray of luminous color refracted through each and every point on the anterior and posterior surfaces of cornea to each and every point on the lens. Otherwise, there is no way for it to tell that the impressions at B' and D' belong to D but not to E, or that the impression at E' belongs to E and not to D. And all these determinations and calculations must occur simultaneously at each instant of perception!

As a result, Alhacen's effort to explain peripheral vision places an extraordinarily heavy burden on the visual faculty, forcing it to make an infinitude of infinitely complex judgments and calculations at every instant of seeing. It also exempts vision from the physical laws of optics. Almost everything that is seen is seen by means of oblique rays that are refracted at the anterior surface of the lens so as to bypass the center of sight, which is the cardinal viewpoint for all visual judgment. But according to Alhacen, everything seen aslant through a refractive interface is seen along the refracted ray. To make matters worse, everything we see outside the cone of radiation is seen by means of imaginary rays converging at the center of sight, and almost everything we see lies outside that cone. For the most part, then, visual perception is based on the illusion that we are seeing things directly according to determinate optical principles, when in fact we are not.[74]

Proposition 17 and Radiation through Glass Spheres: At the end of the discussion of proposition 17 on pp. xxxix-xli above, we noted that Alhacen's analysis of image formation in a transparent sphere has implications for lens theory We also noted that in the Latin version the analysis is misrepresented in the diagram accompanying the proposition and that this misrepresentation also has implications. The source of the misrepresentation seems to lie in a faulty interpretation of proposition 9 in chapter 5. As we saw in our brief account of that proposition on pp. xxxiii-xxxiv, center of sight A in figure 28, p. 426, is assumed to lie on axis AGZD, facing convex surface GE behind which lies a denser medium, presumably glass, that extends indefinitely beyond D. Point B on the axis radiates its light to point E on the refractive interface, where it forms the largest possible angle of incidence BEZ that will allow the light to refract into the air to point A outside the sphere. Angle of refraction HEA must therefore be less than 90° by some infinitesimal amount. It follows, then, that there will be some other point B' on normal BD that radiates its light to some point E' on arc EG at an angle of incidence less than BEZ such that the resulting angle of refraction is commensurately less than 90° and the light is diverted to A. Or, to put it in somewhat different terms, for refracted ray E'A there is an appropriate incident ray that can be traced back to some point B' on the axis to form an angle of incidence less than BEZ. If there is one such point, there must be a multitude, each radiating light to an associated point on arc GE from which it will refract to A.

Since E is the limiting point on the axis for light to refract to A, all the rest of the points on the axis whose light will reach A from arc GE must lie on one or the other side of B. In short, they must lie either on BD or on the rest of the axis beyond B. Unfortunately, Alhacen does not specify which side, and the diagram given in the manuscripts is no help because it shows only B and its incident and refracted rays BE and EA, so it is up to the reader to figure out which is the right alternative. Evidently the translator or the draftsman who composed the original diagram for the Latin version of the text chose the alternative pictured in figure 28a, p. 427, assuming incorrectly that the second point B' lies on BD, the resulting incident ray yielding angle of incidence B'E'Z < BEZ, as mandated by the proposition. Accordingly, all the remaining points whose light will reach A through arc GE must lie on BD, and B will be the outermost point on line BD whose light can refract to A.

Turning to figure 29, p. 428, we can see how this interpretation dictated the way in which the analysis in proposition 17 would be represented diagrammatically in the Latin version. In that version L and H are taken as two points on the axis that will radiate directly to T and G if the medium within which they lie is the same as the medium filling the glass sphere. T must either be the limiting point of refraction for the light from L, or it must lie between that limiting point and B, as mandated by the analysis in proposi-

tion 9 of chapter 5. At T and G the light from L and H will be refracted in
the air outside the convex surface of the sphere along TA and GA, so if we
reverse the direction of radiation, with A as a source, the light from it will
refract toward the normal at T and G to arrive at L and H along rays TZL and
GMH. Then, if we assume that the glass sphere is completely suspended in
air, the light refracted toward the normal along TZ and GM in the glass will
refract away from the normal when it enters the air at Z and M to follow
rays ZO and MK. Reversing the direction of radiation once more, we see
that the light from K and O will follow KM and OZ to arc MZ on the refractive interface, refract toward the normal along MG and ZT to arc GT, and
refract again away from the normal along GA and TA. By the principle of
reciprocity, moreover, the initial angles of incidence ATX and AGY will be
equal to the final angles of refraction FZO and CMK, respectively. Thus, the
entire line segment KO will be seen by A on arc GT, and after a full rotation
about axis AL, that arc will form a ring on the sphere's surface. Since the
chord forming the cross-section of arc TG > object-line KO, the image will
be magnified, as Alhacen claims, and it will be distorted in shape according
to the convexity of the surface on which it appears to lie. All told, then, the
faulty diagram in the Latin text appears to support the analysis in proposition 17 and the conclusions drawn from it.

Now let us choose the second alternative in proposition 9 and assume
(correctly) that point B' lies on the other side of B, as depicted in figure
29a, p. 429, where rays BE and B'E' intersect before reaching arc GE on the
refractive interface. If we redraw the diagram accompanying proposition
17 according to this understanding—which is in fact how the diagram appears in the original Arabic version of the theorem—then both the tenor and
implications of the analysis will change dramatically, even though its word-
ing and logic remain precisely the same.[75] So let us again choose points L
and H on the axis, as represented in figure 29b, p. 430, and let H lie beyond
rather than in front of L. Hence, if the entire space behind the convex surface of the sphere is filled with glass, the light from nearer point L will pass
straight through Z to T and will be refracted in the air along TA, and by the
same token, the light from farther point H will pass straight through M to
G and will be refracted in the air along GA. Rays LT and HT will therefore
intersect before reaching arc GT.

If, as before, we reverse the direction of radiation, with A as a radiative
source, the light from it passing along AT and AG will refract at T and G so
as to reach L and H along TZL and GMH, respectively. Then if we assume
that the sphere is suspended in air, the light passing along refracted ray TZ
will refract along ZO to reach point O on the axis, whereas the light passing
along ray GM will refract along ZK to reach point K on the axis. Finally,
reversing the direction of radiation yet again, we see that the light from O

follows OZ to the refractive interface, is then refracted along ZT, and is refracted once more along TA. Likewise, the light from K follows KM to the
refractive interface, is refracted along MG, and is then refracted along GA.
The form of all the points on line KO will thus be refracted to A from arc
GT, and if the entire figure is rotated about axis AH, that arc will form a ring
on the surface of the sphere facing A. This ring will constitute not only the
locus of refraction for all the points on KO but also KO's image as viewed
from A. The image will thus be larger than its object, it will be distorted
in shape according to the surface upon which it appears to lie, and it will
be inverted because all the rays from KO that produce it will have crossed
before reaching arc GT. Surprisingly enough, Alhacen never mentions this
latter point, critical though it is to a proper interpretation of the analysis.

As far as its implications for lens theory are concerned, proposition 17
takes on new meaning when A is treated not as a center of sight but as a
source of light radiation. In that case, as we just saw, the light incident on arc
BT at the greater angle—i.e., the light from A striking T at angle of incidence
ATY—intersects the axis at point O nearer to D than point K, where the light
striking at the smaller angle of incidence AGX reaches the axis. It therefore
follows that the smaller the angle of incidence becomes as the light from
A strikes points nearer B, the farther out from D it will reach the axis after
double refraction through the sphere. If we add the radiation to the limiting
point of refraction from proposition 9 of chapter 5, which is represented by
point X in figure 29c, p. 431, then it is evident that, when ray AX is refracted
into the sphere along XY, it will be refracted out of the sphere to point X' on
the axis. That point will lie above O and, moreover, will be the *innermost*
point on the axis at which the light from A will reach the axis after refraction
through the sphere. These things can all be inferred from the Arabic version
of the proposition on the basis of the correct figure accompanying it, and
it is obvious from that analysis that the light from A refracted through the
sphere undergoes spherical aberration.[76] In other words, if all possible rays
passing from A through the sphere are taken into account, their points of
intersection on the axis will have an inner boundary defined by the limiting
point of refraction on the surface of the sphere facing A, and no two rays
will have the same intersection point on that axis.

The implications of proposition 17 become even clearer in light of
Alhacen's analysis of the focal property of glass spheres in his *Treatise on the
Burning Sphere*, which was written sometime after the *De aspectibus*.[77] In the
fifth and culminating proposition of this work, Alhacen isolates the region
where parallel solar rays entering a glass sphere will be refracted to the
axis in a dense enough aggregation to cause combustion. His analysis can
be summarized as follows according to figure 30, p. 432. Let the large circle
in the diagram at the upper left represent a plane cutting the sphere of

glass centered on D, let ADC be an axis, and let line LDL' be perpendicular to that axis. Let a ray of sunlight parallel to axis ADC strike point B on the sphere at angle of incidence EBH = 50°, and let it be refracted at angle DBK = 30°, as given by Ptolemy in his tabulations for refraction from air to glass in book 5 of the *Optics* (see Smith, *Ptolemy's Theory*, p. 236). According to those same tabulations, if a parallel ray strikes point B_1 at i = 40°, the resulting angle of refraction will be 25°, so it follows that the refracted rays from both B and B_1 will intersect at point K on the other side of the sphere, leaving arc KC = 10°. BK will then refract on the other side of the sphere to point N on the axis, and B_1K to point N_1 beyond it. The same holds for their counterparts on the other side, which will intersect at K' and then be refracted, respectively, to N and N_1.

If a ray parallel to the axis strikes the sphere at point O above point B on arc BL, it will be refracted to some point J between points K and C on arc KC, and it will then be refracted to some point on the axis between N and C. Hence, every possible parallel ray striking the sphere between B and L will be refracted from arc KC to some point on the axis between N and C. Likewise, if a parallel ray strikes some point F between B_1 and A, it will be refracted to some point U between K and J, whence it will be refracted to some point beyond N_1. Consequently, all the rays that strike the sphere within arc FA = < 40° will be refracted from arc KC to some point beyond N_1.

Now, let us rotate the entire figure about ACD as an axis to produce the outer circle represented in the diagram to the lower right of figure 30 (this circle is described by L at the equator of the sphere) and the inner circle described by B_1 nearer to pole A of the sphere according to the size of arc B_1A, which = 40°. The area of the surface between the two circles, shaded in gray, will be larger than that of the inner circle (somewhat more than 1.4 times as large, in fact), so it will be exposed to more solar rays, and, as has been shown, those rays will be refracted between C and N_1 on the axis. Furthermore, it can be demonstrated that, no matter how close to axis ADC a parallel ray strikes the sphere, when it is refracted through the sphere and then into the air, the line of refraction in the air will be less than a radius of the sphere in length, so it will strike a point on that axis less than a radius of the sphere away from C. Let V be the last such point of intersection, so CV < radius CD of the sphere. Let us bisect CV at S. Thus, SC < one-half CD. It can be shown that CN < one-fifth CD, and it can also be shown (although Alhacen does not do so) that CN_1 is considerably less than one-half CD. Since, therefore, the majority of solar rays reaching the sphere—i.e., those striking arc LB_1 and even beyond toward F—are refracted to points on CS, which is less than half the radius of the sphere, it follows that the area of combustion will lie less than half a radius away from the sphere. It also follows that the outer limit of any parallel radiation through the glass sphere lies less than a radius away from the sphere's outer edge. In short, no ray entering the sphere parallel to the axis will intersect the axis beyond that point.

The connection between this analysis and that in proposition 17 becomes evident if we imagine source-point A in figure 29b to lie so far from the sphere that rays AT, AG, and AB become virtually parallel. Given this connection, we can infer that, wherever A lies in relation to the sphere, all of the rays reaching from it to arc TB will intersect behind the refractive interface and will therefore fall at different points on the axis. We can also infer that, no matter how far A lies from the sphere, there will be two limiting points on the axis beyond which light will not reach it, one on the side of D and the other beyond that point toward H. And we can infer that there is a particular sheaf of rays whose intersections on the axis cluster closely enough to form a focal area on it.

Unlike the *De aspectibus* and the *De speculis comburentibus*, Alhacen's *Treatise on the Burning Sphere* was never translated into Latin. Nor, as far as we know, was it available to Western scholars in any other language until Rashed's recent Arabic edition and French translation. Consequently, without the analysis of parallel radiation through spheres in that treatise to guide them, and relying on a faulty diagram to illustrate the point of proposition 17, Alhacen's Latin readers were bound to take the proposition at face value as an analysis of a particular case of image formation in refraction. Because of the way it was represented in the accompanying figure, moreover, they were bound to misunderstand the proposition and thereby miss its underlying import for the analysis of lenses. This misunderstanding was perpetuated by Alhacen's most authoritative derivative source, Witelo, who relied upon the same faulty diagram to illustrate his version of the proposition in book 10, proposition 43 of the *Perspectiva*, and Friedrich Risner enshrined it some three centuries later in his magisterial tandem edition of Alhacen's *De aspectibus* and Witelo's *Perspectiva*.[78] The resulting failure by later medieval and Renaissance optical thinkers to grasp the deeper import of proposition 17 is evident from the assumption common among them that all the rays from a luminous point facing a glass sphere are refracted through it so as to converge at a single point on the axis.[79] Had they apprehended proposition 17 and its implications properly, those thinkers—or at least the more astute among them—could not have failed to realize that such a mass convergence of refracted rays at a single point is impossible because of spherical aberration. And they might have been helped in that realization had Alhacen mentioned the inversion of KO's image at the end of proposition 17.

As far as we know—and evidently as far as he knew—Johannes Kepler was the first Western scholar to recognize the underlying significance of proposition 17 as transmitted either by Alhacen or by Witelo through Risner's

1572 tandem edition of the *De aspectibus* and the *Perspectiva*. That realization came to him in the course of the optical research that led to the publication of his *Ad Vitellionem Paralipomena* ("Emendations to Witelo") of 1604. As the title suggests, this work was based on Witelo's *Perspectiva*, which is so closely modeled after Alhacen's *De aspectibus* that it amounts to little more than a redaction. The specific context of Kepler's work on lenses is chapter 5 of the *Paralipomena*, where he undertook a close investigation of vision according to the optics of the crystalline lens. For that purpose he used the radiation of light through a water-filled sphere as a model.[80]

The gist of the resulting analysis is as follows, starting with figure 31, p. 433, where LA, MB, NC, and OD are parallel rays striking the surface of the water-filled sphere according to equal arcs—i.e., arc AB = arc BC = arc CD = 30°. D is thus the limiting point of refraction, which means that angle of incidence ODX must be infinitesimally smaller than 90° if the light is actually to enter the sphere rather than graze it along the tangent. Kepler demonstrates that, if all these rays, except axial ray LA, are refracted at arc AD and continue straight through the other side of the sphere without further refraction—which implies that the space beyond the sphere is filled with water—then the farther the incoming ray lies from the axis, the closer to G it will intersect the axis after refraction. Hence, MB will be refracted to E, which is farthest from G, NC will be refracted to intermediate point P, and OD will be refracted to Q, which is closest to G.

When we suspend the sphere in air, it follows that rays BB', CC' and DD' will be refracted away from the normal when they emerge from the sphere at points B', C', and D', respectively. The light entering the sphere along the outermost path OD will therefore intersect the axis at point Q' after double refraction through the sphere. That point is closer to G than is point P', where the refracted path followed by NC through the sphere intersects the axis, and point P' in turn is closer to G than is point E', where the refracted path followed by MB through the sphere intersects the axis. Clearly, then, the light will suffer spherical aberration after double refraction through the sphere.

The similarity between Kepler's approach here and Alhacen's approach in proposition 17 is obvious, the main difference between the two being that Kepler locates the luminous source point infinitely far from the sphere so that all the rays from it are effectively parallel. Both start by determining where the rays refracted at the front surface of the sphere will intersect the axis if the entire space behind that surface consists of the same medium. Kepler, moreover, includes the limiting refraction at point D in figure 30, which is the equivalent of the limiting refraction specified by Alhacen in proposition 9 of chapter 5, i.e., point E in figure 28. Both then complete the analysis by suspending the sphere in the rarer medium and then determining where the rays will intersect the axis after the second refraction.

Having established that light undergoes spherical aberration in its passage through the water-filled globe, Kepler goes on to show that the limiting points of intersection for parallel rays entering the sphere can be defined as shown in figure 32, p. 433. The inner limit I is where the tangent rays LG and L'G' would end up after double refraction through the sphere at points G and G' and H and H', but since rays LG and L'G' cannot enter the sphere, actual refraction through the sphere is limited to rays that strike the sphere above points G and G' on arc GDG'.[81] The outer limit of intersection lies slightly beyond point F, which is where rays KB and K'B' ultimately intersect the axis after striking the sphere at an angle of 10° (i.e., angle of incidence KBE = 10°). Arc BD is thus 10°, and all parallel rays that strike the sphere within that arc and its counterpart B'D on the other side will intersect in a cluster at a spot slightly beyond F. That spot, according to Kepler, lies "little more than the semidiameter of the globe, no farther" from point N.[82] The remaining parallel rays that strike points between B and G on arc BG will intersect the axis between I and F after refraction through the sphere, as will those parallel rays that strike points between B' and G' on arc B'G'.

Conversely, Kepler argues, if I is a source of radiation, none of its rays will be refracted through the sphere so as to intersect the axis beyond D because the outermost rays CG and C'G' in the sphere will refract along lines parallel to the axis, and any rays from I striking arc HH' inside points H and H', e.g., IC, will eventually emerge from the sphere along lines that diverge away from the axis on the other side. For the same reason no point of radiation between I and N will be refracted toward the axis on the other side of the sphere. On the other hand, if some point O between I and F is taken as the source of illumination, then all the rays from it that strike the sphere between point N and tangent-point X at the limit of its possible refraction will be diverted through the sphere to intersect the axis beyond D. If, however, F or or any of the points beyond it is taken as source of radiation, none of the rays extending from it to arc CC' will intersect the axis beyond D because they will all emerge from arc BDB' along lines parallel to or diverging from the axis. The remainder of the rays between FC and tangent FY, which is the limit of refraction, will be refracted through the sphere toward the axis beyond D. Therefore, as Kepler eventually sums it up, as the source point of radiation is drawn increasingly closer to the sphere from a point infinitely far away, the innermost and outermost limits of intersection on the axis beyond D will shift increasingly far both from point D and from each other.

Kepler then observes that if the luminous source lies infinitely far away, and if all the parallel rays from it that strike the sphere on arc MDG are taken into account, the resulting lines of refraction in the air on the other side of the sphere will intersect each other to form a sort of cone defined by all the intersection points. Thus, as illustrated in figure 33, p. 434, when they emerge into the air from the sphere, rays XY, AB, CD EF, GH, KL, MN, X′Y′, A′B′, C′D′ E′F′, G′H′, and K′L′ will intersect by respective pairs at points a and a′, b and b′, c and c′, d and d′, and e and e′, until the last pair converges at f. These points, Kepler concludes, will lie on conjoining hyperbolic sections.[83] The closer to vertex F of the cone formed by these intersection points the refracted radiation gets, the more densely clustered the rays become, until they converge most acutely at that vertex.

This mathematical analysis can be verified empirically, Kepler explains, if we take a glass flask with a spherical base, fill it with water, and, using the sun as an infinitely distant light source, allow parallel rays from it to pass through the base. Let XMX′ in figure 33 represent the edge of the sphere through which select rays are refracted at points X through X′. Holding a sheet of paper right against surface XMX′ of the sphere and keeping it perpendicular to axis MN at position 1, where it passes through intersection points a and a′, we will see a circle of light with a bright ring at its periphery projected onto the paper. As we move the paper downward along axis MN to position 2 through intersection points b and b′, we will see this circle and its bright peripheral ring diminish rapidly in size, and it will continue to diminish in size, but less rapidly, as we move the paper to position 3 through intersection points c and c′. We can thus conclude that the bright peripheral ring is produced at the intersections of adjacent rays, such as XY and AB, AB and CD, and CD and EF. At position 4, where the outermost rays XY and X′Y′ intersect on the axis, we will see a dim dot appear in the center of the circle, and as we draw the paper downward to position 5 through points d and d′, we will notice that the circle of light with its bright peripheral ring diminishes ever more slowly while the inner spot of light becomes ever brighter. Drawing the paper further downward to position 6, we will see the peripheral ring and central dot begin to coalesce until, finally, the paper reaches position 7, where the the two actually do coalesce to render the light as perfectly concentrated as possible. Aside from confirming the spherical aberration of light passing through the flask's base, this test also makes it possible to physically measure the distance between point M and point F of maximum concentration. That distance, of course, is the focal length of the sphere, which Kepler had already deduced mathematically to be little more than the radius of the sphere.[84]

So far Kepler has covered much the same ground as Alhacen did in his *Treatise on the Burning Sphere*, although he has followed a markedly different analytic tack, using a different refracting medium, and arrived at a more definitive result. But unlike Alhacen, who was only interested in the burning power of focused light, Kepler was interested in its capacity to form real images, or "pictures," on a screen. Accordingly, he imagined a luminous surface facing the sphere at a finite but significant distance and standing perpendicular to the sphere's axis. From each point on that surface an axis can be drawn through the sphere, and the light from that point will radiate through the sphere to a focus on the axis on the other side. If a screen is placed at the central focal point, an inverted image of the entire surface will be projected onto the screen. Because of the tangle of light rays reaching the screen, though, this image will be quite strong at the center and increasingly weak at the periphery. In order to clarify the image, Kepler suggests restricting the incoming light by placing an opaque tablet with a narrow aperture directly in front of the sphere. Illustrated in figure 34, p. 434, this arrangement has line KL on some object surface facing the sphere in front of which is opaque tablet DG with aperture EF, which allows a narrow sheaf of rays to pass from points I and H to the sphere. The sheaf from H will converge toward central focal point K and the sheaf from I toward focal point L to the right, so if a screen is placed at K perpendicular to axis HK, the images of I and H will be projected at K and L, but the image at L will be weaker than that at K because the rays reaching L have already intersected at N and M. Consequently, the clearest image of I will be in the region of M and N. This, Kepler concludes, is why the retina is curved rather than flat, so that it can capture sharply defined images of all points in the visual field at the intersections of converging rays, the sum total of those images forming a relatively sharp, inverted depiction of the visual field.

With this analysis of retinal imaging Kepler not only revolutionized visual theory by overturning Alhacen's model of lenticular image selection, but he also took the crucial first step in the development of modern lens theory. Moreover, he did so in a work whose very title, *Ad Vitellionem paralipomena*, is an open acknowledgement of Alhacen (through his proxy Witelo) as the primary source. The obvious question, then, is what role, if any, did that source play in the development of Kepler's theory of spherical lenses? The answer to that question can be addressed at three levels.

At the most general level, Kepler may have learned the fundamental principles governing refraction from Alhacen, but as we have seen, those principles had already been clearly articulated by Ptolemy in his *Optics*, which was a cardinal source for Alhacen. Kepler also may have adopted— in fact probably did adopt—Alhacen's model of light radiation based on the propagation of light in all directions from every point on the surface of any luminous object. Here Kepler's debt to Alhacen appears to be direct and unequivocal, insofar as this radiative model seems to have originated with Alhacen.[85]

At a more specific level, Kepler may have learned the technique of ray analysis from Alhacen, but, again, that technique harks back to the Greek visual-ray theorists, Ptolemy in particular. He may also have have gotten from Alhacen the technique of imagining the space beyond the sphere filled with the same medium as the sphere, tracing the radiation through it to the axis, then suspending the sphere in air and following the rays refracted out of it to the axis. What he clearly did not learn from Alhacen's analysis of refraction in book 7 of the *De aspectibus* was the methodology of theoretical or empirical ray-tracing because nothing like it is to be found there or anywhere else in the *De aspectibus.*

We have already encountered empirical ray-tracing in Kepler's test of sunlight passing through the spherical base of a water-filled glass flask, and we have seen how critical this test was to Kepler's understanding of how the water-filled sphere focuses light. Theoretical ray-tracing comes into play in the analysis illustrated in figure 31, where ray MB strikes the sphere's surface at an angle of 30°, NC at an angle of 60°, and ray OD at an angle of 90°. Drawing on Witelo's tabulations for refraction from air to water, which come directly from Ptolemy's tabulations in book 5 of the *Optics,* Kepler assumed the appropriate angle of refraction from those tabulations in order to calculate precisely where the refracted rays would strike the axis.[86] On that basis, Kepler not only confirmed the spherical aberration of light passing through the water-filled globe but also managed to locate the focal point of that globe fairly accurately. This procedure is reminiscent of Alhacen's approach in the *Treatise on the Burning Sphere,* but since Kepler had no access to that treatise, he could not have learned it there.

At the most specific level is the possible connection between book 7, chapter 7, proposition 17, of the *De aspectibus* and Kepler's account of spherical lenses in the *Paralipomena.* That there is such a connection seems to be borne out by the fact that, unlike his Perspectivist predecessors, Kepler managed to see the underlying implications of that proposition despite the misleading diagram that accompanies it in the Latin version. Likewise, the manifest similarities between Kepler's explicit analysis of spherical aberration and the implicit analysis of spherical aberration in proposition 17 suggest a link between the two.

There are, however, good reasons to doubt such a link, or at least to doubt that proposition 17 was instrumental in the development of Kepler's understanding of spherical lenses and their focal property. First, as we have already seen, the ostensible purpose of proposition 17 is to account for the formation of a particular image on a convex spherical surface, not to explain the radiation of light from a point source through the sphere. Granted, such radiation does enter into Alhacen's account, but he uses it to establish the path that light would follow to center of sight A from the endpoints of object-line KO. Second, the emphasis on image formation is strongly reinforced by the empirical verification based on the small ball of black wax viewed through a glass sphere that Alhacen suggests at the end of proposition 17. At best, then, the implications of that proposition for lens theory are fairly well masked, and at worst, when the proposition is based on the faulty diagram provided in the Latin version, they are buried almost beyond recognition. Third, Kepler did not have Alhacen's *Treatise on the Burning Sphere* to provide a context within which to interpret proposition 17 as an implicit analysis of spherical aberration.

Yet despite all these impediments, Kepler somehow understood the underlying import of the proposition. How? Was it simply a flash of insight that led him to see what none of his Latin predecessors had? Perhaps, but it is at least as likely that Kepler saw through the faulty diagram to the implications of proposition 17 because he already understood those implications from his own analysis based on theoretical and empirical ray-tracing. In other words, he was able to grasp the implications of proposition 17 because he knew what to expect.

Furthermore, even if Kepler *had* first learned of spherical aberration from proposition 17, that knowledge did not dictate the specific direction of his subsequent analysis. This becomes clear if we compare the conclusions he drew about the focusing of parallel light rays through spherical lenses with those of Alhacen in his *Treatise on the Burning Sphere*. Kepler, on the one hand, was able to determine the focal point of his lens at very nearly the radius of the sphere according to the outer limit of intersection, and, in addition, he was able to limit the radiation effectively brought to focus to a relatively narrow shaft of rays surrounding the axis of the sphere. Alhacen, on the other hand, could only define the area of focus vaguely to within less than half the radius of the sphere, and he did so by supposing that the pertinent radiation occurs toward the outer edge of the sphere rather than around the axis. Both men addressed the same phenomemon, both based their analysis on spherical aberration, yet both followed fundamentally different analytic paths to fundamentally different conclusions. Ironically enough, the analytic path Kepler followed to the more successful conclusion included an empirical component that was entirely missing in Alhacen's treatment of parallel radiation through a glass sphere.

As far as original methodological and theoretical elements are concerned, then, it appears that Alhacen's contribution to Kepler's analysis of spherical lenses was fairly limited. In fact, Kepler viewed much of Alhacen's treatment of refraction and its underlying principles in a negative rather than a positive light. For instance, he took Alhacen to task for his causal analysis of refraction on the basis of "easier passage," observing that in order to seek such passage the light must somehow be "endowed with mind [so that] it might of itself perform its own refraction."[87] Nor was he any better disposed toward Alhacen's supposition that the crystalline lens is so exquisitely sensitive that it can distinguish perpendicular impingements

from even the closest neighboring oblique ones.[88] And Alhacen adds insult
to injury, Kepler observes, by then allowing the lens to sense oblique rays,
thus opening sight to a complete confusion of impressions.[89] Perhaps most
significant, however, was Kepler's outright rejection of the cathetus rule of
image-location for curved reflective interfaces and all refractive interfaces,
a rule that was absolutely central to Alhacen's analysis of both reflection
and refraction.[90] As a source, therefore, Alhacen's account of refraction in
book 7 of the *De aspectibus* seems to have functioned more as a *contra quod*
than as a *de quo* for Kepler's lens theory.

Ibn Sahl and Alhacen: According to Roshdi Rashed, Ibn Sahl occupies a special
place in the history of optics for at least three reasons. First, his analysis of
parabolic and elliptical burning mirrors puts him in company with those
who kept a long-standing tradition that harks back at least to Archimedes
and Diocles. This tradition continued during late antiquity with Anthemius
of Tralles, the anonymous author of the Bobbio fragment, Didymus, and a
certain "Dtrums," and it passed into the Arabic ambit with al-Kindī, Qusṭā
ibn Lūqā, and Aḥmad Ibn ʿĪsā.[91] Second, as far as we know, Ibn Sahl was
the first to extend the analysis beyond burning mirrors to burning lenses,
thus inaugurating the study of dioptrics as an outgrowth of catoptrics.
This he did in his *Treatise on Burning Instruments*, which has already been
discussed at some length in the previous section. And third, Ibn Sahl appears to be the first among Arabic optical thinkers before Alhacen to have
studied Ptolemy's analysis of refraction in book 5 of the *Optics*. Hence, as
Rashed sees it, Ibn Sahl assumes a crucial intermediate position between
Ptolemy and Alhacen.
     Rashed, moreover, is convinced that Alhacen was thoroughly acquainted
with Ibn Sahl's *Treatise on Burning Instruments* when he wrote his *Treatise on
the Burning Sphere*. That conviction is based on four considerations. First is
geographical proximity. Ibn Sahl worked in Baghdad, Alhacen in Cairo, so
Alhacen would have had relatively easy access to Ibn Sahl's works through
the lines of communication between the two cities. Second, there is no
question whatever that Alhacen knew of Ibn Sahl's *A Proof that the Celestial
Sphere is not Perfectly Transparent*, because he cites it in his *Discourse on Light*,
so it is not improbable that he also knew of the *Treatise on Burning Instruments*. Third, Alhacen's analysis of paraboloidal mirrors in his *Treatise on
the Parabolic Burning Mirror* is so similar to Ibn Sahl's treatment of the same
mirrors in the *Treatise on Burning Instruments* that there is little doubt that the
one was based on the other. Fourth and finally, the very fact that Alhacen
pursued the study of spherical lenses and their focal property in his *Treatise
on the Burning Sphere* suggests that he was inspired to do so by having read
Ibn Sahl's study of hyperboloidal burning lenses, particularly the biconvex

ones. "When these points are taken into account," Rashed concludes, "it is not too much to suppose that Alhacen indeed read those parts of Ibn Sahl's 'Treatise' devoted to lenses and to refraction."[92]

Although there is no way of determining when Alhacen began to investigate spherical lenses and their focal property, Rashed has no doubt that this investigation was well underway by the time Alhacen composed book 7 of the *De aspectibus*, which was finished before the *Treatise on the Burning Sphere*. The proof, for Rashed, lies in Alhacen's analysis of spherical interfaces in chapters 5 and 7 of that book: specifically propositions 6-11 of chapter 5, and proposition 17 of chapter 7. Accordingly, in his *Géométrie et dioptrique,* he provides the Arabic texts and French translations of all these propositions before supplying the text and translation of Alhacen's *Treatise on the Burning Sphere* to mark not only the chronological priority of those propositions but also their priority in the evolution of Alhacen's thought about spherical lenses. Given, therefore, that the same analytic thread runs through these propositions and through the *Treatise on the Burning Sphere*, it follows that Alhacen knew of Ibn Sahl's treatment of lenses by the time he wrote book 7 of the *De aspectibus*. Ibn Sahl, in short, was probably a source for Alhacen's refraction analysis in that book.

Rashed's case for supposing that Alhacen had read Ibn Sahl's analysis of lenses in *On Burning Instruments* by the time he undertook his refraction analysis in book 7 of the *De aspectibus* is not unreasonable, but it is nonetheless shaky at several points. For a start, Alhacen's failure to incorporate Ibn Sahl's "sine law" into his refraction analysis is an issue. As we have seen, this failure is difficult to explain without assuming either that Alhacen failed to grasp the significance of that law or that he understood it but nonetheless dismissed it as inapplicable to his own analysis.[93] Furthermore, the notion of a focal point, which is central to Ibn Sahl's analysis of lenses, never enters into Alhacen's treatment of spherical interfaces in book 7, either explicitly or implicitly. And finally, with the exception of proposition 17 in chapter 7, Alhacen's analysis of spherical refractive interfaces involves refraction through a single interface, not through an entire sphere—and this includes his analysis of the magnification or diminution of objects seen through such interfaces, which Rashed does not include in his own study.[94] The reason for this limitation is clear. Alhacen's concern with spherical refractive interfaces in book 7 was motivated by the need to set the background for analyzing the apparent displacement or size distortion (i.e., magnification or diminution) of celestial bodies caused by atmospheric refraction, which occurs at the spherical interface between the celestial realm and the atmospheric shell. That, moreover, is why in all cases, except for proposition 9 of chapter 5 and proposition 17 of chapter 7, Alhacen's analysis of spherical refractive interfaces is based on having either the object point or the center of sight lie between the interface and the center of curvature of the sphere containing it—because the visible effects of atmospheric refraction are seen from points on the Earth's surface facing celestial bodies, not from the Earth's center or through the Earth. Within that context, the two exceptional propositions are truly anomalous because they actually do have implications for lens theory, and in that regard they stand pretty much isolated from the rest of Alhacen's refraction analysis.[95] Nevertheless, like the rest of that analysis, these propositions focus upon the way objects are seen through such interfaces, not the way light is radiated through them. To sum up, then, the case for supposing that Alhacen had read Ibn Sahl's analysis of burning lenses by the time he wrote book 7 is not implausible, but neither is it very compelling. Furthermore, even if we do concede this point, there is nothing definitive in Alhacen's refraction analysis in book 7 of the *De aspectibus* to tie it to Ibn Sahl's treatment of lenses in *On Burning Instruments*, which is effectively to deny Ibn Sahl any meaningful role as a source for that analysis.

4. *The Fate of Alhacen's Refraction Analysis in the Latin West*

The Early Perspectivists: Bacon, Witelo, and Pecham: That all three of these late-thirteenth-century authors based their analyses of refraction solely or at least primarily on Alhacen's analysis in book 7 of the *De aspectibus* should come as no surprise, given their reliance on his analysis of direct and reflected vision. Each in his own way, therefore, not only sanctioned but also reinforced various analytic conventions and conclusions proposed by Alhacen in the course of his account of refraction in book 7. Likewise, each in his own way sanctioned and reinforced most of the problematic aspects of that account. In some cases, in fact, they compounded these problems by oversimplifying or confusing Alhacen's analysis, or by adding to it in inappropriate ways.

Take, for example, Alhacen's explanation of the apparent magnification of objects seen in denser media through a plane interface. As discussed earlier (see pp. xxxviii-xxxix), that explanation depends upon two factors: the enlargement of the visual angle caused by the image's being raised in the denser medium and the weakening of the form due to refraction. This latter factor causes the visual faculty to perceive the image as if it lay farther away than it actually does by geometry. Hence, despite the image's being the same absolute size as its object, it appears magnified because it is perceived to lie farther away than it does according to its actual geometrical location. While Witelo cites both factors in his explanation, Bacon ignores the apparent distancing of the image by the weaking due to refraction, citing only the enlargement of the visual angle as the cause of magnification. Hence, his account represents an oversimplification. Pecham, for his part, cites both the enlargement of the visual angle and the perception of distance but concludes that the apparent magnification is due to the image's apparent proximity to the center of sight, "for a larger angle combined with an equal or smaller distance causes the object to be judged larger."[96] However, according to the size-distance invariance principle, which is central to Alhacen's analysis of size perception, the apparent nearness of the image should make it appear diminished rather than magnified. Pecham therefore confuses Alhacen's explanation of magnification.

Another case of oversimplification can be found in Bacon's and Pecham's recapitulation of Alhacen's account of peripheral vision and the visual effect of oblique radiation on the anterior surface of the lens. Bacon, for instance, admits that things can be seen by means of such radiation, albeit weakly, and that within the visual cone oblique radiation from any given luminous point on the viewed object reinforces the direct, orthogonal radiation emanating from it to provide a clearer impression than would be gotten by either radiation alone. He goes on, finally, to repeat Alhacen's claim that, even when blocked with a needle or straw, the eye will see what lies behind it in an unclear fashion.[97] He makes no effort, however, to explain this phenomenon geometrically, nor does he attempt to explain the visual faculty's ability to locate an object seen peripherally by means of virtual radial lines. Even sketchier than Bacon's account, Pecham's includes nothing about the eye's ability to see "through" a needle or straw.[98]

In their very brevity and generality these two accounts gloss over the problematic nature of Alhacen's account of peripheral vision, reducing the phenomenon from a true anomaly (in the Kuhnian sense) to an interesting side issue within the primary account of vision based on orthogonal rays within the cone of radiation. Witelo, as well, glosses over the problem. Admittedly more detailed than either Bacon's or Pecham's, but still considerably less so than Alhacen's, Witelo's discussion of vision by means of oblique radiation occurs not in book 10 of the *Perspectiva*, where he deals explicitly with refraction, but in book 3, where he deals with the eye's reception of light and color.[99] More or less buried in that context, Witelo's account of peripheral vision emphasizes the difference between vision within the cone of radiation and vision by means of oblique radiation from outside it according to the clarity or distinctness of the resulting visual impression. He therefore concludes that "distinct vision takes place only along perpendicular lines, . . . while indistinct vision takes place along non-perpendicular lines, and in this manner indistinct vision assists the distinct."[100] So far Witelo has said nothing at variance with Alhacen, but it becomes clear in the course of his analysis that, for all practical purposes, oblique radiation is irrelevant; virtually all of the radiation required for normal vision lies within the cone of radiation. To support this position, Witelo assumes that the vertex angle of the cone at the center of sight is so large (slightly less than 90°) that it provides an adequately broad field of view to account for what we see and how we see it under normal and natural conditions.[101] Peripheral vision thus becomes a sort of departure from the norm in much the same way as diplopia is a departure from the norm of binocular vision.

These are just a few examples of how the early Perspectivist authors drew upon and mediated Alhacen's analysis of refraction. Other examples abound. All three authors, for instance, repeat Alhacen's explanation of refraction on the basis of easier or stronger passage in the direction of the normal, although Witelo is the only one to follow Alhacen's vectorial analysis according to horizontal and vertical components.[102] All three deal with atmospheric refraction and its effect on celestial observation in terms of both the apparent position and size of celestial objects, and all three discuss the Moon Illusion and its cause in the misperception of distance at zenith and horizon.[103] Moreover, both Bacon and Witelo follow Alhacen's lead in claiming that celestial objects are sometimes magnified by vapor in the atmosphere.[104]

Yet, despite their basic adherence to Alhacen's analysis, all three extend it beyond the limits set in book 7 of the *De aspectibus*. For example, whereas Alhacen offers no quantitative results from his experiments to determine angles of refraction, Witelo provides a slightly (and oddly) modified set of values drawn directly from Ptolemy's tabulations for refraction from air to water, air to glass, and water to glass in book 5 of the *Optics*.[105] On that basis, he extrapolates a corresponding set of values for refraction from water to air, glass to air, and glass to water by applying the principle of reciprocity. In doing so, he runs roughshod over the issue of critical angle, adducing a sizeable number of refractions and resulting angles that are impossible.[106] Another departure from Alhacen's refraction analysis is the inclusion by both Witelo and Pecham of a discussion of the rainbow and its formation, a phenomenon not even mentioned in the *De aspectibus*.[107]

Among the remaining instances in which the early Perspectivist authors extend the analysis of refraction beyond the limits set by Alhacen, perhaps the most telling is found in the way all three authors explain the radiation of light from a luminous point through a glass sphere suspended in the air containing the luminous point. This, of course, is the situation implicit in proposition 17 of book 7 (see pp. xxxix-xli). Pecham and Bacon make no bones about it. All the radiation from that point will converge at a single point on the opposite side of the sphere, and if the luminous point lies on the surface of the sun, the light refracted to the point of convergence will be intense enough to cause burning.[108] Witelo's analysis is somewhat more oblique because it is posed hypothetically, but it at least implies that all radiation from a point in the air outside a glass sphere and passing through it will converge at a single point in the air on the opposite side.[109] Whatever the case, it is clear that none of the three authors understood the implications of proposition 17 for spherical aberration and, therefore, that they were not only grappling in the dark with the problem of spherical lenses but doing so ineptly.

The Later Perspectivist Tradition: From a modern standpoint the shortcomings of the early Perspectivist refraction analysis based on book 7 of the *De aspectibus* are manifest: an inadequate grasp of the quantitative relationship between angles of incidence and refraction, a deficient understanding of magnification, a problematic physical theory of refraction, and an incoherent theory of peripheral vision by means of rays refracted through the cornea. All, or at least most of these deficiencies can be traced to a failure on the part of the early Perspectivists to properly understand how light is refracted through spherical lenses and brought to approximate rather than to point focus. Still, the effort to analyze this phenomenon, however abortive, at least brought the problem of lenses into the open for subsequent analysis. The importance of this point cannot be overstated because it was in resolving the lens problem at the turn of the seventeenth century that Kepler laid the foundations of modern optics.

That it took some three centuries for the problem of lenses to be resolved is perplexing enough. Even more perplexing is that the problem itself seems to have been all but ignored by optical theorists until the mid-sixteenth century.[110] What makes this so perplexing is that it was precisely during the period between roughly 1300 and the mid-sixteenth century that the manufacture of eyeglasses became increasingly widespread and sophisticated. The earliest eyeglasses were relatively simple, consisting of biconvex or plano-convex lenses with a fairly large radius of curvature designed for the correction of presbyopia or farsightedness.[111] Even before the invention of eyeglasses in the very late thirteenth century, plano-convex lenses were fairly commonly used to aid in reading. Bacon mentions these devices in part III of his *Perspectiva*, pointing out that, in order to work properly according to optical principles, they must be less than hemispherical.[112] It was thus a fairly short conceptual and technological leap from lenses of this sort, with a short radius of curvature, to lenses of a longer radius of curvature suitable for eyeglasses.

By the mid-fifteenth century a fairly clear picture of the manufacture, diffusion, and use of eyeglasses emerges. First, there was a crucial change in the base of manufacture from monasteries and nunneries in the fourteenth century to urban centers in the fifteenth. Florence seems to have been predominant in that regard. Second, the system of manufacture changed dramatically, from sporadic and scattered local production to mass production under the control of craft guilds. Not only that, but the production itself was systematized according to key components. Glass blanks might be produced in Pisa, Venice, or Nuremberg and shipped to Florence for grinding and finishing. Likewise, frames might be fashioned elsewhere and then sent to Florence so that the finished lenses could be inserted into them and the final, assembled product distributed along established trade networks. And they were traded in surprisingly large numbers at surprisingly affordable prices. Third, and concomitantly, by no later than the mid-fifteenth century eyeglasses with concave lenses for the correction of myopia or nearsightedness were in production. Furthermore, both types of lens were fabricated more or less according to prescription. This is especially the case for convex lenses, which were graded by age level at five-year intervals from 30 to 70, each stage defined by a specific radius of curvature that shortened incrementally with every five-year increase in age. Concave lenses, on the other hand, were graded for only two stages of myopia: medium and distant vision of the young. All told, this system of manufacture and distribution had become so sophisticated and efficient by the end of the sixteenth century that, according to Vincent Ilardi, "eyeglasses were perhaps almost as common [in Western Europe] as desktop computers are in developed countries today."[113] Though perhaps somewhat hyperbolic, the analogy is apt because, like desktop computers, eyeglasses transformed the working habits of scholars, merchants, and fine craftsmen in fundamental, even revolutionary ways.

Given the prevalence of eyeglasses during the later Middle Ages and early Renaissance, it is highly unlikely that they passed unnoticed by optical theorists, some of whom must have actually worn them. Yet, as far as we know, the earliest attempt to make sense of the lenticular correction of faulty vision was undertaken by Francesco Maurolyco in the mid-sixteenth century. That no one before him addressed the issue is especially bewildering because Perspectivist refraction analysis would have been perfectly adequate to explain the correction of presbyopia by convex lenses on the basis of the apparent magnification of what is seen through them. Such an explanation would have been based on the principles to which Roger Bacon appealed in his account in *Perspectiva*, III, 2, 4, of the plano-convex lenses used by the elderly and those of weak eyesight to make reading and close work easier.[114]

Why the long silence on eyeglasses in particular and lenses in general? Several explanations have been proposed. The most far-fetched, at least on the face of it, is that of Vasco Ronchi, who suggests that, because refraction through lenses was understood by medieval thinkers to distort rather than correct vision, the study of eyeglasses and lenses was in essence prohibited.[115] Lindberg argues more plausibly that by the time eyeglasses appeared the study of mathematical optics had lost its creative impulse, so no one during the later Middle Ages was likely to tackle new issues. The study of optics, in short, had become ossified. He also suggests that the problem of lenses would have been ignored because it had no new theoretical implications. The Perspectivists already understood the principles of refraction through a single interface, and the study of refraction through a second interface would contribute nothing new to that understanding. Furthermore, Lindberg adds, the analysis of lenses would have been an exercise in practical application rather than theory, and, being focused on the latter, the later Perspectivists had little or no interest in the former.[116]

While Lindberg's rationale is both cogent and compelling, Ronchi's argument from conspiracy seems absurd. But it should not be summarily dismissed. The Perspectivists did in fact view refraction as a source of illusion not only because it causes objects to appear displaced from their actual location, but also because it causes them to appear distorted by magnification or diminution, as well as by image inversion. This point is crucial because it means that, rather than correct weak eyesight, eyeglass lenses actually distort it and, in the process, delude the viewer into thinking he sees things more clearly and correctly. Eyeglasses, in other words, cause an unnatural modification of natural vision. Furthermore, the Perspectivist theory of visual apprehension by the crystalline lens is not really optical; it is physiological. The burden of vison is thus borne by the visual spirit enlivening the lens, and, as we have seen, that burden becomes extraordinarily heavy when peripheral vision is taken into account. Not being an optical problem, therefore, weak eyesight is not properly subject to optical correction. On the contrary, weak eyesight is generally due to an inadequate infusion of visual spirit; the weaker the one, the weaker the other.[117] That, evidently, is why late-medieval and Renaissance physicians were skeptical about the use of eyeglasses because eyeglasses only address the symptom, not the root problem. The true corrective could only be found in medical treatments designed to improve the flow of visual spirit to the lens.[118] In a sense, therefore, Ronchi is right. During the later Middle Ages and early Renaissance certain cultural and intellectual factors did conspire to repress the study of lenses by making that study irrelevant.

It is easy to deprecate the later Perspectivists for their conservative and inflexible approach to optical theory and practice. Bear in mind, though, that taken as a whole, the Perspectivist theory of sight and light is remarkably coherent and, as such, has broad explanatory power. Still, it does have rough edges, one of which is its inability to account for the lenticular correction of myopia by concave lenses. Francesco Maurolyco's account of how eyeglasses work represents an attempt to smooth this edge out. Dated to 1554,[119] his explanation is based on two fundamental Perspectivist principles:

that the lens is sensitive and that it funnels the rays reaching its anterior surface into the hollow optic nerve in proper fashion. This latter point is key. If the rays emerging out of the lens at its concave posterior surface are refracted too steeply or not steeply enough, the entire sheaf of rays will tend toward convergence too soon or too late to enter the optic nerve in the appropriate way and, presumably, at the appropriate size.[120] The result in both cases will be indistinct vision.

The basic cause in both cases is an improper curvature of the lens's anterior surface. Under normal conditions, Maurolyco claims, this surface is curved in such a way that all the rays responsible for proper vision, except the axial ray, will reach it obliquely rather than along the perpendicular. Each object point in the visual field will therefore be seen along a ray that strikes the lens's anterior surface at a particular angle of incidence, starting with the axial ray, which impinges orthogonally at the surface's centerpoint, and increasing as the surrounding rays strike the lens ever farther from that point. Being direct, therefore, the axial ray yields the most distinct visual impression, whereas the surrounding rays yield increasingly indistinct impressions on account of their growing obliquity. That, in a nutshell, is why visual clarity decreases continually from the center to the edges of the visual field.

In myopia, the anterior surface of the lens is too sharply curved, so all the rays responsible for vision, except the axial one, impinge more obliquely than they should and are thus refracted more severely than they would be under normal conditions. The upshot is that they are refracted out of the lens more obliquely and steeply than they should be and are brought toward convergence too soon as they are funneled toward the optic nerve. Placing a concave lens in front of the eye corrects this problem by forcing the rays that reach the lens obliquely to impinge at a somewhat gentler angle of incidence and, therefore, to emerge from the lens at a correspondingly gentler angle of refraction. Presbyopia, on the other hand, is due to the lens's not being sharply curved enough. Hence, in not striking the lens's surface at a steep enough angle, the non-axial rays emerge at an inadequately steep angle of refraction and are brought toward convergence too late. By sharpening the angle of incidence of the oblique rays reaching the lens's anterior surface, a convex lens causes the corresponding rays that emerge from the lens to be more sharply refracted and thus to be brought toward convergence sooner.[121]

Maurolyco's account of the lenticular correction of presbyopia and myopia does not represent a frontal assault on Perspectivist visual analysis and its fundamental principles. True, in order to make his explanation work, he was forced to deny the primacy of orthogonal rays within the cone of radiation, but that primacy had already been compromised by the Perspectivists' inclusion of oblique radiation as a contributing factor both within and outside the cone of radiation. Maurolyco was therefore bending the rules of Perspectivist analysis somewhat in order to accommodate them to his explanation.

Several other optical phenomena came to the fore during the sixteenth century either because they were first noticed then or because, although long-recognized, they were looked at in new ways by scholars of the time. The *camera obscura* is one such phenomenon. Well before the sixteenth century this device had been used for the observation of solar and lunar eclipses,[122] and all the early Perspectivists had tried their hand at explaining why, when projected through a small, triangular hole to a screen placed well beyond on the other side, sunlight or moonlight is cast on the screen in the shape not of the hole but of the luminous object.[123] But all these explanations treated the phenomenon as a case of light- and shadow-casting, not of image formation. By the later sixteenth century, though, the capacity of a small aperture to project real, inverted images into a darkened room (hence, *camera obscura*) was well known, and Giambattista della Porta suggested improving the device by adding a convex spherical lens at the aperture and, in addition, positioning a concave spherical mirror beyond the aperture and directing it toward a screen in order to enhance and enliven the projected image.[124]

This modification of the camera obscura was no accident. By the early sixteenth century the optics of both convex spherical lenses and concave spherical mirrors were being seriously studied. In Theorem XXXI of the *Photismi*, for instance, Maurolyco demonstrates geometrically that rays from a single luminous point facing a concave spherical mirror at a finite distance will reflect in such a way as to intersect the axis at various points depending on their distance from the axial ray. The farther from that axial ray the incident ray strikes the mirror's surface, the closer to the surface the intersection of the reflected ray with the axis will be.[125] In short, light undergoes spherical aberration when reflected from concave spherical mirrors. More than a decade earlier, Leonardo da Vinci undertook a close geometrical study of reflection from concave spherical mirrors based on parallel incident rays. His drawings indicate that he, too, understood spherical aberration and knew that the farther from the axis the incident ray is, the closer to the mirror's surface its reflected ray will intersect the axis. He also concluded that the point of combustion will lie halfway on the axis between the mirrors' center of curvature and its surface.[126]

Both Maurolyco and Leonardo undertook these studies in order to establish both that and how concave spherical mirrors cause combustion at a particular point. They had little or no interest in the capacity of such mirrors to produce images, whether by the projection of real images or by the formation of virtual images.[127] By around 1560, however, the two capacities were analyzed in tandem by Ettore Ausonio in his brief *Theorica speculi concavi sphaerici*, and both were shown by him to be associated with the focal point, which he called the "point of inversion" (*punctum inversionis*). This point can be defined according to two stipulations: that the center of sight face the reflecting surface from beyond the mirror's center of curvature and that the object lie between the mirror's surface and the center of curvature. Hence, if the object is placed right next to the mirror's surface and then moved along the axis toward the center of curvature, it will reach a point where, just after its image disintegrates into a chaotic sprawl filling all or most of the reflecting surface, it reappears inverted. This is Ausonio's *punctum inversionis*, and it coincides with the mirror's focal point, both lying midway on the axis between the mirror's surface and its center of curvature.[128]

Although never published, Ausonio's *Theorica* circulated in manuscript form and was influential within Italian circles during the later sixteenth century, in part because of Ausonio's reputation as an expert, and in part because of a growing interest in the subject matter. Galileo, for instance, owned a copy of Ausonio's treatise, and Giambattista della Porta's familiarity with it is indicated by the opening of book XVII, chapter 4 of his *Magiae Naturalis*, where he describes a method for determining the *punctum inversionis* of any concave spherical mirror. This determination, he assures us, is necessary because "you will not get full use of such [a mirror] unless you first know the point of inversion."[129] From this quotation it is clear that Porta's primary concern with the *punctum inversionis* was practical. He was more interested in the sorts of optical effects that can be produced with concave spherical mirrors than in the theoretical principles underlying those effects.

As we have seen, one of those effects—the capacity of concave spherical mirrors to focus sunlight to a burning-point—was recognized and analyzed long before Porta's day. Another—the ability of such mirrors to magnify images seen in them—was also recognized before his day. But this latter ability had become of paramount interest in the later sixteenth century because several thinkers of a practical bent were convinced that concave spherical mirrors could be put to effective use for telescopy. This idea was far from new at the time. Tales of mirrors with extraordinary magnifying (and burning) power had been circulating from late Antiquity, and they were evidently accepted as credible by thinkers who were considered authoritative.[130] By the second half of the sixteenth century, however, improvements in the quality of mirrors and a better understanding of how they work, coupled with the proliferation of old and new accounts of their wondrous powers, inspired various thinkers to put the idea into actual practice.[131] Some concentrated on perfecting the mirrors themselves by lengthening their focal length while at the same time refining the curvature of the reflecting surface. Others combined lenses and mirrors in the hope of improving both magnification and image resolution. Neither expedient worked as well as hoped because of technical and optical limitations, but success always seemed to lurk around the corner.[132]

The inclusion of convex spherical lenses in this telescopic mix makes sense. After all, convex spherical lenses and concave spherical mirrors are optically analogous. Both focus sunlight to a point of combustion, both magnify images seen in or through them, both cause image inversion at a certain point, and both project real inverted images onto a screen. Why not, then, exploit the analogy and extend the analysis of the one to the other? This is apparently what Maurolyco had in mind in theorem XVIII of part 1 of his *Diaphaneon*, where he demonstrates that, "when two parallel rays are travelling through a transparent sphere at unequal distances from the center, the one which is more remote from the axis parallel to them will intersect the axis at a point nearer to the sphere than will the other."[133] The analogy between this proposition and Theorem XXXI of the *Photismi*, briefly discussed on p. xci, is clear. The latter shows that light reflecting from a concave spherical surface undergoes spherical aberration and, moreover, that the farther the ray from the center of curvature, the closer to the reflecting surface it intersects the axis.[134] The former shows precisely the same thing for light refracting out of the concave, inner surface of a glass sphere. Granted, Maurolyco's analysis of refraction out of the sphere is based on rays passing parallel to the axis within the sphere, whereas his analysis of reflection inside a sphere is based on radiation from a single point. But that point can easily be conceived to lie so far from the reflecting surface that the rays are virtually parallel to the axis.

In his *De Refractione Optices Parte: Libri Novem* (Naples, 1593) Giambattista della Porta takes the analogy between concave spherical mirrors and convex spherical lenses even further than Maurolyco.[135] He starts in book 2, proposition 1, pp. 36-41, with an analysis of concave spherical mirrors. This analysis can be summarized on the basis of figure 37, p. 436, which is a composite of several diagrams in Porta's text. Let arc MP represent a section of the reflecting surface, and let X be its center of curvature, through which axis XP passes. Mark off chord MP equal to the side of an inscribed hexagon, chord LP equal to the side of an inscribed octagon, IP equal to the side of an inscribed dodecagon, and HP equal to the side of an inscribed hexadecagon. Then draw lines ME, LD, IC, and HB parallel to the axis, and let them represent rays striking the mirror. Connect normal XM, bisect it, and drop a perpendicular from the point of bisection to the mirror. It will intersect the axis at point P, where the axis and the reflecting surface meet, and MP will be the reflected ray. The same method applied to normals XL, XI and XH will yield intersection points Q', R' and S' on the axis where reflected rays LQ', IR' and HS' meet it.[136] According to this construction, then, angle of incidence EMX = angle of reflection XMP, angle of incidence

DLX = angle of reflection XLQ', and so forth for the remaining points. Also, according to the construction, the angles of incidence and reflection are 60° at M, 45° at L, 30° at I, and 22.5° at H. A couple of important points flow from this construction. First, no matter how close to point P the incident ray may strike, its reflected ray will never intersect the axis beyond point F, which is the bisection point of radius XP of the mirror and constitutes the *punctum inversionis* or focal point. Second, the reflected ray is equal in length to the part of the axis between the center of curvature and the point of intersection, so LQ' = Q'X, IR' = R'X, etc.

With this analysis in mind, Porta turns in book 2, proposition 2, pp. 41-43, to the refraction of light through a glass sphere. Let VMP in figure 38, p. 436, represent half the sphere centered on X, through which axis VS passes. Let VM', VL', VI', and VH' be sides, respectively, of an inscribed hexagon, an inscribed octagon, an inscribed dodecagon, and an inscribed hexadecagon. Finally, let EM', DL', CI', and BH' be parallel rays striking the sphere's surface at angles of incidence, respectively, of 60°, 45°, 30°, and 22.5°. The refraction of light at those angles, Porta contends, will mirror the reflection of light at those angles within the concave spherical mirror. Accordingly, the light arriving along EM' will refract along M'P to intersect the axis at the sphere's surface. The light arriving along DL', CI', and BH' will in turn refract to points Q, R, and S on the axis, and those points will lie the same distance below P as their counterparts Q', R', and S' in reflection lie above point P.

Porta did not end with this geometrical analysis, though. He devised a simple method for confirming the theoretical results of that analysis by empirical ray-tracing. The requisite apparatus consists of four components: a cylindrical disk of polished glass around half a foot (*semipes*) in radius and somewhat less than a digit (*digitus*) thick, a semicylindrical glass disk of the same radius and thickness, a rectangular glass plate one foot long and half a foot wide of the same thickness but with a semicylindrical section cut out, and a strip of wood or copper sufficently long and wide to block sunlight from shining on the edge of the glass pieces when properly placed. The various pieces of glass are then attached to an iron plate, and the strip is attached above it so that, when it is posed toward the sun, the strip blocks the sun. Small holes are bored into the strip at specific intervals according to the parameters of the theoretical analysis described above. Figure 39, p. 437, illustrates the resulting arrangement for the glass disk, the strip being pierced at E, D, C, B, and A so that rays of sunlight can shine through those holes to points M, L, I, H, and V. We should then be able to see that ray EM is refracted to point P, that ray DL is refracted either to or toward point Q, and so on.

As described, this experiment is somewhat confusing, because all of the rays except MP should be refracted when they emerge into the air on the other side of the disk, but Porta's diagram does not show this, so it is unclear from the diagram itself whether Porta took lines MP, LQ, IR, and HS to be the actual refracted rays or the imaginary lines connecting the points of incidence and the points at which the light will eventually intersect the axis after passing through the disk. The subsequent experiments with the glass piece out of which a semicylindrical section is cut indicate that Porta meant those lines to represent the actual rays. For instance, in book 2, proposition 3, pp. 43-45, Porta describes the experiment illustrated in figure 40, p. 438, in which the rays strike the concave surface of the the glass along EM, DL, CI, and BH at angles of incidence, respectively, of 60°, 45°, 30°, and 22.5°. Circular sections of the same radius as the concave hollow of the glass piece are then placed tangent to the hollow's circumference at points M, L, I, and H. Each of those sections represents an equivalent section on the glass disk in figures 38 and 39. Accordingly, ray EM will refract along MP to intersect axis VP at point P, ray DL will refract along LQ to point Q on the extension of axis VP, and, the same holds for rays CI and BH, which will refract to points R and S, respectively.

This same analytic model underlies Porta's experimental accounts in book 2, propositions 4 and 5, pp. 45-47. Proposition 4 is devoted to the experiment illustrated in figure 41, p. 439. According to the analysis in proposition 2, if the space below the curve were filled with glass and the space above it with air, rays EM, DL, CI, and BH would refract to points P, Q, R, and S on axis AS. But since the rays are refracted out of glass into air, the rays are refracted symmetrically in the opposite direction along MP', LQ', IR', and HS'. Hence, angle P'MM' = angle M'MP, and so forth. In proposition 5 the light passes through the semicylindrical disk as illustrated in figure 42, p. 439. If the space below arc MH were filled with glass and the space above with air, the rays would refract along MP, LQ, IR, and HS. However, since the rays are passing from glass into air, their refracted paths will be symmetrical and opposite, along MP', LQ', IR', and HS' such that angle PMM' = angle M'MP', etc.

Porta's refraction model was clearly wrong at the quantitative level because it took the analogy between convex spherical lenses and concave spherical mirrors too literally and because it depended too heavily on geometrical a priorism.[137] Nevertheless, his effort to understand lenticular refraction was not fruitless. For one thing, Porta established beyond question, on both geometrical and empirical grounds, that light refracted through spherical interfaces undergoes spherical aberration. In this regard, he was plowing a furrow already turned by Maurolyco, but his analysis was considerably more sophisticated and, moreover, was published well before Maurolyco's. Equally important, Porta's analysis demonstrates that the rays brought to focus in a spherical lens consist of a fairly narrow beam of incident rays surrounding the sphere's axis. Perhaps even more important, Porta showed the potential value of empirical ray-tracing as a tool in both the qualitative and quantitative analysis of refraction. Granted, his empirical results were fairly far off the mark because his observations were so badly skewed by theoretical expectations, but the experimental technique he claims to have followed is both simple and feasible.[138]

Neither Maurolyco nor Porta made significant theoretical advances beyond their Perspectivist sources. They both accepted the crystalline lens as the visually sensitive component of the eye, they both followed the cathetus rule in their analyses of image location in mirrors and lenses, and they both failed to derive a satisfactory law governing the relationship between angles of incidence and refraction. Indeed, Maurolyco's assumption that the angles of incidence and deviation are constantly proportional represents a significant step backward, not only with respect to Perspectivist analysis, but also with respect to Ptolemy's.[139] Yet, for all their basic conservatism, these two authors were symptomatic of a fundamental change in optics during the sixteenth century. The overarching symptom was the emergence of a pragmatic approach to optics and a concomitant focus on phenomena that had been ignored by the Perspectivists and that turned out to defy adequate analysis according to Perspectivist principles. The correction of refractive visual disorders by convex and concave lenses was one such phenomenon. The projection of real images onto a screen by concave spherical mirrors and convex spherical lenses was another. And the application of mirrors and/or lenses to telescopic devices was yet another. In all these respects, opticians during the second half of the sixteenth century laid the groundwork for Kepler not only by isolating certain phenomena and associated problems but also by suggesting ways to interpret and resolve them.[140]

We should also bear in mind that it was during the sixteenth century that the publishing industry came into its own. Especially important within this context was the publication of Alhacen's *De aspectibus* and Witelo's *Perspectiva* by Friedrich Risner in his *Opticae Thesaurus* of 1572. Not only did Risner make these works available in an eminently readable font, but he also organized them systematically according to cross-references within and between the two texts. In addition, he went to great lengths to break the *De aspectibus* into propositional elements, adding his own enunciations to summarize the points made in them. He also provided numerous source citations at spots in the text where they were either missing or vague, and he imported his own commentary at key points to explain steps or conclusions that are not immediately obvious. Risner thus made both texts far more accessible than they had been in manuscript form, and he did so in three crucial ways: first, by making many more copies available; second, by standardizing both texts; and third, by restructuring them, particularly the *De aspectibus*, so that they could be more easily assimilated in piecemeal fashion, as reference texts rather than as continuous treatises. Based as it was on Risner's edition, Kepler's extensive critique of Alhacen's refraction analysis in the *Paralipomena* was doubtless informed in critical ways by how the texts of Alhacen's *De aspectibus* and Witelo's *Perspectiva* were presented in that edition.

5. *Putting Alhacen in His Proper Place*

Over the past few decades, a consensus has formed among historians of science that Alhacen's optical synthesis marked not just a turning point, but a revolutionary turning point in the history of optics. Gérard Simon, for example, locates this turning point in Alhacen's rejection of visual rays in favor of light rays because the adoption of the resulting intromissionist model of sight had revolutionary implications at all levels of analysis, from the physics and mathematics of sight to the physiology, psychology, and epistemology of visual perception.[141] For David Lindberg, as well, Alhacen's focus on light rays as the primary cause of vision is of cardinal importance, but of equal importance is his reliance on a punctiform model of light radiation from all spots on the surfaces of visible objects.[142] Roshdi Rashed, meantime, sees two features of Alhacen's visual theory as critically important in the break with ancient optics: the treatment of light as a subject of analysis independent of vision and the development of a new approach to mathematical physics.[143] Saleh Omar finds this approach manifested in the creation of a new experimental method based on carefully controlled observation using precise instrumentation.[144]

Not only are these scholars in general agreement that Alhacen's analysis of light and vision was innovative in crucial respects and thus marked a radical departure from ancient optics. They are also in general agreement that this analysis was instrumental in the development of modern optics. On the face of it there is nothing exceptionable about either of these positions. Alhacen's theory of light and the intromissionist visual model based on it did have significant ramifications in physics insofar as Alhacen made it possible to treat light as a subject of study in its own right apart from its visual manifestations. Alhacen's intromissionist model did call for a profound and broad revision of the physics, physiology, psychology, and epistemology of visual perception. Nor is there any question that Alhacen's approach to optics was more empirical than that of his Greek predecessors, Ptolemy in particular, or that his experiments were designed for different purposes and with more elaborate equipment than Ptolemy's. There is also no question that Alhacen's model of light and sight was instrumental in the development of modern optics. Lindberg has done a masterful job of showing how

influential and pervasive that model was among European opticians of the
Middle Ages and Renaissance up to the time of Kepler.

In great part because of the interpretive slant of these scholars, the
terms "revolutionary" and "modern" are now applied almost reflexively
to Alhacen. Thus, in a fairly recent review article, Hélène Bellosta speaks
confidently of "the development of [optics] from its Greek origins to the eve
of the *revolution* carried out by Ibn al-Haytham. . . , who, breaking definitively
with the theory of the 'visual ray' will make optics enter *modernity*."[145] A.
I. Sabra makes much the same point in a piece entitled "Ibn al-Haytham's
Revolutionary Project in Optics: The Achievement and the Obstacle."[146]
On the one hand, he argues in that article, Alhacen's effort to revise the
science of optics from top to bottom on the basis of intromissionism was
revolutionary in its import. On the other, he continues, Alhacen failed to
bring the revolution to fruition because he insisted on treating the lens as
a sensitive rather than a purely optical instrument. Implicit in this argument is that, were it not for this one theoretical and conceptual impediment,
Alhacen could have, indeed would have, brought optics to what Bellosta
refers to as modernity. In other words, Alhacen not only laid the foundations of modern optics, but he also erected most of the superstructure. It
was Kepler, presumably, who put the final touches on this superstructure
with his theory of retinal imaging.[147]

The underlying logic of this argument seems to be as follows. The restructuring of optical theory undertaken by Alhacen could have resulted
in a revolution if followed to its logical conclusion. While Alhacen failed
to carry it to that conclusion, Kepler did, so he brought Alhacen's potential
revolution to actuality. But what of the optical revolution that followed—the
one carried out by Galileo, Descartes, Huygens, and Newton? Either this
revolution represented an overturning of the revolution completed by Kepler,
or it was no revolution at all, just a tying up of loose ends left by Kepler.
Neither conclusion is plausible in light of what we know about the development of optics from Kepler to Newton. Kepler's work played a positive and
crucial role in that development, and the end result in Newton's theory of
light and color represented at least as radical a break with Alhacenian optics
as Alhacenian optics did with ancient visual-ray optics.

The problem with the current evaluation of Alhacen's optical work lies
in the facile use of "revolutionary" and "modern." One major source of that
problem is the recent historiography of medieval Arabic science. Let us go
back some seventy years to this statement by Henry Crew in the introduction to his translation of Maurolyco's Photismi:

> Alhazen's most important role, like that of many another Arabian
> scholar, appears to have been that of a translator and preserver.
> We are deeply indebted to them, as Professor Lynn Thorndike

puts it, for having "kept in cold storage" so much of ancient learn-
ing.[148]

To be sure, Crew was badly informed about medieval optics in general
and Alhacen in particular, but his somewhat dim view of medieval Arabic
science reflected the general understanding of his time, at least among non-
specialists.

No one today with even a superficial knowledge of medieval Arabic
mathematics, medicine, astronomy, or optics would concur with Crew that
Arabic science was static and uncreative. This sea-change in attitude has
evolved over the past few generations, as a cadre of eminent researchers,
Rashed and Sabra among them, has shown in detail how medieval Arabic
thinkers not only contributed to a range of scientific disciplines in innovative
and sophisticated ways but also took those disciplines well beyond the level
reached by such ancient luminaries as Galen, Ptolemy, Euclid, and Aristotle.
Part of the process has involved uncovering and publishing new texts in
order to establish the breadth and depth of interest in science among Arab
thinkers of the Middle Ages. Rashed has been particularly active in this
regard. Part of the process has involved a close, technical analysis of those
texts and a concomitant effort to achieve a synthesis. And part of it has
involved presenting the results in an accessible way to an educated public
with little or no real knowledge of the subject.

In many cases, Alhacen's included, a true appreciation of those results
and their implications rests on a firm grasp of the technical details of analysis underlying them. But in order to present those results and implications
in an accessible way, scholars have been forced to simplify their accounts
by glossing over what they take to be unnecessary details or by distilling
the necessary ones to what they think are their essentials. Unfortunately,
simplification can easily lead to oversimplification. Furthermore, in an
effort to provide a meaningful interpretive evaluation of those results for
non-specialists, scholars have been forced to sacrifice nuance for the sake of
simplicity, clarity, and effect. Here, too, simplification can lead to oversimplification. Add to this the fact that scholars have differing perspectives on core
significance, and it is no wonder that they sometimes lapse into hyperbolic
characterizations, such as "revolutionary," in order to highlight what they see
as important, innovative, or ingenious about a given thinker's achievement.
No wonder, as well, that they cast this achievement in a modern light so as
to emphasize both its significance and its positive, progressive nature. In
addition, we, as readers, are all too apt to infer such characterizations, even
when they are not explicitly given in the accounts themselves. Nowhere,
for instance, does Lindberg refer to Alhacen as revolutionary, yet if one is
predisposed to view Alhacen as such, he can easily find corroboration of
that view in Lindberg's account.

All of this, I suggest, has played a role in the reflexive application of "revolutionary" and "modern" to Alhacen. That tendency is exemplified in the way Alhacen's optical work is presented in popular accounts. Take the current Wikipedia article on "Alhazen."[149] In it we learn that Alhacen "is regarded as 'the father of optics.'" We also learn that he "proved the intromission theory of vision and refined it into its essentially modern form." In the process, the article continues, Alhacen "discovered a result similar to Snell's Law of sines" in refraction; "correctly hinted at the retina being involved" in the formation of visual images; argued that light "is made of particles traveling in straight lines"; and devised "the rudiments of what may be designated as a hypothetico-deductive procedure in scientific research." In short, he anticipated Kepler, Galileo, Snell, and Newton by a good six hundred years or more.

Then there is the recent popular biography by Bradley Steffens, *Ibn al-Haytham: First Scientist*,[150] whose title says it all: Alhacen invented modern science, and he did so, according to Steffens, by importing the modern experimental method and the attendant emphasis on empiricism into his optical analysis. Alhacen's modernity is echoed by Rosanna Gorini in a summary article entitled "Al-Haytham the Man of Experience, First Steps in the Science of Vision," where she points to the crater on the moon named after Alhacen (i.e., "Alhazen") as "proof of the relevance of his studies."[151] One wonders what sort of proof that is.

As absurd or problematic as these claims may appear to be at first glance, most of them are based on scholarly sources and represent either what those sources actually say or what the authors infer from them. By whom, precisely, is Alhacen regarded as "the father of optics?" One particular author, as it turns out, but the implication is that this view is relatively widespread and uncontroversial. Did Alhacen actually conceive of light as consisting of particles zipping through space in rectilinear trajectories? As we have seen, he treated light *as if* it consisted of such particles (with problematic consequences), but this treatment was based on an analogy that Alhacen never meant to be taken literally. Did Alhacen actually discover a result similar to the sine law of refraction? That he could not have done so is evident from the fact that he never compiled the quantitative observational data from which to derive any result, much less one similar to the sine law. Anyway, it is clear from earlier discussion in this introduction that Alhacen either did not know or failed to accept the sine law. Was Alhacen in fact following the hypothetico-deductive method in his experiments dealing with the equal-angles law of reflection and the rules governing refraction? It is difficult to imagine that he was, since all evidence suggests that he never conducted those experiments at all, or at least not as described. In addition, those experiments were designed not to test but to confirm hypotheses already accepted *a priori*.

Whether intentionally or not, these secondary accounts, as well as many of the primary scholarly studies on which they are based, convey the message that, as an optical theorist and practitioner, Alhacen had no genuine antecedents. If indeed he was the *first* scientist, as Steffens claims (meaning, presumably, the first modern scientist), then Alhacen holds a unique position in the history of optics. No one before him, including Ptolemy, was doing anything remotely similar. Likewise, if it was Alhacen who took the *first* steps in the science of vision, as Gorini implies, then it follows that Euclid, Ptolemy, and Al-Kindī were involved with something else. While dissociating Alhacen from the past, moreover, these accounts link him firmly to the future according to the modernity of his achievement and his anticipation of ideas and discoveries usually credited to later thinkers. Uprooted entirely from his past and thrust into the future, Alhacen thus emerges as lone genius, a thinker not of his time but of ours.

As far as the history of optics is concerned, this portrayal of Alhacen has some pernicious implications. For one thing, if Alhacen was entirely uprooted from the past, then it is pointless to bring that past into any account of his achievement, unless, of course, to mine it for all the errors he corrected. For another, to assume that Alhacen brought optics to the verge of modernity and that Kepler carried it over that verge is to concede that whatever occurred in the Latin West between the appearance of the *De aspectibus* and the advent of Kepler played no meaningful role in the development of modern optics—other, perhaps, than to keep Alhacen's optical analysis "in cold storage" for over three centuries. Here of course the tables are turned on Crew; instead of medieval Arabic science, it is medieval Latin science that is static and uncreative. And for yet another thing, the resulting "history" is regressive, a return to the anachronistic and discredited "great man" historiography of science that prevailed a century or more ago.

Granted, I have taken the implications of Alhacen's portrayal as "revolutionary" and "modern" to the extreme here, but that they can be taken to this extreme indicates just how questionable those terms can be. Properly speaking, revolutions entail a more or less abrupt toppling or overthrow of what went before and its replacement with something fundamentally different in kind. True, in certain respects Alhacen's optical synthesis fulfills both criteria. His rejection of visual rays in favor of light rays represents just such an overthrow, as does his analysis of distance perception on the basis of ambient clues. But as radical as they were, these changes formed only part of a complex and extensive revision of Ptolemaic visual theory that in the end yielded something not so much different in kind as in degree.

As we have seen throughout this critical edition of the *De aspectibus*, elements of Ptolemaic analysis are evident throughout Alhacen's optical synthesis. Ptolemy's visual cone persists both mathematically and analytically in Alhacen's cone of radiation. The rays within that cone still maintain a point-to-point correspondence between external objects and their visual representations. The center of sight at the vertex of that cone remains the cardinal reference point for visual analysis. Luminous color continues to be the proper object of vision. The "common sensibles"—albeit much expanded by Alhacen—still constitute the proper objects of perception. Vision is still understood to unfold in stages, starting with brute sensation and ending with intellectual judgment. Image formation in reflection and refraction is still defined by the cathetus rule, and the images themselves are still treated as psychic rather than physical entities. These are not loose ends dangling from a radically new core of physical light theory and waiting to be tied up by Kepler; they are systemic, integral parts of an essentially coherent whole.

Furthermore, the revolutionary consequences of revolutions are supposed to be fairly immediate. Take the so-called Copernican Revolution as an example. According to standard interpretation, what made that revolution revolutionary was Copernicus's having put the Earth in orbit about the sun, thereby overturning not only the old geocentric Aristotelian-Ptolemaic cosmology but also Ptolemy's cumbersome mathematical system of planetary and solar motion based on eccentrics and equants. But as far as we know, Aristarchus of Samos proposed the heliocentric system some eighteen centuries before Copernicus. Why not, then, call it the "Aristarchan Revolution" and concede that it took almost two milennia to come to fruition with Copernicus?[152]

Consider also that when we pass from the conceptual discussion in the first book of the *De revolutionibus* to the technical account of planetary motion in the subsequent books, we find Copernicus's model, with its compounding of circular motions, to be conceptually identical to, and no less cumbersome than, Ptolemy's. At the technical level, therefore, Copernicus's heliocentric model represents not an overturning but a revision and refinement of Ptolemy's geocentric one.[153] The real overthrow of Ptolemaic astronomy had to await Kepler and his rejection of circular deferents and epicycles in favor of single elliptical orbits for all the planets. Recognition of this point by N. R. Hanson nearly fifty years ago led him to credit the revolution to Kepler, granting Copernicus the distinction of having created a mere "disturbance."[154] Viewed in this light, Alhacen might well be seen as having created the disturbance that led to a Keplerian revolution in optics.

Apropos of this point, it bears noting that Alhacen's intromissionist model of sight by no means superseded the extramissionist alternative in the Latin West. On the contrary, both were viewed as viable theories throughout the Middle Ages and Renaissance. Roger Bacon, for instance, insisted that the visual act requires radiation from the eye in order to be completed, but he was not followed in this supposition by his Perspectivist colleagues Pecham or Witelo, who remained staunch intromissionists. Yet despite its Perspectivist imprimatur (and its obvious superiority by modern lights), intromissionism did not sweep the field. In the *Della pittura* of 1436 Leon Battista Alberti bases his system of artistic perspective on visual rays, although he seems to be aware of the intromissionist alternative.[155] Indeed, among Renaissance artists the visual-ray model was often used as an analytic device.[156] And the same holds for physicians; some adhered to an extramissionist account of sight, and most followed Ḥunayn ibn Isḥāq in locating the lens at the very center of the eye rather than toward the front as Perspectivist theory requires.[157]

A glance at the manuscript transmission of Euclid's *Optics* and *Catoptrics*, as well as Ptolemy's *Optics*, during the later Middle Ages and Renaissance reveals a continuing and presumably lively interest in the visual-ray theory. Twenty-five extant copies of Euclid's *Optics*, in one form or another, are datable to the fourteenth, fifteenth, and sixteenth centuries; thirty-four extant copies of his *Catoptrics* trace back to those centuries; and thirteen copies of Ptolemy's *Optics* are from the same period, two of them in fact dating from the seventeenth century.[158] As late as the end of the sixteenth century the debate over extramissionism and intromissionism still apparently raged. When composing his *De Refractione* of 1593, Porta felt compelled to devote a significant portion of book 4 to disproving the extramissionist theory, and ten years later Kepler lauded him for having finally put the "quarrel" between extramissionists and intromissionists to rest in favor of the latter.[159]

These examples should be adequate to make my argument crystal clear, if not entirely compelling: characterizing Alhacen and his optical work as revolutionary or modern is both simplistic and ahistorical, even antihistorical. Still, to deny Alhacen revolutionary status is not to deny him his place in the sun. Bound by the conceptual, technical, and analytic constraints of his time, Alhacen, like Plato's Demiurge, had to make the best of what he had at hand. And like that of Plato's Demiurge, Alhacen's best was exceptionally good. But not perfect. As we have seen, Alhacen's optical synthesis was logically vulnerable at certain points, and that in part is what makes his synthesis so important. For it was in exposing and exploiting those vulnerabilities that subsequent thinkers were finally able to bring the science of optics toward Bellosta's modernity.

So there is no doubt that Alhacen's work holds an important, perhaps even a pivotal, place in the history of optics. But "important" and "pivotal" cannot and should not be confused with "revolutionary." Nor can the difference in connotation here be obviated by characterizing Alhacen's achievement as almost revolutionary or not quite revolutionary, as Sabra seems to do. That, ironically enough, is to demean the achievement by implying that it was a failure, that it should have been revolutionary. By the same token, to ignore the distinction in connotation and confuse "important" or "pivotal" with "revolutionary" is to devalue the real historical significance of that achievement. The history of optics deserves a more nuanced approach than that. And so, I might add, does Alhacen.

NOTES

[1]Whenever Alhacen refers to "glass" without specification in book 7, he means clear glass rather than colored versions, such as cobalt glass.

[2]As will become clear later on in our discussion of Alhacen's physical explanation of refraction (see pp. lvii-lxii), Alhacen ties refractivity to the relative transparency of various media: the more transparent the given medium, the less refractive it is. He also ties refractivity to relative density, so that the denser the given medium, the more refractive it is. Given the implicit connection between relative density and relative transparency, therefore, Alhacen uses the two terms interchangeably when describing the refractive quality of media such as air, water, and glass.

[3]See notes 7 and 8 to paragraphs 7, 2.2-3, pp. 333-334, for a discussion of the measures used by Alhacen in constructing the apparatus.

[4]The circle on the water's surface will of course be distended into an ellipse according to the angle at which the light beam strikes the water: The greater the angle, the greater the ellipticity of the circle.

[5]Saleh Omar, *Ibn al-Haytham's Optics: A Study of the Origins of Experimental Science* (Minneapolis: Bibliotheca Islamica, 1977), 136-138.

[6]More specifically, at any point on Earth above or below the tropics, which lie at a latitude of 23.5° north or south of the equator, the sun never reaches true zenith. At the tropics it reaches zenith only once, at summer solstice for the northern tropic and at winter solstice for the southern one. At the equator it reaches zenith twice, at vernal and autumn equinox. At any other point between the tropics it also reaches zenith twice, once as the sun ascends to summer solstice, and once as it descends back down to winter solstice.

[7]Alhacen in fact does this later in chapter 4, pp. 265-270, where the point of the experiment is to show that an object placed on the normal dropped from the center of sight through the water's surface is seen along that normal where it actually lies.

[8]See note 43 to paragraph 7, 2.77, p. 347, for an explanation of Alhacen's distinction between essential and accidental light.

[9]Implicit in this analysis is that any angle greater than BEZ will either be the critical angle or be larger than the critical angle, in which case the light from B will be internally reflected at the refracted interface. Why Alhacen did not apply this understanding to his description of the refraction of light from glass to air is therefore something of a mystery (see pp. xxvi-xxvii). Since BEZ is the largest possible angle of incidence at which the light will pass through the interface, then HEA will be the largest possible resulting angle of refraction in the modern sense, and it must be less than 90° by some infinitesimal amount. This follows from the principle of reciprocity, according to which BEZ is the angle of refraction that

results if A is a point source of radiation and HE is an incoming ray forming angle of incidence AEH. If AEH = 90°, then ray HE will be tangent to the sphere and will thus not penetrate its surface, so it must be less than 90° by some amount, no matter how tiny.

[10]Alhacen has already claimed in book 5, paragraph 2.313, that in reflection, when the image is located at the center of sight, it will appear at the point of reflection on the mirror's surface, so he is merely extending that claim here to refractive interfaces; see A. Mark Smith, ed. and trans., *Alhacen on the Principles of Reflection*, Transactions of the American Philosophical Society, 96.3 (Philadelphia: American Philosophical Society, 2006), 448.

[11]See note 126 to paragraph 7, 5.72, p. 378.

[12]For Alhacen's detailed explanation of visual selection by the lens on the basis of the cone of radiation, see A. Mark Smith, ed. and trans., *Alhacen's Theory of Visual Perception*, Transactions of the American Philosophical Society, 91.5 (Philadelphia: American Philosophical Society, 2001), 355-376.

[13]Smith, *Alhacen's Theory*, 348-355.

[14]In taking such radial lines "transumptively," we are taking them as if they were real radial lines, even though they are not, in much the same way we can take light rays as trajectories for light particles, even though light is not actually particulate according to Alhacen's theory.

[15]For a brief explanation of Alhacen's account of variable visual acuity in binocular vision, see Smith, *Alhacen's Theory*, lxxiii-lxxvi.

[16]Alhacen's theory of size perception in fact depends upon this principle because it is based on a correlation of distance and visual angle so that anything subtending a given visual angle at a given distance will be perceived as of a given size and will be perceived to be of the same size when it subtends a different visual angle at a different given distance. Hence, as Alhacen puts it in book 2, paragraph 3.137, "when a visible object lies one cubit from the eye and is then moved farther away . . . until it lies two cubits from [the eye], there will be a significant difference in the two angles subtended by that object at the center of sight. Still, sight does not perceive the object lying two cubits away as any smaller than the object lying one cubit away" (Smith, *Alhacen's Theory*, 475).

[17]See book 2, chapter 3, paragraph 3.159, and book 3, chapter 7, paragraphs 7.250-7.252, in Smith, *Alhacen's Theory*, pp. 486 and 625.

[18]As pointed out in the reference provided in note 11 above, the largest possible image formed on arc EG in figure 14 will fall slightly short of point G, so it will not occupy the whole of the surface between E and G but will form a ring centered on G, a ring that occupies most of that surface.

[19]If, in fact, the atmosphere becomes continuously denser as it approaches the Earth, then the light passing through it should be continuously refracted through the gradient so as to follow a curved line. That it follows a straight line instead, Alhacen argues, is because, being continuous, the gradient has no lines of demarcation within it and, thus, no surfaces within it at which the light can be refracted.

[20]The effect of atmospheric refraction in the apparent displacement of celestial bodies at or near the horizon depends upon two factors: the height of the atmosphere

and the density differential between the aither above the atmospheric shell and the fire at the top of the shell. In the case of the moon, which lies relatively close to the Earth, the height of the atmosphere is a particularly important factor because the closer to the moon the top of the atmospheric shell lies, the greater the density differential between aither and fire must be. For an explanation of this point, see note 92 to paragraph 7, 4.30, pp. 368-370. That Alhacen located the aither-fire interface quite close to the Earth is indicated by his analysis in proposition 21, pp. 324-325; see esp. note 181 to that proposition on p. 392.

[21]For Alhacen's detailed account of distance-perception, see book 2, paragraphs 3.67-93, in Smith, *Alhacen's Theory*, 448-457.

[22]For Alhacen's detailed account of size-perception, see book 2, paragraphs 3.3.135-170, in Smith, *Alhacen's Theory*, 475-493.

[23]For Alhacen's detailed account of shape-perception, see book 2, paragraphs 3.127-134, in Smith, *Alhacen's Theory*, 471-474.

[24]For further details, see note 193 to paragraph 7, 7.71, pp. 394-395.

[25]Roshdi Rashed suggests Ibn Sahl as another potential source, but we will defer the discussion of that issue to a later section (see pp. lxxxii-lxxxiv). It is always possible that Arabic sources or Greek sources known only in Arabic translation will emerge in the course of close study of the Arabic version of the *De aspectibus*, but even so, such sources would have gone unrecognized by readers of the Latin version.

[26]See Smith, *Alhacen on the Principles*, lxxiv-lxxvii.

[27]For Ptolemy's description of the apparatus and the experimental derivation of the angles of refraction based on it, see A. Mark Smith, *Ptolemy's Theory of Visual Perception*, Transactions of the American Philosophical Society 86.2 (Philadelphia: American Philosophical Society, 1996), 231-238.

[28]Ptolemy is vague about the sighting device, simply calling it a "diopter" (*dioptra*), which may or may not consist of a sighting slit. Whatever the form of this device, its purpose is obvious: to confine the experimenter's view to a single line of sight. Ptolemy is equally vague about how (or even whether) the diopter and the marker at H are attached to the disk. Since both are clearly meant to slide along their respective arcs AD and BG, however, it is reasonable to suppose that each lies at the end of a thin rod attached to an axle, preferably on the back of the bronze disk, so as to allow it to pivot.

[29]To my knowledge no one has ever suggested that Ptolemy did not actually perform the refraction experiments described in the *Optics*, despite his rather vague description of the apparatus and its construction, particularly with regard to the sighting device and the marker and their attachment to the bronze disk. The relative accuracy of Ptolemy's results is compelling evidence that he did conduct the experiments. That Ptolemy's experimental results were not "raw" as given is obvious from the underlying pattern of his tabulations on the basis of constant second difference, so there is no question that he adjusted his results according to that algorithm. In *Ibn al-Haytham's Optics*, Saleh Omar characterizes this adjustment procedure as "doctoring"; see esp. pp. 126-130. By contrast, of course, Alhacen's experimental results were completely undoctored because he gave none. For a detailed (and more sympathetic) analysis of Ptolemy's results and their adjustment, see A. Mark

Smith, "Ptolemy's Search for a Law of Refraction: A Case-Study in the Classical Methodology of 'Saving the Appearances' and its Limitations," *Archive for History of Exact Sciences*, 26 (1982): 221-240.

[30]This failure on Alhacen's part is all the more puzzling in that, being significantly larger that Ptolemy's, Alhacen's experimental apparatus should have yielded commensurately more precise results, certainly beyond the half-degree tolerance of Ptolemy's.

[31]See Ptolemy, *Optics*, 5, 1-3 and 5, 17, in Smith, *Ptolemy's Theory*, 229-231 and 235-236.

[32]Ptolemy, of course, is analyzing the refraction of visual rays, not light rays, but since the only effective difference between the two types of rays lies in the direction of radiation, what holds for the one holds reciprocally for the other.

[33]Ptolemy in fact offers an experimental verification that light (understood as the reciprocal of visual radiation) passing through a refractive interface along the normal passes straight through without refraction; that light inclines toward the normal when it passes at a slant into a denser medium and away from the normal when it passes at a slant into a rarer medium; that the denser the medium the light enters, the more severe the refraction; and that refraction is reciprocal; see experiment V.4, in Smith, *Ptolemy's Theory*, 242-244. At the end of this experiment Ptolemy adduces as a general rule that, when light passes from a rarer to a denser medium, if $i > i'$, and if i yields r, while i' yields r', then $i{:}i' > r{:}r'$, or, by alternation, $i{:}r > i'{:}r'$. From this it follows that, $i - r > i' - r'$, which is to say that, as i increases, $i - r$ also increases. By implication the converse holds for refraction from a denser to a rarer medium, i.e., $i{:}r < i'{:}r'$, so that as i increases, $r - i$ also increases. No matter the direction of radiation, then, it holds generally that, as i increases, the difference between i and r also increases. Alhacen expands on this weak generalization with his set of seven rules at the end of chapter 3, but, as we will see later, on pp. lxiii-lxv, in doing so he overgeneralizes in certain cases.

[34]See Ptolemy, *Optics*, 3, 1-2, and 5, 1-2, in Smith, *Ptolemy's Theory*, 130-131 and 229.

[35]Smith, *Ptolemy's Theory*, 244; Latin: Albert Lejeune, *L'Optique de Claude Ptolémée* (Leiden: Brill, 1989), 246.

[36]For instance, both use the "floating coin" experiment to demonstrate the fact of refraction (cf. Alhacen, paragraph 5.3, pp. 274-275, and Ptolemy, paragraphs V, 5-6 [Smith, *Ptolemy's Theory*, 230-231]); both describe the same experiment showing that, when a rod is stood upright in water, its base under water will appear magnified (cf. Alhacen, paragraph 7.32, p. 317, and Ptolemy, paragraph V, 68 [Smith, *Ptolemy's Theory*, 255]); both attribute the magnification or diminution of object lines seen through a plane refractive interface to the enlargement or diminution of the visual angle under which the images of those lines are viewed (cf. Alhacen, propositions 12-14, pp. 309-314, and Ptolemy, theorems V.13 and V.14 [Smith, *Ptolemy's Theory*, 257-258]); and both assume that in the case of refraction through a convex spherical interface the image location can vary according to where and whether the cathetus of incidence meets the extension of the refracted ray (cf. Alhacen, proposition 7, pp. 289-290, and Ptolemy, theorem V.10 and paragraphs V.62-63 [Smith, *Ptolemy's Theory*, 252-253]).

[37]G. J. Toomer, trans., *Ptolemy's Almagest* (Princeton, NJ: Princeton University Press, 1998), 39.

[38]According to Sabra, this issue was addressed by Alhacen in at least four other works: "Treatise on the Solution of Difficulties in the First Book of the Almagest," "Commentary and Summary of the Almagest," "On the Appearance of the Stars," and "Doubts on Ptolemy"; see A. I. Sabra, *The Optics of Ibn al-Haytham*, vol. 2 (London: Warburg Institute, 1989), xxxiii-xl.

[39]Demonstrating that the refraction occurs in a single plane perpendicular to the refracting interface does, however, demand utmost precision in both the construction and deployment of the pan so that the axial ray of the sunbeam passing through the two holes and reaching the pan's rim on the other side lies within the plane of the middle circle on the inner wall of the rim.

[40]See note 6, p. cv.

[41]Gérard Simon, "L'expérimentation sur la réflexion et la réfraction chez Ptolémée et Ibn al-Haytham," Régis Morelon and Ahmad Hasnawi, eds., *De Zénon d'Élée à Poincaré* (Louvain, Paris: Peeters, 2004), 355-375, esp. 373-374.

[42]Actually, this two-hole system for channeling sunlight into the apparatus allows for a slight bit of error in lining it up with the sun; see note 17 to paragraph 7, 2.18, pp. 339-340.

[43]Simon, "L'experimentation," 373. Simon points out quite rightly that the enlargement of the circle of light cast on the inner rim of the pan by diffraction will not only skew the quantitative determinations of the angles of refraction but will also throw off the tagging experiments, which depend on placing the needle-point at the very center of the shaft of light as it passes through each hole as well as beyond both holes. He is mistaken, however, in claiming that, in order to achieve the level of accuracy that Ptolemy reached in his empirical determinations of the angles of refraction, Alhacen would have had to be capable of measuring those angles to within a half-minute of arc. As far as we know, Ptolemy's apparatus was accurate only to within half a degree of arc.

[44]See note 60 to paragraph 7, 3.20, pp. 354-356.

[45]See Smith, *Alhacen on the Principles*, xxv-xxvi; see also A. Mark Smith, "Le De aspectibus d'Alhacen: Révolutionnaire ou réformiste?" *Revue d'histoire des sciences*, 60 (2007): 65-81.

[46]See Smith, *Alhacen on the Principles*, xxiii.

[47]Actually, Alhacen offers two somewhat different explanations of light absorption by visible objects. The first, which occurs in book 1, chapter 8, traces such absorption to opacity as a function of the object's inherent color. Color, however, cannot be seen on its own. In order to become visible, it requires the addition of light, which is thereby "fixed" (*figatur*) to the object's surface, where it mingles with its color to create a secondary form (*forma secunda*) consisting of luminous color. It is through this secondary form, which radiates from the object to the eye, that the object is visually perceived; see esp. paragraphs 8.9 and 8.10 in Smith, *Alhacen's Theory*, 393-394. The second explanation, which is only hinted at in chapter 3 of book 4, appeals to the physical structure or texture of the visible object's surface. Perfectly continuous and smooth surfaces are perfectly reflective. Being interrupted by pores and crevices, rough surfaces are only partially reflective

according to the portions that are continuous and smooth and the porous portions
into which the light is diverted and presumably trapped. One might infer that
such surfaces are seen according to the light scattered by them, but Alhacen never
adverts to such scattering explicitly. Rather, he explains the perception of rough
surfaces in book 2 on the basis of how uniform or non-uniform the light on them
appears; see book 2, chapter 3, paragraphs 3.189-191, in Smith, *Alhacen's Theory*,
500-501.

[48]Whether it should be taken as a model at all or whether Alhacen intended it
as a mere analogy in the loosest and most flexible sense is an open question. Both
positions are supportable, but the evidence in favor of taking it as a model is fairly
compelling. Two things are particularly telling in this regard. First is Alhacen's
supposition that light takes time to travel, which necessitates that it has some
velocity, however unimaginably great; see book 2, chapter 3, paragraphs 3.60-3.62,
in Smith, *Alhacen's Theory*, 445-447. Second is Alhacen's explanation of vision on
the basis of the sensible impression incoming light makes on the anterior surface
of the crystalline lens. If that impression is strong enough, Alhacen contends at the
beginning of book 1, the lens will actually suffer pain (*dolor*), and the dynamic effect
may linger in the afterimage (see Smith, *Alhacen's Theory*, 343-344). In explaining
vision this way Alhacen is in a sense literalizing Aristotle's seal-and-wax analogy
in *De anima* and, by extension, literalizing the physical stamping of the object's
form on the eye (see Smith, *Alhacen's Theory*, note 95, 409-410).

[49]In that paragraph Alhacen contends that "the tunics of the eye . . . are also
transparent and denser (*spissiores*) than the air" (p. 302; for the Latin, see p. 107,
lines 291-292). Alhacen in fact cites thickness or *spissitudo* as one of the twenty-two
visible intentions in book 1, paragraph 1.44, and by it he clearly means "opacity";
for the Latin, see Smith, *Alhacen's Theory*, 111, line 94. Alhacen also uses the term
grossities ("crassness" or "thickness") to designate density; see book 7, paragraphs
2.81 and 2.86, pp. 244-245 and 246 (for the Latin, see pp. 33-34 and 35).

[50]For instance, in book 3, chapter 5, paragraph 5.10, Alhacen informs us that
"opacity (*soliditas*) is a cause of error in sensation if the opacity (*soliditas*) is scant, as
[it is] in the case of glass. . . "; Smith, *Alhacen's Theory*, 596; see ibid., 295 for the Latin.

[51]See, e.g., book 3, chapter 3, paragraph 3.10, in Smith, *Alhacen's Theory*, pp.
590-591; Latin: pp. 288-289, esp. p. 289, line 129-130.

[52]Alhacen, for instance, asserts explictly that, in order for vision to occur, the
air between the visible object and the eye must be continuous; see, e.g., book 1,
paragraphs ó.28, 6.63, and 6.81, in Smith, *Alhacen's Theory*, pp. 364, 375, and 378-
379. This of course can be taken in at least two ways: either the air itself is perfectly
continuous or there is no opaque body in it to interrupt the line of sight between
object and eye. In favor of the former interpretation is Alhacen's contention that
light radiation is continuous. Because it is a formal effect, its continuity would
therefore depend upon the continuity of the transparent medium supporting it, be
that medium air, water, or glass.

[53]In the first paragraph of chapter 3 in book 4 Alhacen characterizes a polished
body as having "an extremely smooth surface," the smoothness consisting "in the
parts of the surface being continuous without many pores" (Smith, *Alhacen on the
Principles*, 300).

[54]In fact, there are also terminological variants in the Arabic text; see, e.g., *densus, spissitudo, and spissus* in the Latin-Arabic glossary in Sabra, *Optics*, vol. 2, pp. 179 and 198-199. The spectrum of Latin terms would no doubt be narrower than it is had only one translator been responsible for the entire text of the *De aspectibus*. For instance, the terms *soliditas/solidus* for "opacity/opaque" appear for the first time with the change of translators in book 3 and recur a few times in book 5, but they disappear from use in book 7, which belongs to the portion of text beginning in book 6, chapter 6 that was taken over by a new translator; see "Manuscripts and Editing," pp. cxxv-cxxvi. Likewise, the term *grossities* for "density" is unique to book 7.

[55]For amplification and clarification, see note 50 to paragraph 2.87, pp. 348-351.

[56]See pp. l-li above.

[57]See Omar, *Ibn al-Haytham's Optics*, 145 (listed there as rule 5), and Roshdi Rashed, *Géométrie et dioptrique au Xe siècle* (Paris: Les Belles Lettres, 1993), lxiv (listed there as rule 4). Evidently, then, the rendering ". . . semper erit minor angulo . . ." in the Latin text of rule 4 (see p. 51, line 220) should have been ". . . semper erit minor medietate anguli . . ."

[58]Stated thus, the rule makes no sense whatever in light of the experimental results Alhacen should have obtained or the results Ptolemy did obtain in book 5 of the *Optics*. For instance, according to Ptolemy, if light passes from air to water at an angle of incidence of 10°, the resulting angle of refraction will be 7.5°, which translates to an angle of refraction/deviation of 2.5° according to Alhacen's measure. Clearly, half of 10° + 2.5° (i.e., 6.25°) is far greater than the angle of refraction/deviation of 2.5°. Cf. Ptolemy's tabulations for refraction from air to water in Smith, *Ptolemy's Theory*, 233.

[59]Omar, *Ibn al-Haytham's Optics*, 145 (listed there as rule 6), and Rashed, *Géométrie et dioptrique*, lxiv (included there in rule 4). As in the case of rule 4, so in this case, the Latin version seems to veer off from the Arabic at one key point: instead of "tunc angulus reflexionis erit medietas coniuncti duorum angulorum," (p. 51, lines 223-224), the phrase should have been rendered ". . . tunc angulus reflexionis erit minor medietate coniuncti duorum angulorum . . ."

[60]Nor would Ptolemy's tabulations for refraction from air to glass have invalidated the rule, no matter whether taken directly or reciprocally. According to those tabulations, when $i = 80°$, $r = 42°$, which translates to $r/d = 38°$ by Alhacen's measure, so $r/d < .5i$. On the other hand, if we take the same values reciprocally, so that i and r are interchanged, then when $i = 42°$, $r = 80°$, which translates to $r/d = 38°$ by Alhacen's measure, so $r/d < .5(i + r/d)$—i.e., $38° < 40°$. Cf. Ptolemy's tabulations for refraction from air to glass in Smith, *Ptolemy's Theory*, 236.

[61]This point can be easily understood by recourse to the table in figure 25, p. 425. According to Ptolemy's tabulations for air to glass (in Smith, *Ptolemy's Theory*, 236), when $i = 80°$, $r = 42°$; when $i = 70°$, $r = 38.5°$; when $i = 60°$, $r = 34.5°$; when $i = 50°$, $r = 30°$; and when $i = 40°$, $r = 25°$. The two tables in figure 25 reverse the direction of radiation so that r for refraction from air to glass becomes i for refraction from glass to air, and vice-versa. For the sake of comparison with

Ptolemy's values in the top table, I have given the corresponding modern values derived by the sine law in the bottom table. Accordingly, in the top table, when i in column 1 = 42°, r in column 3 = 80°, which translates to r/d in column 4 = 38°. By the same token, when i in column 1 = 38.5°, r in column 3 = 70°, which translates to r/d in column 4 = 31.5°. Consequently, the difference between the two angles of incidence in column 3—i.e., 3.5°—is less than the difference between the two angles of refraction/deviation $r/d - r/d'$ in column 5—i.e., 6.5°. Continuing down the table from i = 38.5° to i = 25°, we see that $r/d - r/d' > i - i'$—i.e. 6° > 4° and 5.5° > 4.5°, until i = 25°, when the difference between it and i = 30° (5.0°) is equal to the difference between the respective angles of refraction/deviation r/d (5.0°). In the bottom table we see the same general pattern among the modern values derived by the sine law.

[62]For an Arabic edition and French translation of the entire treatise, see Rashed, *Géométrie et optique*, pp. 1-52.

[63]This list takes the tradition up to the sixth century, and Rashed has extended it to the tenth century according to its inclusion of, among others, "Dtrums" and al-Kindī; see *Géométrie et dioptrique*, xv-xviii.

[64]Anthemius did in fact deal with elliptical mirrors, but as far as we know he was unique in doing so until Ibn Sahl took them up; see Rashed, *Géométrie et dioptrique*, xxvi. For Ibn Sahl's analysis of the two kinds of lenses, see ibid., pp. 23-52.

[65]In order to simplify the subsequent analysis and make it somewhat easier to follow, I am taking liberties with Ibn Sahl's construction by putting all the relevant elements in a single plane. He has the plane of refraction oblique to arc SBP on the lens's hyperboloidal surface, so his analysis is three dimensional, whereas mine is two dimensional. I am also recasting his figures in slightly different form and orientation. In neither case, however, am I doing violence to his overall intent in the analysis. For his full demonstration, see Rashed, *Géométrie et dioptrique*, pp. 23-30, and for Rashed's résumé of that demonstration, see ibid., pp. xxix-xxxvi. For an analysis of Ibn Sahl's overall treatment of burning mirrors and lenses in English, see Rashed, "A Pioneer in Anaclastics: Ibn Sahl on Burning Mirrors and Lenses," *Isis*, 81 (1990): 464-491. Most of the analysis in that article is drawn directly from chapter 1 of *Géométrie et dioptrique*, xv-xlii.

[66]The proof for the biconvex hyperboloidal lens follows directly and easily from the proof for the plano-convex hyperboloidal lens. Let STBP in figure 26b, p. 424, be the face of the plano-convex lens in the previous proof, and let ST'B'P be its mirror image so that B'L' = BL, latus rectum B'N' = latus rectum BN, A'L' = AL, A'B' = AB, etc. By the principle of reciprocity, then, if A' is a point of radiation, ray A'T' will be refracted at T' along T'T parallel to axis A'A, so it will be refracted again at T to A. Nor does it matter whether the two intersecting hyperbolic sections are congruent. ST'B'P can be any other section with A' as the focus of its opposite branch, and any ray from A' to that section will refract in a line parallel to axis AA' so as to refract again to A.

[67]The derivation of the sine law is as follows according to the lower diagram of figure 26, p. 422. Drop normal QCR through point C of refraction parallel to normal GEH. Hence, angle of incidence DCQ = corresponding angle CHG, and

angle of refraction RCE = alternate angle CEG. GC/CH is the sine of angle CHG, which = angle of incidence DCQ; and GC/CE is the sine of angle CEG , which = angle of refraction RCE. Hence, (GC/CH [= sine angle CHG])/(GC/CE [= sine angle CEG]) = (GC/CH)(CE/GC) = CE/CH. It therefore follows that CE/CH = sine *r*/sine *i*. If the terms are inverted, we end up with CH/CE = sine *i*/sine *r*, which translates to the ratio CH:CE = sine *i*:sine *r*. Another, slightly different form of the law follows from conceiving of E as an object point and D as a center of sight, in which case EG will be the cathetus of incidence and H the image of E. The resulting law would therefore be something like "in all refractions, the ratio of the distance from the object point to the point of refraction and the distance from the image point to the point of refraction is constant." For Rashed's commentary on Ibn Sahl and the sine law, see *Géométrie et dioptrique*, pp. xxix-xxxiv.

[68]*Géométrie et dioptrique*, xxiv and lxx-lxxi.

[69]The point of Ibn Sahl's proof that the celestial sphere is not perfectly transparent is to show that, like all other transparent media, the *aither* filling that sphere has some opacity or density. In the brief prologue to the treatise he alludes explicitly to book 5 of Ptolemy's *Optics*. As far as we know, Ibn Sahl was the first Arabic scholar to read, or at least use, Ptolemy's *Optics*. For the Arabic text and a French translation of Ibn Sahl's "Proof," see Rashed, *Géométrie et dioptrique*, pp. 53-56. On Ibn Sahl as the first Arabic scholar to use Ptolemy's *Optics*, see ibid., p. lxix, note 30, and Sabra, *Optics*, vol. 2, p. lix. Ibn Sahl's denial of the perfect transparency of the celestial sphere may be reflected in Alhacen's claim in book 7, paragraph 2.81, pp. 244-245, that "when light passes through any transparent body, [that body] resists the light to some extent according to how dense it is because in every physical body there must be some density, since slight transparency has no limit in the imagination when it conceives of translucency, and what is [imaginable is] that all physical bodies [can] reach a limit [of rarity] they cannot transgress." It would be rash, however, to take this as compelling evidence of Ibn Sahl's influence.

[70]This of course begs the question of what can be taken as a trustworthy angle pair. For that purpose, the experiments described in chapter 4 would have been better than the ones in chapter 3. In the former, the experimenter is to insert a reed through the two holes in the apparatus and drill a narrow channel through it to constrict the line of sight as much as possible. To derive the appropriate angle pair for air to water (actually water to air), he could then fill the carrying vessel with water up to the center of the register plate and rotate the pan to the desired angle of "incidence" (actually refraction). Sighting through the narrow opening in the reed, he could then move the needle point along the middle circle under water and opposite the sighting hole until he could see it and mark the spot. That would give the appropriate angle of "refraction" while obviating the problem of diffraction plaguing the experiments in chapter 3. The equivalent could be done *mutatis mutandis* with the glass quarter-sphere. Both determinations, of course, should be made at moderate angles of "incidence" rather than at the extremes. Hence, although the actual determinations would be based on refraction from water to air or glass to air, the results are interchangeable because of the law of reciprocity.

[71]*Géométrie et dioptrique*, lxviii-lxxv.

[72]Alhacen never actually specifies the relative density of the cornea, albugineous humor, glacial humor, and vitreous humor, but there is strong reason to suppose that he thought the vitreous humor to be denser and more refractive than the glacial humor (see Smith, *Alhacen's Theory*, note 15, pp. 531-534). As to the glacial humor, Alhacen's description of it in book 1, chapter 8, paragraph 7.5 stresses its somewhat limited transparency "so that light and color can pass through it, but . . . so that the forms of light and color can persist in it for awhile in order to let the form of the light and color impressed on it be seen by the sensitive faculty" (ibid., 388). The transparency of the albugineous humor, on the other hand, seems to be such as only to let the light and color radiate through it without having any physical effect on it, so presumably it is rarer than the glacial humor (see paragraph 7.3 in ibid., 387). It is not unreasonable to suppose, therefore, that the range of densities from the cornea through the three successive humors follows the order of their placement in the eye.

[73]See note 144 to paragraph 7, 6.25, p. 383.

[74]For a more detailed critique of Alhacen's account of peripheral vision and its implications, see A. Mark Smith, "The Measure of Vision in the Middle Ages," forthcoming in *Micrologus*.

[75]For the the figure accompanying the Arabic text, see Rashed, *Géométrie et dioptrique*, p. 106.

[76]"Notons que, au cours de son étude de la lentille sphérique, Ibn al-Haytham utilise l'aberration sphérique d'un point à distance finie dans le cas du dioptre, pour étudier l'image d'un segment qui est une portion du segment définie par l'aberration sphérique," ibid., liii.

[77]For the Arabic text and a French translation of this work see Rashed, *Géométrie et dioptrique*, 111-132; see also pp. liii-lx for a critical evaluation. It should be noted that this work has come down to us in a redaction by Kamāl al-Dīn al-Fārisī, the brilliant early-fourteenth-century commentator on Alhacen's optical work who is best known for his remarkable explanation of the rainbow.

[78]See, e.g., Cambridge, Emmanual College, MS 20, f. 186r, for an example of the faulty diagram accompanying book 10, proposition 43 in Witelo's *Perspectiva*. As was his wont, Friedrich Risner used the same diagram for book 7, proposition 17 of the *De aspectibus* and book 10, proposition 43 of the *Perspectiva*; see his *Opticae thesaurus. Alhazeni arabis libri septem, nunc primum editi. Eiusdem liber De crepusculis et Nubium ascensionibus. Item Vitellonis thuringopoloni libri X.* (Basel, 1572; reprint, New York: Johnson Reprint, 1972), 221 (Alhacen) and 441 (Witelo).

[79]See Roger Bacon, *De multiplicatione specierum*, II, 3, in David C. Lindberg, ed., *Roger Bacon's Philosophy of Nature* (Oxford: Clarendon Press, 1983), p. 117; Witelo, *Perspectiva*, book 10, proposition 48, in Risner, *Opticae Thesaurus*, pp. 443-444; Pecham, *Perspectiva communis*, III, proposition 16, in David C. Lindberg, *John Pecham and the Science of Optics* (Madison: University of Wisconsin Press, 1970), pp. 228-231. In book 2 of his *De refractione opticae partes libri novem* (Naples, 1593), however, Giambattista della Porta shows that solar rays striking a sphere along parallel rays are refracted to different points on the axis; for further details on Porta's analysis of spherical aberration, see pp. xciv-xcv.

[80]The discussion that follows is based on chapter 3, section 3, pp. 177-203 of the original Latin edition; for an English translation, see William H. Donahue, *Johannes Kepler, Optics* (Santa Fe: Green Lion Press, 2000), 191-218.

[81]It is in the course of defining this inner limiting point that Kepler cites Witelo, *Perspectiva*, 10, 43 (i.e., Alhacen's prop. 17); see Donahue, *Kepler*, p. 202.

[82]Donahue, *Kepler*, 207.

[83]This hyperbolic section thus forms the so-called caustic curve.

[84]That Kepler in fact did some measuring, however rough, is indicated by his claim that the central dot of light first appears in the circle of light at a distance of about one-twentieth the sphere's diameter.

[85]David Lindberg, however, sees a foreshadowing of Alhacen's model of "punctiform analysis" in Al-Kindī's account of visual radiation in a work that has come down to us in Latin under the title *De aspectibus*; see *Theories of Vision from Al-Kindi to Kepler* (Chicago: University of Chicago Press, 1976), 26-30.

[86]It should be noted that, although Witelo took his tabulations for the refraction of light from air to water directly from Ptolemy, he made a slight change in Ptolemy's value for the angle of refraction when the angle of incidence is 10°. Instead of adopting Ptolemy's figure of $r = 7.5°$ for $i = 10°$, he unaccountably altered it to 7.25°, the value upon which Kepler based his conclusion that a parallel ray of light incident on a water-filled sphere at 10° will refract to the axis at a point very near the focal point of the sphere.

[87]Donahue, *Kepler*, 100.

[88]Ibid., 200.

[89]Ibid.

[90]See *Paralipomena*, chapter 3, propositions 18 and 19, and chapter 4, sections 3-5, in Donahue, *Kepler*, 86-91 and 104-123.

[91]For a useful summary of this tradition and Rashed's role in exposing it through publication of the actual texts, most of them extant only in Arabic, see Hélène Bellosta's review essay, "Burning Instruments from Diocles to Ibn Sahl," *Arabic Sciences and Philosophy*, 12 (2002): 285-303, esp. 285-286.

[92]"Compte tenu de ces observations, il n'est pas excessif d'admettre qu'Ibn al-Haytham a bien lu les parties du *Traité* d'Ibn Sahl consacrées aux lentilles et à la réfraction," *Géométrie et dioptrique*, lxxiii.

[93]See pp. lxv-lxvii.

[94]These include propositions 15-16 and 19-21, pp. 314-318 and 321-325.

[95]It should be noted, however, that toward the end of paragraph 6.8, p. 299, Alhacen alludes to, but does not analyze, the size distortion of objects seen through a sphere and the variation in that distortion depending on how far or near the object is to the back edge of the sphere.

[96]*Perspectiva communis*, III, 10, in Lindberg, *Pecham*, 221. For Bacon's account of such magnification, see *Perspectiva*, III.iii.2, in David C. Lindberg, ed. and trans., *Roger Bacon and the Origins of* Perspectiva *in the Middle Ages* (Oxford: Clarendon Press, 1996), 292-297; and for Witelo's account see *Perspectiva*, 10, prop. 31, in Risner, *Opticae Thesaurus*, 431-432.

[97]See Bacon, *Perspectiva*, III, iii, 1, in Lindberg, *Bacon and the Origins*, 286-293. In *Perspectiva*, I, vi, 2, Bacon maintains that vision occurs principally through

the cone of radiation consisting of perpendicular rays striking the eye's surface orthogonally and that by its relative intensity each such ray "conceals all of the oblique rays [reaching the same point on the eye], just as a greater and stronger light conceals many weak lights" (see ibid., p. 75).

[98]See Pecham, *Perspectiva communis*, I, prop. 42 and III, prop. 10, in Lindberg, *Pecham*, pp. 124-127 and pp. 220-221.

[99]Witelo's overall account of the eye's reception of luminous radiation can be found in *Perspectiva*, 3, props. 17 and 18, in Sabetai Unguru, ed. and trans., *Witelonis Perspectivae liber secundus et liber tertius*, Studia Copernicana, XXVIII (Wrocław, Warsawa, Kraców: Ossolineum, 1991), 122-127; his account of vision by means of oblique radiation can be found in ibid., 124-125.

[100]Ibid., 125.

[101]Witelo's claim that the vertex angle of the cone of radiation is just under 90° and therefore that the field of view is "virtually a fourth of the great circle of the celestial sphere" is found in *Perspectiva*, 4, prop. 3, in Carl J. Kelso, Jr., ed. and trans., "Witelonis *Perspectivae* liber quartus," (Ph.D. diss., University of Missouri, 2003), 73-74.

[102]See Bacon, *De multiplicatone specierum*, part II, chapter 3, in Lindberg, *Bacon's Philosophy*, pp. 104-119, esp. pp. 110-117; Pecham, *Perspectiva communis*, I, 15 and 16, in Lindberg, *Pecham*, pp. 88-93; and Witelo, *Perspectiva*, 2, prop. 47, in Unguru, *Witelonis* Perspectiva *liber secundus et liber tertius*, pp. 94-98.

[103]Bacon, *Perspectiva*, II, iii, 6 and III, ii, 4, in Lindberg, *Bacon and the Origins*, 226-229 and 313-317; Pecham, *Perspectiva communis*, I, prop. 82, and III, props. 12 and 13, in Lindberg, *Pecham*, 153 and 222-229; Witelo, *Perspectiva*, 10, props. 49-54, in Risner, *Opticae Thesaurus*, 444-449.

[104]Bacon, *Perspectiva*, III, ii, 4, in Lindberg, *Bacon and the Origins*, 313-315, and Witelo, *Perspectiva*, 10, prop. 54, in Risner, *Opticae Thesaurus*, 448-449.

[105]Witelo gives these tabulations in *Perspectiva*, 10, prop. 8, in Risner, *Opticae Thesaurus*, 412, arranging his tabulations according to both the modern angle of refraction—which he calls the *angulus refractus*—and Alhacen's version (i.e., the modern angle of deviation)—which he calls the *angulus refractionis*. The slight (and only) modification comes at the very first value for refraction from air to water at 10° incidence. Instead of Ptolemy's value of 8°, Witelo unaccountably gives 7° 45' for the *angulus refractus*. Even more unaccountable is the corresponding value he gives for the *angulus refractionis* (i.e., deviation), which is the angle of incidence minus the refracted angle, or 10° minus 7° 45'. Instead of the expected 2° 15', the value given is 2° 5'. Otherwise, Witelo's values are identical to those given by Ptolemy in book V, experiment V.1 of the *Optics* (cf. Smith, *Ptolemy's Theory*, 233). Likewise his values for refraction from air to glass and water to glass are precisely the same as those given by Ptolemy in experiments V.2 and V.3 (Smith, *Ptolemy's Theory*, pp. 236 and 238).

[106]Applying the principle of reciprocity, Witelo adduces the values for the *anguli refracti* (i.e., the modern angles of refraction) from water to air, glass to air, and glass to water simply by adding the *anguli refractionis* (i.e., deviation) already given for refraction from air to water, air to glass, and water to glass to the angles of incidence. Thus, when the angle of incidence is 10° for a ray of light passing from water to air,

the resulting *angulus refractus* is 12° 5′, which equals 10° plus the 2° 5′ Witelo gives for the angle of refraction for light passing from air to water. This technique leads him to adduce values for refraction from water to air and glass to air for angles of incidence of 50° to 80°, which is impossible, since they exceed the respective critical angles of somewhat less than 42° and 49°. Likewise, his values of refraction for light passing from glass to water at angles of incidence of 70° and 80° are spurious because those angles exceed the critical angle, which is around 61°. Clearly, then, despite his detailed instructions, which repeat Alhacen's almost verbatim, Witelo did not actually do the experiments to determine any of the refractions; for his description of the construction of the apparatus and the resulting experimental determinations of the angles of refraction, see *Perspectiva*, 2, prop. 1 (in Unguru, *Witelonis* Perspectiva *liber secundus et liber tertius*, 40-45), and *Perspectiva*, 10, props. 4-7 (in Risner, *Opticae Thesaurus*, 407-412).

[107]Pecham, *Perspectiva communis*, III, props. 18-20, in Lindberg, *Pecham*, 232-235, and Witelo, *Perspectiva*, 10, props. 65-84, in Risner, *Opticae Thesaurus*, 457-474. Although Bacon does not broach the subject of rainbows in his *Perspectiva*, which is part 5 of his *Opus majus*, he does take it up in part 6 of the *Opus majus* on experimental science. For the classic account of medieval rainbow theory in the Latin West, see Carl B. Boyer, *The Rainbow: From Myth to Mathematics* (New York: Yoseloff, 1959), pp. 85-142. For a more recent general account based heavily on secondary sources, see Raymond Lee and Alistair Fraser, *The Rainbow Bridge: Rainbows in Art, Myth, and Science* (University Park: Pennsylvania State Press, 2001), pp. 148-166. See also David C. Lindberg, "Roger Bacon's Theory of the Rainbow: Progress or Regress?" *Isis*, 57 (1966): 235-248.

[108]Bacon, *De multiplicatione specierum*, II, 3, in Lindberg, *Bacon's Philosophy*, 117-119, and Pecham, *Perspectiva communis*, III, prop. 16, in Lindberg, *Pecham*, 228-231. The diagram that accompanies this latter proposition makes the point explicitly by showing several rays from the luminous point passing through the sphere and converging at a point on the other side of the sphere. Interestingly enough, Bacon's discussion is based on the examples of sunlight passing through a solid glass sphere or a water-filled urinal flask.

[109]In *Perspectiva*, 2, prop. 50, Witelo establishes by supposition that, if a luminous point faces a convex spherical surface beyond which lies a denser transparent medium, all the rays from that point refracted at that surface will congregate at a single point beyond the center of curvature; see Unguru, *Witelonis* Perspectiva *liber secundus et liber tertius*, 99-101. Later, in *Perspectiva*, 10, prop. 48, Witelo attempts to show that sunlight passing through a sphere of glass will bring that light to a point on the other side of the sphere, where combustion will occur; see Risner, *Opticae Thesaurus*, pp. 443-444. Combined with proposition 50 of book 2, this analysis implies that all the rays from a point on the sun that are refracted through the sphere will congregate at a single point.

[110]Theoderic of Freiberg stands as a mild exception in his account of the rainbow in *De iride et radialibus impressionibus*, which was composed sometime in the first decade of the fourteenth century That account was based on the geometrical analysis of rays of sunlight refracted into myriad individual raindrops, reflected internally, and then refracted out of them at specific angles tied to specific colors

in the spectrum of the rainbow's arc. The primary bow involves a single internal reflection, the secondary bow two internal reflections, hence the reversal of colors and their weakening by added refraction. Although conceptually similar to various thirteenth-century accounts, Theodoric's was exceptional in two respects: first, he attempted to explain the rainbow on clear geometrical principles according to Perspectivist ray theory, and second, he experimented with sunlight passing into a water-filled urinal flask to derive and test his model. See Boyer, *Rainbow*, pp. 110-124; William A. Wallace, *The Scientific Methodology of Theodoric of Freiberg*, Studia Friburgensia, n.s. 26 (Fribourg: Fribourg University Press, 1959); and Lee and Fraser, *Rainbow Bridge*, pp. 161-166.

[111]According to the best documentary evidence so far available, eyeglasses were invented in Pisa in the mid-1280's. Their diffusion from that point throughout Italy during the fourteenth century was relatively swift, although spotty, and archaeological evidence indicates that they were being used in northern Europe by at least the later fourteenth century, if not earlier. For details, see Vincent Ilardi, *Renaissance Vision from Spectacles to Telescopes* (Philadelphia: American Philosophical Society, 2007), 3-73.

[112]*Perspectiva*, III, 2, 4, in Lindberg, *Bacon on the Origins*, 317-318.

[113]Ilardi, *Renaissance Vision*, 152. For details on the manufacture and the development of a widespread commercial trade in eyeglasses from the mid-fifteenth century, see ibid., pp. 75-205. For specifics on the gradation of convex and concave lenses, see ibid, 82-95. For a more technically oriented account of the invention and manufacture of eyeglasses during the Middle Ages and Renaissance, see Rolf Willach, *The Long Route to the Invention of the Telescope* (Philadelphia: American Philosophical Society, 2008).

[114]Although Perspectivist optics was adequately equipped to explain the "correction" of presbyopia by convex lenses on the basis of magnification, it could not account for the correction of myopia by concave lenses because such lenses actually reduce rather than magnify things viewed through them.

[115]Vasco Ronchi, *Optics: The Science of Vision*, trans. Edward Rosen (New York: New York University, 1957), 32-33.

[116]For a full account of his reasoning, see David C. Lindberg, "Lenses and Eyeglasses," in Joseph Strayer, ed., *Dictionary of the Middle Ages*, vol. 7 (New York: Scribner's, 1986), 538-541.

[117]Medieval physicians did recognize that in certain cases visual acuity was compromised not by an inadequate flow of visual spirit but by physical conditions, such as growths on the corneal surface or cataracts, which were assumed to block incoming light. In such cases, of course, the visual problem might well not be associated with the production and flow of visual spirit.

[118]For more details, see A. Mark Smith, "Petrus Hispanus' 'Treatise on the Eyes'," in A. Mark Smith and Arnaldo Pinto Cardoso, *O tratado dos olhos de Pedro Hispano* (Lisbon: Alétheia Editores, 2008), 9-56, esp. 46-54.

[119]This account is given in the last part of a more general tripartite analysis of refraction entitled *Diaphaneon seu Transparentium*, part one dealing with refraction through spherical lenses, and part two with the rainbow. This work is contained in the *Photismi de lumine et umbra ad perspectivam et radiorum incidentiam facientes* (Naples, 1611), which was published thirty-six years after Maurolyco's death.

[120]Maurolyco does not mean that the rays emerging into the eye from the back of the lens converge to an actual focus in the visual process. Rather, he means that the entire sheaf of those rays forms a cone converging toward a vertex. That vertex, however, lies beyond the entrance of the optic nerve into the back of the eyes, so all the rays enter that nerve before converging, just as Alhacen would have it. Indistinct vision occurs when the vertex of the cone lies either too far forward or too far back in the optic nerve. Presumably, then, distinct vision entails a cone that is just the right size for all the rays on its outer surface to enter the optic nerve at its outer edges so that the entering image is precisely the same size as the opening in the nerve.

[121]For Maurolyco's full account of the eye and lens, the apprehension of radiation by the lens, the cause of myopia and presbyopia, and the correction of each by its appropriate kind of lens, see ibid., pp. 69-80. Aside from his fairly detailed illustration of the eye and its components, Maurolyco offers only two diagrams. One of them shows rays of light emanating from a point and shining onto the surface of a biconvex spherical lens, then passing through the lens along parallels, and converging toward a point after refraction on the other side. This is meant to show that biconvex lenses collect (*congregare*) light. The other shows rays of light converging on the surface of a biconcave lens, passing through the lens along parallels, and then diverging after refraction through the other surface. This is meant to show that biconcave lenses spread light out (*disgregare*). The explanation of myopia and presbyopia on the basis of the crystalline lens's curvature is entirely qualitative and vague, and the resulting explanation of how the appropriate lenses compensate for the problems caused by inappropriate curvature of the lens is problematic in several respects. Maurolyco is, however, crystal clear on one point: proper vision is not due to the lens's reception of orthogonal radiation.

[122]The earliest known published representation of a camera obscura is to be found in Gemma Frisius, *De Radio Astronomico et Geometrico* (Antwerp/Louvain, 1545), folio 31v. Clearly pictured in this representation is an inverted image of a solar eclipse projected through the aperture onto a screen beyond it.

[123]Bacon, *De speculis comburentibus*, V, in Lindberg, *Bacon's Philosophy*, 305-323; Pecham, *Perspectiva communis*, I, prop. 5, in Lindberg, *Pecham*, 66-73; Witelo, *Perspectiva*, 10, props. 49-54, in Risner, *Opticae Thesaurus*, 444-449.

[124]See Giambattista della Porta, *Magiae Naturalis libri XX* (Naples, 1589), book 17, chapter 6, pp. 266-267, where he deals with various ways of using the camera obscura and enhancing its effects.

[125]Dated to 1521, Maurolyco's demonstration can be found in Theorem XXXI, in *Photismi*, 25-27: *Si in cavum speculum à signo quopiam tres radij incidant, unus quidem per centrum, duo vero à centro inequaliter remoti, horum qui remotior est, inferius cum eo, qui per centrum concurret* ("If from any point whatever three rays are incident upon a concave mirror, one of these passing through the center and the other two at unequal distances from the center, the more remote of these two rays will [after reflection] intersect the ray through the center at a lower point"—translation from Henry Crew, *The Photismi de lumine of Maurolycus: A Chapter in Late Medieval Optics* [New York: Macmillan, 1940], 39). Maurolyco's geometrical demonstration is based on figure 35, p. 435, where inner ray CF is reflected to H and outer ray CG

is reflected to O. Accordingly, because of the equal-angles law, angle CFA (formed by straight line CF and arcal line FA) = angle HFB (formed by straight line HF and arcal line FB), and, by the same token, angle CGA = angle OGB. Maurolyco then goes on to prove that angle HGB > angle CGA, so it follows that angle OGB < angle HGB, which means that O must lie below H on the axis.

[126]See Sven Dupré, "Optics, Pictures, and Evidence: Leonardo's Drawings of Mirrors and Machinery," *Early Science and Medicine,* 10 (2005): 209-236.

[127]Maurolyco does devote one proposition (Theorem XXXIV) in the *Photismi* to explaining why real images projected by concave spherical mirrors are inverted; see *Photismi,* 28-29. In theorem XXXV, he goes on to claim that, although the sun's rays are subject to spherical aberration after reflection from a concave spherical mirrors, those rays that strike close to the axis will come very close to convergence at a spot on the axis and will thus be concentrated enough to cause combustion, ibid., 29. He does not, however, offer a quantitative determination of that point.

[128]See Sven Dupré, "Ausonio's Mirrors and Galileo's Lenses: The Telescope and Sixteenth-Century Practical Optical Knowledge," *Galileana,* 2 (2005): 145-180. The focal point of concave spherical mirrors had been determined in Antiquity, and its location on the axis halfway between the mirror's center of curvature and its surface was well known to various Arabic thinkers, including Alhacen. Nonetheless, as Dupré points out, Perspectivist writers were confused about the focusing properties of concave spherical mirrors, in part because they were misled by the last proposition of Pseudo-Euclid's *Catoptrics,* where a specious proof is given to show that the point of combustion lies at the center of curvature; see I. L. Heiberg, ed., *Euclidis opera omnia,* vol. 7 (Leipzig: Teubner, 1895), 340-343.

[129]"Sed cum eo nil perfectè operaberis, nisi prius punctum inversionis cognoueris. . . ," *Magiae Naturalis,* 264.

[130]In *Perspectiva,* III, 3, 4, for instance, Roger Bacon cites Caesar's supposed use of huge mirrors (presumably concave spherical ones) to spy on Britain from across the Channel as an example of how useful such mirrors could be; see Lindberg, *Bacon and the Origins,* 332-333. In the prologue to book XVII of his *Magiae Naturalis,* Giambattista della Porta adverts to Archimedes's deployment of mirrors to burn the Roman fleet beseiging Syracuse and recounts how King Ptolemy installed a mirror of such magnifying power atop the *pharos* of Alexandria that it revealed things 600 miles away.

[131]For a good recent account of these tales and their proliferation, see Eileen Reeves, *Galileo's Glassworks: The Telescope and the Mirror* (Cambridge, MA: Harvard University Press, 2008), esp. 15-46.

[132]Ibid., pp. 47-144. Reeves is convinced that Galileo's own telescopic research was focused on a mirror-lens combination until he got wind of the actual convex-concave lens combination of the Dutch telescope.

[133]Crew, *The Photismi,* 65. The Latin enunciation is as follows: *Parallelorum radiorum intra perspicuum orbem à centro inaequaliter distantium, remotior cum axe sibi parallelo propius sphaerae concurret, quam reliquus.* Maurolyco's demonstration is based on figure 36, p. 435, where CD and AB are the rays parallel to axis EF, which passes through center of curvature K. Maurolyco starts with the false premise that the angles of deviation are proportional to the angles of incidence: that is, $i{:}i' = d{:}d'$. Let AB refract along BG and CD along DH. Suppose that AB refracts along BH. It

follows that angle BHF = angle of deviation A'BH for angle of incidence ABK, and angle DHF = angle of deviation C'DH for angle of incidence CDK. But ABK:CDK > BHF:DHF, so BHF is too small to be the appropriate of angle deviation. Hence, refracted ray BG from outer ray AB will intersect the axis above point H. On the basis of his assumption that the angles of incidence are proportional to the angles of deviation, Maurolyco concludes that in refraction from glass to air, the angle of incidence is invariably two and two-thirds the size of the angle of deviation (i.e., in the ratio of 8:3), and he informs us that this figure is based on experiment: *Ergo et angulus inclinationis ad angulum suae fractionis semper unam servat rationem. Estque dupla et duas tertias superpatiens sicut experimentato chrystallina sphera probavimus—Photismi*, 36. Suffice to say, Maurolyco's "law" of constant proportionality between angles of incidence and deviation flies in the face of Alhacen's rules of refraction, and any tabulations derived on its basis will be wildly at odds with those given by Ptolemy and Witelo.

[134]One implication of theorem XVIII, which Maurolyco makes explicit in the second corollary to theorem XX (*Photismi*, 45-46), is that all the parallel rays equidistant from the axial ray will refract from a circle on the surface of the sphere, and the refracted rays will form a cone with its base on that circle and its vertex at the point where they all intersect the axis. Consequently, in figure 36, all the rays parallel to AB and equidistant from EF will refract from a circle formed by the rotation of B around the axis, and the refracted rays will all form a cone with its base on that circle and its vertex at point G. The same holds for the rays parallel to CD and equidistant from EF, and for any other set of rays parallel to EF and equidistant from it. In the scholium to the twenty-fourth, and last, theorem of the first part of the *Diaphaneon*, Maurolyco observes that, if all these rays are taken in the aggregate, they will form a cone whose surface curves constantly inward "according to the successive intersections of these sorts of rays (*propter huiusmodi successivas radiorum sectiones*)"—*Photismi*, 48. The vertex of this cone, he goes on to say, lies at the endpoint of convergence of all these cones. Though he has not defined it mathematically, Maurolyco is obviously aware of the caustic curve, and he is also aware that the resulting curved cone has a limiting point in the vertex. This limiting point, which Maurlyco leaves undefined, is the focus of a plano-convex spherical lens.

[135]Porta also briefly discusses lenses in *Magiae Naturalis*, book 17, chapters 10-13, pp. 269-270.

[136]This method of pinpointing where the reflected ray will intersect the axis by dropping a perpendicular from the midpoint of the normal comes at the very end rather than the beginning of the proposition; see *De Refractione*, pp. 40-41.

[137]Since Porta ties the analyses of reflection from concave spherical mirrors and refraction through glass spheres so closely, it follows that the focal point for rays refracted through such spheres lies half the radius of the sphere below point P, at point F as marked in figure 38, p. 436. This happens to be the actual focal point for crown glass according to modern theory, but that it follows from Porta's analysis is serendipitous. Porta's analysis is essentially qualitative rather than quantitative. Thus, although he locates the points of incidence according to the chords specified by the sides of the inscribed regular polygons, he gives no numerical values for

the resulting angles, which can nonetheless be easily derived from the model. Nor does he give any numerical values for the angles of refraction, which are much more difficult to derive. According to his model, those figures are: for $i = 22.5°$, $r = \sim 13.4°$; for $i = 30°$, $r = \sim 17.6°$; for $i = 45°$, $r = \sim 25.5°$; and for $i = 60°$, $r = 30°$. According to the sine law, those figures should be: for $i = 22.5°$, $r = \sim 14.6°$; for $i = 30°$, $r = \sim 19.2°$; for $i = 45°$, $r = \sim 27.7°$; and for $i = 60°$, $r = \sim 34.7°$. The derived values for the angles of refraction according to Porta's model are therefore off by as much as $4.7°$ (for $i = 60°$) and as little as $1.2°$ (for $i = 22.5°$).

A peculiar twist to Porta's analysis comes with his specification on p. 41 of *De Refractione* that the glass sphere (or cylindrical section) used for the experiments is to be "crystal" (*crystallina pila*), which is a lead glass with a much higher index of refraction than crown glass: up to 1.7 as opposed to 1.52. In that case, ray EM′ in figure 38, will refract at an angle of very nearly 30° (30.63°), which will bring the refracted ray to intersection quite close to point P. However, ray DL′ will refract along a line which, if continued straight through bottom edge HP of the sphere, will intersect the axis at point Q″ well above point Q predicted by the model. The same holds for rays CI′ and BH′, although their points of intersection will increasingly approach the predicted points R and S. The results are even worse if the refraction out of the glass into air is taken into account because in that case the light striking the top edge of the sphere along BH′ at an angle of incidence of 22.5° will intersect the axis below the sphere at a point not far below Q″, more than halfway between P and point S predicted by the model. In that case, of course, the focal length will be far less than half the radius of the sphere, which is what is required by the model.

[138]As an example of how problematic Porta's putative observations are, we need only look at the case in which ray EM′ strikes the glass disk in figure 38 at an angle of incidence of 60°. According to Porta's model, as discussed in the previous note, the refracted ray reaches point P at angle of refraction of XM′P = 30°. But according to the sine law, that angle should be slightly over 34.7°, so the light should actually strike arc HP to the left of P and then be refracted toward the axis so as to intersect it below point P. Also, in the experiment illustrated in figure 41, p. 439, ray DL, which is incident on the interface between the glass and air at an angle of 45°, will not refract out of the glass into the air because its angle of incidence exceeds the critical angle for glass to air. And the same holds *a fortiori* for the light striking at a angle of 60°. These anomalies suggest at least three possibilities: the experiments described by Porta are simply thought experiments; he explained away his failure to get the expected results on the basis of flaws in the glass or in the inadequate narrowness of the rays; or he set the experiments up inaccurately enough that he got results that seemed to confirm his hypothesis.

[139]For a fairly negative appraisal of Maurolyco's and Porta's contributions to optics, see David C. Lindberg, "Optics in Sixteenth-Century Italy," in *Novità celesti et crisi del sapere,* supplement to *Annali dell'Istituto di Storia della Science* (1983): 131-148.

[140]Porta looms especially large in this regard. Not only does Kepler cite his *Magiae Naturalis* several times in the *Paralipomena,* but he mentions an "Optics" to which Porta adverts in *Magiae Naturalis,* book 17, chapter 10, claiming that he

sought but could not find a copy. This work could have been a preliminary draft of the *De Refractione*, which was published four years after the *Magiae Naturalis*. There are, in fact, indications that Kepler had either consulted the *De Refractione* or at least had it summarized for him. In particular, his account of Porta's analysis of the eye and its function suggests familiarity with books 3-7 of the *De Refractione*. It was certainly not based on the *Magiae Naturalis* because Porta says almost nothing about the eye in that work. My thanks to Yaakov Zik for pointing this out to me.

[141]See, e.g., the collection of articles in Simon, *Archéologie de la vision* (Paris: Seuil, 2003), pp. 77-181. See also Simon, "The Gaze in Ibn al-Haytham," *The Medieval History Journal*, 9 (2006): 89-98. In "Optique et perspective: d'Ibn al-Haytham à Alberti," Simon is unequivocal about "the revolution (the word does not strike me as too strong) brought about in optics by Ibn al-Haytham at the beginning of the eleventh century" (. . . *la révolution [le mot ne me paraît pas trop fort] opérée en optique au début du XIᵉ siècle par Ibn al-Haytham. . .*), *Archéologie*, 167.

[142]Lindberg, *Theories*, 58-65.

[143]Rashed, "Lumière et vision: L'Application des mathématiques dans l'optique d'Ibn al Haytham," in René Taton, ed., *Roemer et la vitesse de la lumière* (Paris: Vrin, 1978).

[144]Omar, *Ibn al-Haytham's Optics*.

[145]"Burning Instruments,"—my italics.

[146]"Ibn al-Haytham's Revolutionary Project in Optics: The Achievement and the Obstacle," in Jan P. Hogendijk and A. I. Sabra, eds., *The Enterprise of Science in Islam: New Perspectives* (Cambridge, MA: MIT Press, 2003), 85-118.

[147]This argument is perfectly consonant with Lindberg's contention in *Theories of Vision* that, in taking Alhacenian visual theory to its logical conclusion, Kepler was the last of the Perspectivists, who were of course the Latin heirs of Alhacen. In all fairness, however, I must point out that Lindberg never characterizes Alhacen's optics as revolutionary; indeed, in the so-called continuity debate over whether early-modern science represented a sharp break with medieval science (discontinuity), Lindberg inclines strongly to the side of continuity; see David Lindberg, *The Beginnings of Western Science* (Chicago: University of Chicago Press, 1992), 355-360.

[148]*Photismi*, ix.

[149]http://en.wikipedia.org/wiki/Alhazen.

[150]Greensboro, NC: Morgan Reynolds, 2007.

[151]*Journal of the International Society for the History of Islamic Medicine* (2003): 53-55.

[152]That Aristarchus can be meaningfully viewed as a precursor to Copernicus is implicit in the subtitle of T. L. Heath's study, *Aristarchus of Samos: The Ancient Copernicus* (Clarendon: Oxford University Press, 1913).

[153]George Saliba makes a convincing argument that some of these revisions and refinements—key ones in fact—were borrowed from late-medieval Arabic astronomers whose works, as far as we know, were not translated into Latin; see *Islamic Science and the Making of the European Renaissance* (Cambridge, MA: MIT Press, 2007). Saliba's work on medieval Arabic astronomy exemplifies the revisionist approach to Arabic science mentioned earlier, and it manifests some of the tendencies of that approach toward hyperbolic interpretation.

[154]N. R. Hanson, "The Copernican Disturbance and the Keplerian Revolution," *Journal of the History of Ideas*, 22 (1961): 169–184. In many ways, Peter Barker's recent evaluation of Copernican and Keplerian astronomy reflects this distinction; see, e.g., Peter Barker, Hanne Andersen, and Xiang Chen, *The Cognitive Structure of Scientific Revolutions* (Cambridge: Cambridge University Press, 2006), esp. chapter 6.

[155]See John R. Spencer, trans., *Leon Battista Alberti On Painting* (New Haven, CT: Yale University Press, 1966).

[156]For details, see Smith, *Alhacen's Theory*, pp. cvi-cix.

[157]See Smith, "Petrus Hispanus."

[158]See David C. Lindberg, *A Catalogue of Medieval and Renaissance Optical Manuscripts* (Toronto: Pontifical Institute of Mediaeval Studies, 1975), pp. 46-55 and 74.

[159]*Paralipomena*, 224.

MANUSCRIPTS AND EDITING

Manuscripts and Textual Matters: For this edition I have collated the same manuscripts I used in the previous editions of books 4-5 and 6: i.e., *F*, *P1*, *S*, *E*, *L3*, *O*, and *C1*. A full description of these seven manuscripts, along with the ten others not chosen for inclusion, can be found in my edition of books 1-3, and my reason for choosing these particular manuscripts can be found in my edition of books 4-5.[1] In the case of book 7, manuscript *F* is incomplete, ending abruptly after "maior" on p. 125, line 204. However, since that manuscript contains almost 90% of the text of book 7, I decided to continue using it rather than drop it or replace it.

In my edition of book 6 I concluded that two translators were at work on that book, the second one taking over at the beginning of the sixth chapter. That conclusion was based on terminological and stylistic features typical of the second portion of book 6 but not found, or only rarely found, in the first portion.[2] Many of these terminological and stylistic features are to be found in book 7 as well: the frequent use of "nam" and "tunc" to introduce new clauses; the overwhelming preference for "deceptio" and "fallacia" over "error" to denote "misperception"; the consistent imposition of chapter headings in the text; the designation of most of those chapter headings by "capitulum" rather than "pars"; choppy sentences strung together with the same introductory adverb; and a fairly consistent use of the hortatory subjunctive rather than the jussive.[3] All indications, therefore, are that the translator responsible for the second portion of book 6 was also responsible for book 7. It should be noted, however, that the two texts are rather dissimilar in style and content, the second portion of book 6 consisting almost entirely of geometrical theorems interspersed occasionally with narrative passages, book 7 consisting mainly of narrative passages interspersed occasionally with geometrical theorems. It should also be noted that book 7 has its own idiosyncrasies. For instance, the first two chapters are designated by "differentia" rather than "capitulum," and tenses and moods are randomly mixed in the same passage for no apparent reason.[4]

As I also pointed out in the edition of book 6, this shift in translators is mirrored in book 3, chapter 3, paragraph 13, where the Latin text devolves from a fairly faithful rendering to a highly abbreviated paraphrase of the Arabic original (see the reference in note 2 above). In both cases, moreover, the breaks between textual segments in books 3 and 6 are not clean. On the contrary, there is significant overlap, which is fortunate in that the overlapping texts allow a point-by-point comparison of vocabulary and style that leaves no doubt whatever that the textual segments on both sides of the break are fundamentally, sometimes even radically, different in style.

It is of course possible that the tripartite division of the Latin text into an initial segment (from the beginning of the treatise to book 3, chapter 3), an intermediate one (from book 3, chapter 3, to book 6, chapter 6), and a concluding one (from book 6, chapter 6, to the end of the treatise) is reflected in the Arabic version, or versions, on which the Latin translation was based. In that case, the Latin text could conceivably have been produced by one translator—or possibly more—who changed vocabulary and style according to the dictates of the Arabic original(s). If, however, we assume that the Arabic version was unbroken and consistent throughout so that the segmenting of the Latin text was due to a shift in translators, then it follows that at least two translators must have been at work on the Latin text: one reponsible for the middle segment between book 3, chapter 3, and book 6, chapter 6; another responsible for the segments flanking it. There is, in fact, reason to suppose that no fewer than *three* translators were involved in rendering the Arabic original into Latin. This supposition is based on a striking terminological anomaly in the concluding section that indicates that its translator was not the same as either of the translators responsible for the initial or intermediate segments.[5]

The Critical Text: The topical organization of book 7 is laid out explicitly according to chapter headings, which generally include a brief description of the subject to be covered in the given chapter. With one exception these headings are given in the majority of manuscripts, the one exception being chapter one (*prima differentia*) which is missing from all but *O* and *C1*. I decided to include that chapter heading because, even though it is absent in five of the seven manuscripts, it is consistent with majority usage for the rest of the book. As mentioned earlier, most of book 7 consists of narrative interspersed with geometrical theorems. According to my organization of the text, there are twenty-two such propositions, one in chapter 4, ten in chapter 5, and eleven in chapter 7, all of them labeled in consecutive order from 1 to 22. I have followed previous practice by inserting fairly strong spacing-breaks between propositions and weaker ones between cases. Each proposition is further demarcated by a numerical designation (e.g., [PROPOSITIO 1]). In addition to this particular organizational scheme, I have imposed a paragraph structure on the entire text in order to make it easier to follow, each paragraph being numbered according to its chapter (e.g., [7.1], which designates the first paragraph of chapter 7). This structure is my own; it derives neither from the Latin nor the Arabic text. Likewise, the punctuation of both the Latin text and English translation is entirely of my devising and is meant not so much to reflect the syntax of the Latin text as to ease the modern reader's way through the narrative.

Diagrams: In line with my practice in books 5 and 6, I have included as canonical only those diagrams that are keyed by lettering to a particular proposition. As before, I traced the figures to accompany the Latin text directly from diagrams scanned for the most part from manuscript *O*, although I have relettered them with a modern font and have occasionally reoriented them. I have designated the resulting text diagrams in capital letters according to the format "FIGURE 7.5.9," the number series indicating book (7), chapter (5), and position in the overall sequence of diagrams (9). For a detailed description of the criteria I followed in choosing and representing the appropriate diagrams, see *Alhacen on the Principles*, pp. cxi-cxv.

The Critical Apparatus: For the conventions used in the critical apparatus of this edition, I refer the reader to *Alhacen's Theory*, pp. clxxii-clxxiv.

The Translation and Commentary: The general guidelines I followed in translating book 7 are those discussed in *Alhacen's Theory*, pp. clxxiv-clxxvi and *Alhacen on the Principles*, pp. cxv-cxvii. Unlike the diagrams that accompany the Latin text, those that are matched to the translation are meant to reflect as faithfully as possible the actual conditions specified in the constructions and proofs. In order to distinguish these diagrams from their counterparts in the Latin text, I have designated them according to the lower-case format "figure 7.7.65," the number-series indicating book, chapter, and position in the overall sequence of diagrams. As I did in the editions of books 4-5 and 6, I have placed the figures that go with volume 2 at the end of volume 1 and vice-versa so that the reader can match the logical flow of each propositions with its appropriate diagram(s) without flipping back and forth between proposition and diagram within the same volume. The reference-aids provided in this edition are the same as those provided in the previous two: i.e., a Latin-English index keyed to technical terms in both Latin text and English translation; an English-Latin glossary for cross-referencing to that index; and a general index keyed primarily to the introduction and commentary.

NOTES

[1]See Smith, *Alhacen's Theory*, clv-clxi, and *Alhacen on the Principles*, cvii-cviii.

[2]For the complete argument, see Smith, *Alhacen on Image-Formation*, xlv-xlviii

[3]Cf. ibid., xlv-xlvi.

[4]A good example of this peculiar mixing of tenses and moods—as well as persons—is found in paragraph 2.6, p. 6 (Latin), where the sequence of verbs starts with a future active indicative in the second person plural (*accipiemus*), shifts to a present passive subjunctive in the third person singular (*adequetur*), then shifts to a future passive indicative in the third person singular (*fiet*) and ends with an present active subjunctive in the second person plural (*dividamus*).

[5]See note 104 to paragraph 5.13, pp. 372-373.

ALHACEN'S
DE ASPECTIBUS

LATIN TEXT

SEPTIMUS TRACTATUS
Libri Alhacen filii Alhaycen *De aspectibus*

Et sunt septem differentie. Prima differentia de proemio; secunda quod lux transit per diaffona corpora secundum verticationes line-
5   arum rectarum et reflectitur cum occurrerit corpori cuius diaffonitas fuerit diversa diaffonitati corporis in quo existit; tertia de qualitate reflexionis luminum in diaffonis corporibus; quarta differentia quod quicquid comprehenditur a visu ultra diaffona corpora quorum diaffonitas differt a diaffonitate corporis in quo visus existit cum fuerit
10 declinis a perpendicularibus existentibus super superficies eorum, comprehenditur secundum reflexionem; quinta de fantasmatibus; sexta quomodo visus comprehendit visibilia secundum reflexionem; septima de fallaciis visus que accidunt ex reflexione.

PRIMA DIFFERENTIA

15   [1.1] Predictum est in proemio quarti tractatus huius libri quoniam visus tribus modis comprehendit visibilia—videlicet, secun-
dum rectitudinem, et secundum conversionem a tersis corporibus, et secundum reflexionem ultra diaffona corpora que differunt in diaffonitate a diaffonitate aeris—et quod visus nichil comprehendit ex

---

1 septimus: septimi *R* / septimus . . . *aspectibus* (2) *om. S* / tractatus libri (2) *transp. L3*   2 libri
. . . *aspectibus om. R* / Alhacen: alhaycen *F* / Alhaycen: alhaichen *L3*; alhacchem *C1*; alhaithem
O; alhaicen *E*   3 et *om. R* / differentie: partes *R* / differentia de proemio: pars est proemium
R / proemio: prohemio *P1C1E*   4 transit: transeat *R* / per *om. L3ER* / *post* verticationes *scr. et
del.* lineas *P1*   5 et *inter. O* / et reflectitur *inter. a. m. L3*; *mg. a. m. E* / reflectitur: refringatur
R / occurrerit: occurreret *L3*; occurrit *R*   6 fuerit: fuit *FP1* / diversa *corr. ex* diversaverit *L3* /
diaffonitati: a diaffonitate *R* / corporis: corpori *SL3O* / reflexionis: refractionis *R*   7 diaffonis:
diaffonibus *SFP1C1E* / corporibus *rep. F* / differentia *om. R*   8 a visu *om. C1*   10 declinis:
decline *R* / a *inter. a. m. E* / existentibus: exeuntibus *FP1R*; *alter. in* exeuntibus *a. m. E* / superficies: superficiem *SL3ER* / *post* eorum *add.* quicquid *SO*   11 reflexionem: refractionem *R*
12 sexta *corr. ex* septima *a. m. E* / visus *om. FP1* / comprehendit: comprehendat *R* / reflexionem:
refractionem *R*   13 que: qui *C1* / reflectione: refractione *R*   14 prima differentia *om.*
SFP1L3ER   15 proemio: prohemio *P1L3C1*; *corr. ex* primo *F* / tractatus: tractus *S* / quoniam:
quod *R*   16 comprehendit: comprehendat *R* / *post* secundum *add.* reflexionem *P1*   17 et[1]
om. R / conversionem: reflexionem *R*   18 reflexionem: refractionem *R*   19 a: et *S* / a
diaffonitate *mg. a. m. F*

3

20   visibilibus nisi aliquo istorum trium modorum, et quod quolibet istorum modorum comprehendit visus visibilia et omnes res que sunt in
visibilibus et omnibus modis visionis quorum distinctio declarata est
in ultima differentia secundi tractatus.

[1.2] In precedentibus autem tractatibus declaratum est qualiter
25   visus comprehendit visibilia secundum rectitudinem et secundum
conversionem, et ostendimus diversitatem comprehensionis visus ad
visibilia secundum utrumque istorum modorum. Remanet ergo declarare qualiter visus comprehendit visibilia secundum reflexionem
ultra corpora diaffona. Nos autem in tractatu isto solummodo de
30   reflexione tractabimus; et manifestabimus formam reflexionis, et dis-
tinguemus eius modos, et dividemus proprietates eius, et declarabi-
mus quomodo accidit visui deceptio in huiusmodi visione. Et primo
proponemus quedam fundamenta que certificant quicquid dependet
ab hac re.

SECUNDA DIFFERENTIA
Quod lumen transit per diaffona corpora, et extenditur in
eis secundum lineas rectas, et reflectitur cum occur-
rerit corpori diaffono differenti in diaffonitate
a diaffonitate corporis in quo existit

40   [2.1] Quoniam lumen quidem transit in aera et extenditur secundum rectas lineas declaratum est in tractatu primo huius libri. Aer
autem est unum de corporibus diaffonis; aqua autem, et vitrum, et
diaffoni lapides lumen transit per ipsa, et extenditur secundum lineas rectas. Hoc autem comprehenditur per experientiam.

20 *post* et *scr. et del.* comprehendit visus visibilia E / quod *om.* S / quolibet: quodlibet S; *corr.*
ex quodlibet E / *post* istorum² *add.* trium SFP1   21 in visibilibus (22) *corr. ex* visibiles C1
22 et omnibus *inter. a. m.* L3   25 comprehendit: comprehendat R   26 conversionem:
reflexionem R   27 declarare: declararare R   28 qualiter: quomodo ER / comprehendit: comprehendat R / reflexionem: refractionem R   29 nos: non F / tractatu isto *transp.* R
30 reflexione: refractione R; *corr. ex* flexione O / reflexionis: refractionis R   32 accidit: ac-
cidat R / deceptio *om.* R / huiusmodi: huius O; *corr. ex* huius L3   33 quedam *om.* R / certificant: certificat S   35 secunda differentia *om.* R / secunda . . . existit (39) *om.* S   36 quod:
quoniam E / lumen transit: lux pertranseat R / et . . . eis (37) *om.* R   37 *post* secundum
add. verticationes R / lineas rectas: linearum rectarum R / reflectitur: refringatur R / occurrerit:
concurrerit L3; occurrit R   38 diaffono . . . diaffonitate: cuius diaffanitas fuerit diverse
R / differenti: differenter O; *corr. ex* existenti E   39 a diaffonitate *om.* L3E / diaffonitate:
diaffonitatem O   40 quoniam: quod R / lumen quidem *transp.* FP1 / quidem: quod
SL3O / *ante* transit *add.* quod FP1 / transit: transeat R / aera: aerem R; *corr. ex* aerem C1 / extenditur: extendatur R   41 rectas lineas *transp.* L3R / est *om.* C1 / libri: operis R   42 *post*
diaffonis *add.* per R / et¹ *om.* R / et² *inter.* L3   43 per ipsa *om.* R

45   [2.2] Si quis ergo experiri voluerit, accipiet laminam ex ere rotundam cuius diameter non est minus uno cubito, et sit spissitudo eius aliquantulum fortis. Et habeat horas rotundas perpendiculares super superficiem eius, et sit altitudo horarum eius non minor latitudine duorum digitorum. In medio autem dorsi lamine sit aliquod corpus

50   parvum columpnale rotundum cuius longitudo non minor latitudine trium digitorum, et sit perpendiculare super superficiem lamine. Et ponamus hoc instrumentum in retornativo in quo tornatorii retornant instrumenta cupri, et ponamus alterum dentem tornatorii in medio lamine et reliquum in medio extremitatis corporis quod est in

55   dorso lamine. Et radamus revolvendo hoc instrumentum abrasione vera quousque verificetur rotunditas horarum suarum intus et extra, et adequetur superficies interior et exterior, et fiant due superficies equidistantes. Et abrademus etiam corpus quod est in dorso donec fiat rotundum.

60   [2.3] Cum ergo instrumentum hoc fuerit perfectum per abrasionem, signemus in superficie eius interiori duos diametros secantes se perpendiculariter et sic transeuntes per centrum eius. Deinde signemus punctum in basi hore instrumenti cuius distantia ab extremitate alterius duorum diametrorum secantium se est latitudo unius digiti.

65   Deinde extrahamus ex isto puncto tertium diametrum transeuntem per centrum lamine quod extenditur in tota superficie eius. Deinde extrahemus a duobus extremis huius diametri duas lineas in superficie hore instrumenti perpendiculares super superficiem lamine. Deinde dividemus ex altera istarum duarum linearum tres lineas

70   parvas equales quarum prima sequetur superficiem lamine, et longitudo cuiuslibet earum sit cum quantitate medietatis grani ordeacii. Fient igitur super lineam perpendicularem tria puncta que sunt fines istarum linearum.

45 accipiet: accipiat R / laminam corr. ex laminas E / ere: aere OR   46 est: sit R / minus: minor R
47 aliquantulum: aliquantum P1   48 latitudine corr. ex altitudine C1   50 columpnale:
columpnare C1ER; corr. ex columpne a. m. L3 / latitudine corr. ex longitudine a. m. L3   51 perpendiculare: perpendicularis FC1OE / super om. L3 / lamine: laminum SL3O / post lamine scr. et
del. enim C1   52 retornativo: tornatorio R; corr. ex retornsitorio a. m. E / tornatorii: retornatorii
FP1; om. E / tornatorii retornant: tornant tornarii R   53 ante instrumenta add. retornarii E /
post et add. etiam L3 / post ponamus add. etiam E / tornatorii: tornatori S   54 lamine: laminum
SL3O / est om. E   55 radamus: radiamus S / abrasione: abratione S   57 et³ om. L3O / due
om. C1   58 equidistantes corr. ex equidistans a. m. C1 / abrademus: abredemus E   60 instrumentum hoc transp. R   61 interiori: exteriori O; interiore R / duos: duas R   62 sic: sint
C1E; sunt R   63 basi hore transp. L3 / hore corr. ex ore L3 / distantia: differentia SL3; corr. ex
instantia P1   64 alterius om. FP1 / duorum: duarum R   65 tertium: tertiam R   66 quod:
que R / post quod add. quidem diameter R / extenditur: extendatur R   67 extrahemus: extrahamus FP1ER / post huius add. tertii FP1 / post diametri scr. et del. 1 S   68 super om. S; inter. a.
m. L3   69 duarum linearum transp. S / linearum mg. a. m. L3   70 sequetur: sequitur L3ER
71 earum: harum ER / cum: in R / ordeacii: ordeacei SC1R   72 fient: fiant FP1C1; sunt O / sunt
om. S   73 istarum: illarum ER

[2.4] Et deinde reducamus hoc instrumentum ad tornatorium,
et signemus in ipso tres circulos equidistantes transeuntes per tria
puncta que sunt super lineam perpendicularem super extremitatem
diametri. Secetur igitur alia extremitas que est perpendicularis super
aliam extremitatem huius diametri per istos tres circulos, et fient in
ipsa tria puncta. Et fient in unoquoque trium circulorum duo puncta
opposita que sunt extrema alicuius diametri ex eius diametris.

[2.5] Deinde dividamus medium circulum ex istis tribus circulis
per trecentas sexaginta partes, et si possibile fuerit per minuta. Deinde
perforemus in hora instrumenti foramen rotundum cuius centrum sit
medius punctus trium punctorum que sunt super alteram duarum
linearum perpendicularium super extremitatem diametri lamine,
et sit medietas diametri eius in quantitate distantie que est inter circulos. Perveniet igitur circumferentia foraminis inter duos circulos
equidistantes qui sunt in extremitatibus.

[2.6] Postea accipiemus laminam subtilem quadratam aliquantule
spissitudinis cuius longitudo sit in quantitate altitudinis hore instrumenti et cuius latitudo sit prope hoc. Et adequetur superficies eius
quantum potest, et adequetur spissitudo eius etiam que sequetur alteram extremitatem eius quousque differentia communis inter superficiem faciei eius et inter superficiem spissitudinis eius fiet linea recta,
quam lineam dividamus in duo equalia a cuius medio extrahamus
lineam rectam in superficie faciei eius perpendicularem super illam
rectam lineam que est communis differentia.

[2.7] Deinde dividamus ex hac linea perpendiculari ex parte extremitatis que est super communem differentiam tres lineas equales
inter se et equales unicuique parvarum linearum que distincte sunt
super perpendicularem lineam in hora lamine. Fient igitur super lineam perpendicularem in facie lamine parve tria puncta. Deinde
perforabimus hanc parvam lineam foramine rotundo cuius centrum
sit medius punctus punctorum que distingunt lineas que sunt in ea,

74 et *om.* ER / tornatorium: tornatorum S   75 signemus: assignemus L3 / equidistantes *corr.*
ex estentes *a. m.* L3   77 diametri . . . extremitatem (78) *om.* S / secetur: secabitur R / igitur:
etiam L3 / *post* alia *scr. et del.* a P1 / extremitas: perpendicularis R   78 fient: fiant C1O
79 et . . . puncta[2] *mg. a. m.* E / fient: fiunt C1 / trium *inter.* O   80 *post* diametri *scr. et del.* eius
F / eius: ipsorum R   81 circulum: circuli S; *om.* O   82 per[1,2]: in R / trecentas: trescentas
S / sexaginta: sexagintas L3O / *post* si *scr. et del.* in F   84 medius punctus: medium punctum R
87 foraminis *om.* FP1; *corr. ex* foramis C1   88 qui: que SOC1   89 postea accipiemus
transp. FP1 / accipiemus: accipiamus L3ER   91 et[1]: vel S / adequetur: ad FP1   92 etiam:
et F; *om.* P1 / sequetur: sequitur P1E   94 faciei eius *mg.* F / superficiem *om.* P1 / spissitudinis
corr. ex spissitudinem P1 / fiet: sit ER   95 dividamus: dividemus R / a cuius *corr. ex* alicuius L3
96 *post* lineam *scr. et del.* extra FP1 / illam *om.* ER / illam . . . lineam (97): rectam lineam illam L3
97 rectam lineam *transp.* OER / *post* rectam *add.* scilicet C1 / *post* est *add.* linea S   101 super[1]
om. O / hora: ora E   103 lineam: laminam C1E; *om.* P1   104 medius punctus: medium
punctum R / punctorum *inter. a. m.* L3 / distingunt *alter. in* distinxerunt O

105   et sit medietas diametri eius equalis alicui uni linearum parvarum.
       Erit ergo hoc foramen equale foramini quod est in hora instrumenti.
       [2.8] Deinde signabimus super diametrum lamine super cuius ex-
       tremitates sunt due linee perpendiculares punctum in medio linee
       que est inter centrum lamine et extremitatem diametri que est in par-
110   te foraminis, et faciamus transire super hoc punctum lineam perpen-
       dicularem super diametrum. Deinde ponamus basim lamine parve
       super hanc lineam quousque differentia communis que est in parva
       lamina superponatur huic linee perpendiculari super diametrum, et
       erit punctus qui dividit differentiam communem que est in parva
115   lamina in duo equalia superpositus super punctum signatum in dia-
       metro lamine.
       [2.9] Hoc autem toto facto, applicetur parva lamina cum maiori
       completa applicatione et consolidatione. Tunc ergo foramen quod est
       in parva lamina erit oppositum foramini quod est in hora instrumen-
120   ti, et erit linea recta intellecta que copulat centra duorum foraminum
       in superficie circuli medii trium circulorum qui sunt in interiori hore
       instrumenti, et erit equidistans diametro lamine, et erit lamina parva
       que applicabitur puncto quasi hore astrolabii.
       [2.10] Hoc autem completo, secetur de hora instrumenti quarta
125   que est que sequitur quartam in qua est foramen ex quatuor quartis
       distinctis per duos primos diametros se perpendiculariter secantes,
       quarta propinqua tornatorio extra quod sunt hore, et adequetur locus
       sectionis donec fiat unum cum superficie lamine.
       [2.11] Deinde accipiamus regulam eris cuius longitudo non sit mi-
130   nor sed maior uno cubito, et sit quadrate figure quam circumdent
       quatuor superficies equales in latitudine duorum digitorum, et ade-
       quentur superficies eius quantum possunt donec fiant equales et
       habentes angulos rectos. Deinde perforetur in medio alicuius super-
       ficiei eius foramen rotundum cuius amplitudo sit tanta quanta pos-
135   sit recipere corpus quod est in dorso instrumenti quod revolvatur in

---

105 eius *om. S* / equalis *corr. ex* essentialis *a. m. L3*   107 signabimus *corr. ex* significabimus
F / cuius extremitates *transp. E*   110 et faciamus *inter. a. m. L3*   112 *post* hanc *scr. et
del.* parvam *E*   113 superponatur: supponatur *S*   114 punctus qui: punctum quod
R / *post* in *scr. et del.* para *F* / parva lamina (115) *transp. S*   115 superpositus: suppositus
S; positus *E*; positum *R*   116 lamine *corr. ex* linee *L*   117 toto *om. R* / maiori: maiore *R*
118 consolidatione *corr. ex* solidatione *O*   120 recta *om. R* / *post* que *scr. et del.* cent *O* /
centra: centrum *O*   121 circuli: tertii *FP1* / circuli medii *transp. L3* / interiori hore: interiore
hora *R*   123 hore: hora *R*   124 de *inter. E*   125 que¹ *om. SP1C1ER* / est¹ *om. P1C1ER*
126 duos primos: duas primas *R* / se perpendiculariter *transp. R*   127 quarta . . . hore *om.*
R / extra quod: ex quo *E*   128 fiat *corr. ex* faciat *C1* / unum: unus *R; om. O*   129 sit *om.*
SL3O   130 sed maior *om. O* / uno cubito *transp. FP1* / sit *om. R*   131 digitorum *corr. ex*
angulorum *E*   132 *post* eius *add.* in *ER* / possunt: possent *C1*; potest *R*   134 quanta:
ut *R*   135 quod: ut *R*

ipso non levi revolutione sed difficili, et sit foramen perpendiculare super superficiem regule et transiens in regulam ad aliam partem.

[2.12] Deinde ponamus instrumentum super regulam, et mittamus corpus quod est in dorso instrumenti in foramine quod est in medio
140   regule donec superponatur superficies instrumenti superficiei regule. Hoc autem facto, secetur illud quod superfluit ex extremitatibus regule super diametrum lamine, nam regula longior est quam diameter lamine, quia sic posuimus eam. Cum ergo secaverimus duas superfluitates ex duabus extremitatibus regule, reducemus has duas su-
145   perfluitates, et ponemus illas super duas extremitates regule ita quod ponemus duas extremitates superfluitatum super duas extremitates illius quod remansit de regula. Et applicabimus superficiem extremitatum cum superficie dorsi instrumenti, et erit illud quod ponetur ex utraque duarum superfluitatum super residuum regule equale latitu-
150   dini unius digiti. Hac autem positione considerate, eminebuntur due superfluitates super duas extremitates regule, et si perforatum fuerit illud quod superfluit ex corpore, et missum fuerit in foramine eius stilus cupreus qui ipsum prohibeat exire erit melius. Hoc autem perfecto, perficietur instrumentum, et hec est forma dorsi instrumenti.

155   [2.13] Deinde accipiat experimentator regulam cupream parve latitudinis cuius latitudo sit duplum diametri foraminis quod est in hora instrumenti, et cuius spissitudo sit equalis diametro foraminis, et cuius longitudo non sit minor medietate cubiti. Et verificabitur ista regula donec fiat valde recta et vera, et fiant superficies eius
160   equales et equidistantes. Deinde oblique secabimus alteram latitudinem eius quousque finis longitudinis eius contineat cum fine latitudinis eius angulum acutum ut possit homo declinare et movere eam

136 perpendiculare *corr. ex* perpendiculamen *F*   137 et *inter. O* /in regulam *om. R* /*post* partem *add.* regule *R*   139 *post* corpus *add.* quod est in instrumento *E* /dorso instrumenti *transp. R* /instrumenti . . . superficies (140) *om. S* /foramine: foramen *R*   140 regule² *corr. ex* linee *E* 141 *post* secetur *scr. et del.* e *P1*; *add.* ergo *S* /illud *corr. ex* illum *F* /*post* quod *scr. et del.* in *S*   142 *post* super *add.* id est ultra *E* /longior est: longiorem *S* /*post* est *rep. et del.* longior est *E*   143 sic *corr. ex* si *C1* /secaverimus: secuerimus *R*   144 *ante* ex *scr. et del.* et ponemus illas super duas *S* /reducemus *corr. ex* reducamus *E*   145 quod: ut *R*   146 ponemus: ponamus *R* /*post* extremitates¹ *add.* superfluitates *P1* /superfluitatum: superfluitatem *S* /superfluitatum . . . extremitates² *mg. F* /extremitates²: extremites *S*   147 illius . . . regula *om. P1*   148 ponetur: ponemus *FP1*; ponitur *O*   149 superfluitatum *corr. ex* fluitatum *O*   150 considerate: considerata *R* /eminebuntur: eminebunt *FP1R*   152 ex: de *R*; *inter. a. m. L3* /*post* corpore *add.* in dorso instrumenti *R* /missum: missus *E*; immissus *R*; *alter. in* missus *C1* /fuerit: fuit *S* /foramine: foramen *R* /eius: acus *O*; *om. S*   153 cupreus: ferreus *R* /hoc autem *transp. S* /perfecto: profecto *S*   154 perficietur: perfectum erit *R* /et . . . instrumenti *om. R* /dorsi instrumenti *transp. L3*   156 *post* sit *add.* duplex *O* /duplum: dupla *R*   157 *post* instrumenti *rep. et del.* foraminis (156) . . . instrumenti *E*   158 cubiti: cupri *O*; *corr. ex* cubito *E*   159 ista regula *transp. R* /*post* fiat *scr. et del.* regula *F* /eius equales (160) *transp.* deinde *corr. C1*   160 et *inter. C1* /deinde oblique *alter. in* de obliquo *a. m. E* /*post* deinde *scr. et del.* de *L3* /alteram: altera parte *R* /latitudinem: longitudinem *S*   162 *post* possit *add.* sic facilius *R* /homo *om. ER*

quocumque voluerit, et ponet latitudinem eius ex alia extremitate
perpendicularem super finem longitudinis eius. Deinde dividemus
165 hanc latitudinem in duo equalia, et extrahemus a loco divisionis lineam in superficie faciei regule que extenditur in longitudine eius, et
erit perpendicularis super latitudinem eius.

[2.14] Cum ergo hec regula fuerit superposita superficiei lamine,
erit superficies eius superior in superficie circuli medii trium circulo-
170 rum signatorum in interiori hore instrumenti, nam spissitudo huius
regule est equalis diametro foraminis, et diameter foraminis equalis perpendiculari exeunti de centro foraminis quod est in hora instrumenti ad superficiem lamine, quia diameter foraminis est equalis
duabus lineis trium linearum parvarum que distincte sunt de linea
175 perpendiculari in interiori hore instrumenti. Cum ergo hec regula
fuerit erecta super horam ipsius et fuerit superficies latitudinis eius
super superficiem lamine, tunc linea descripta in medio eius erit in
superficie medii circuli predicti, quia perpendicularis que egreditur
a quolibet puncto huius linee ad finem longitudinis regule est equa-
180 lis perpendiculari que egreditur a centro foraminis ad superficiem
lamine, nam utraque istarum perpendicularium est equalis diametro
foraminis.

[2.15] Cum ergo experimentator voluerit experiri transitum luminis in aqua per hoc instrumentum, accipiet vas rectarum horarum, ut
185 cadum cupri, aut ollam figuli, aut consimile. Et sit altitudo horarum
eius non minor medietate cubiti, et sit diameter circumferentie eius
non minor diametro instrumenti. Et adequentur hore eius donec superficies que transit per horas eius sit superficies equalis, et ponamus
in fundamento eius corpus diversarum partium aut diversorum colo-
190 rum, ut anulus aut argentum depictum, aut depingatur in fundamento aque pictura manifesta.

163 *post* quocumque *add.* quis *R* / *post* alia *add.* parte *C1*   165 latitudinem: longitudinem
O / extrahemus: extrahamus *SC1*; *corr. ex* extrahamus *OE* / a: ex *R*   166 regule *corr. ex* linee *a.
m. L3* / extenditur: extendatur *R*   170 signatorum: significatorum *S*; figuratorum *R* / interiori
hore: interiore hora *R*   171 diameter: diametrum *FP1* / *post* foraminis² *add.* est *R*; *scr. et del.* erit
P1 / equalis² . . . foraminis (172) *mg. F*   172 exeunti: eunti *FP1* / de: e *R*   173 *post* equalis
scr. et del. dia *F*   175 interiori hore: interiore hora *R* / ergo *inter. E*   176 erecta: recta *P1*
177 *post* superficiem *scr. et del.* eius *L3*   178 medii *inter. a. m. L3*   179 longitudinis:
latitudinis *C1*   180 perpendiculari: perpendicularis *L3*   183 *post* transitum *scr. et del.* la
F / luminis: liminis *S*   184 *post* in *add.* quacumque hora *P1* / aqua: qua *F*; *corr. ex* qua *O*; *om.*
P1 / accipiet: accipiat *FP1L3* / *post* horarum *scr. et del.* eius non minor medietate *S*   185 cupri:
cupreum *R* / figuli: figulinam *R*; *corr. ex* singuli *L3*   186 diameter: diametrum *C1*   187 adequentur: adequantur *P1*   188 *ante* que *add.* eius *C1* / horas: hora *O*   189 fundamento:
fundo *ER*   190 anulus: anulum *C1ER* / aut¹ *inter. P1* / *post* depictum *scr. et del.* de ex B per
lineam BE quod reflectitur per lineam AB si ergo forma A reflectatur ad B ex alio puncto quam
ex E sequitur quod forma B reflectatur ad A ex illo puncto sed iam declaratum est quod cum
forma extensa fuerit per lineam BE et reflexa per lineam *L3* / depingatur *corr. ex* depinguatur
L3 / fundamento: fundo *R*   191 aque: aliqua *P1*; eius *ER*

[2.16] Deinde fundatur in vas aqua clara donec impleatur, et expectetur donec motus eius quiescat. Cum ergo motus eius quieverit, erigatur aspiciens aut sedeat erectus, et aspiciat ad vas, et apponat visum suum corpori quod est in fundo aque aut picture que est in fundo aque donec linea inter visum et medium illius corporis aut illius picture sit perpendicularis super superficiem aque quoad sensum, et aspiciat corpus quod est in fundo aque aut picturam. Tunc inveniet illud eo modo quo est, et inveniet ordinationem suarum partium inter se adinvicem eo modo quo ordinarentur si aspiceret illud cum vas esset vacuum. Hoc autem declarato, certificabitur quod illud quod comprehenditur in fundo aque, cum aspexerit illud eadem positione qua aspexit corpus quod est in fundo aque et picturam, comprehenditur secundum ordinationem suarum partium.

[2.17] Hoc autem certificato, si quis voluerit experiri transitum lucis, eligat locum super quem oritur lux solis in quo ponat vas, et preservet se ut superficies circumferentie vasis sit equidistans orizonti. Hoc autem potest observari hoc modo quod sit circumferentia superficiei aque equidistans circumferentie vasis: et si intus in vase aut prope circumferentiam eius fuerit signatus circulus equidistans circumferentie vasis, erit melius ad hoc quod circumferentia superficiei aque comparetur ad circumferentiam circuli.

[2.18] Deinde experimentor debet imponere instrumentum rotundum intra hoc vas ita quod due regule parve posite super duo extrema regule maioris superponantur hore vasis ex utraque parte. Tunc medietas instrumenti cum regula extensa in longitudine instrumenti erunt intra vas. Deinde addatur aqua aut diminuatur de ea donec fiat in superficie aque unum cum centro instrumenti, et sit aqua clara. Deinde revolvetur instrumentum in circuitu vasis donec obumbretur illud quod est intra aquam ex horis eius ab illo quod est supra aquam ex horis eius. Tunc teneatur regula altera manuum

192 fundatur: infundatur *R*/fundatur in *transp. O*/in vas *mg. a. m. E*/in . . . clara: aqua clara in vas *ER*/*post* vas *scr. et del.* qua clara *P1*/aqua: aliqua *F*/expectetur: exspectetur *S* 194 aspiciat *corr. ex* respiciat *C1* 195 *post* in¹ *rep. et del.* est in *P1* 196 illius picture (197) *transp. ER* 197 *post* super *scr. et del.* f *E* 198 aque *om. ER* 199 illud: illam *L3ER*/inveniet *corr. ex* inveniret *F* 200 adinvicem *om. R*/ordinarentur:ordinantur *FP1O*/illud: illum *F* 201 certificabitur: certificatur *R*/quod² *om. FE* 202 illud: illum *F*/positione: ratione *L3* 203 qua: quam *E*/aspexit: aspexerit *SP1C1E*/et: aut *R*/picturam: pictura *E* 206 ponat: ponet *F* 207 preservet: observet *R*/ se *om. R* 208 quod: ut *R* 209 superficiei: superficie *S*/equidistans: equidistantis *S*/et . . . vasis (211) *om. S* 210 aut *om. R* 211 vasis: basis *E*/quod: ut *R* 214 intra: ad *O*/quod: ut *R*/parve *corr. ex* ma *F* 216 cum: et *R* 218 in *om. C1ER*/superficie: superficies *C1R*/*post* superficie *scr. et del.* unu *P1*/unum: una *R*; *corr. ex* vinum *L3*/cum *inter. O* 219 revolvetur: revolvatur *ER* 220 *post* obumbretur *scr. et del.* illud *C1*/est . . . aquam: intra aquam est *L3*/*post* eius *scr. et del.* ad quod est super aquam ex horis eius *E*/ab . . . eius (221) *om. ER* 221 manuum: manu *L3ER*

et revolvatur instrumentum reliqua manu super se in circuitu centri
eius donec foramen quod est in hora instrumenti sit oppositum corpori solis, et transeat lumen solis in foramen, et perveniat ad alterum
225   foramen, et transeat per aliud foramen. Cum ergo pertransiverit forma in duobus foraminibus, perveniet ad fundum aque. Tunc experimentator preservabit quod situs lucis in regula de secundo foramine
sit situs equalis.

[2.19] Hoc ergo situ preservato et luce preveniente ad superfi-
230   ciem aque, auferet experimentator manus suas ab instrumento, et
stet erectus vel sedeat erectus, et inspiciat ad fundum aque ex quarta
cuius hore sunt scisse, et preservet positionem quam preservaverat
cum aspexerit corpus quod erat in fundo aque ut sit certus quod illud
quod videt est secundum quod est. Tunc ergo cum intuebitur illud
235   quod est intra aquam de hora instrumenti, inveniet lumen pertransiens ex duobus foraminibus super anterius hore instrumenti quod
est intra aquam.

[2.20] Et inveniet lumen inter duos circulos equidistantes extre-
mos de tribus circulis signatis in anteriori parte hore instrumenti,
240   aut addetur super distantiam que est inter circulos modicum, et erit
additio eius ex duobus lateribus circulorum equalis. Sequitur ergo
ex positione quod punctus qui est in medio luminis apparentis intra aquam quod est super interiorem partem hore instrumenti sit per
medium circulum trium circulorum equidistantium qui sunt in in-
245   teriori parte hore instrumenti. Et hoc lumen quod est intra aquam
erit manifestius, quia hora superior instrumenti que circumdat superius foramen obumbrat interiorem partem hore instrumenti que
circumdat lumen quod est in interiori parte hore instrumenti, et sic in

222 instrumentum *om.* E / instrumentum . . . manu: reliqua manu instrumentum R / super se
corr. ex superficie L3   224 in: per R / *post* foramen *add.* hore instrumenti R / *post* alterum *scr.
et del.* foraminum L3   225 *post* foramen[1] *add.* tabule parve R / aliud: illud R / foramen[2] *om.*
R / pertransiverit: pertransierit ER / forma: foramen S; foramina C1; formam O; *corr. ex* formam L3
226 *ante* in *add.* lucis ER / in . . . foraminibus: per duo foramina R / perveniet: perveniat L3; *corr.
ex* perveniat P1 / fundum: foramen SFP1O; *corr. ex* foramen *a. m.* L3   227 preservabit: ob-
servabit R / quod: ut R   229 hoc ergo *transp.* L3 / ergo: autem ER / situ *inter.* E   230 auferet:
auferat R; aufert L3; *corr. ex* aufert O   231 stet *corr. ex* fiat *a. m.* L3 / erectus[1] *om.* R / vel . .
. erectus[2] *om.* L3 / ex *corr. ex* et L3   232 scisse: fixe O; cisse E; abscisse R / preservet: servet
R / preservaverat: servaverat R   233 aspexerit: aspexerat ER   234 cum *om.* FP1 / illud
om. FP1   235 de: ad S / lumen *inter.* O / pertransiens: transiens L3   236 anterius: interius
P1; superficiem R / quod: que R   238 inveniet *corr. ex* veniet C1   239 anteriori: interiore
R / hore instrumenti *transp.* ER   240 aut *corr. ex* ut E   242 punctus qui: punctum quod
R / est *inter. a. m.* L3 / luminis *mg.* C1   243 est *inter. a. m.* L3; *om.* P1 / hore instrumenti *transp.
deinde corr.* C1   244 circulum: circuli SC1; circulorum FP1L3E / equidistantium *inter. a. m.*
E / interiori: interiore R   245 *post* instrumenti *scr. et del.* et sic in illo loco non erit ex interiori
parte L3 / et . . . instrumenti (246) *om.* S / quod: que L3   246 manifestius: manifestum R / quia:
quod ER; *corr. ex* qua L3   247 interiorem: inferiorem FP1   248 in[1] *om.* L3 / interiori:
interiore R

illo loco non erit ex interiori parte hore instrumenti aliquid de lumine
250   solis nisi lumen quod exit ex duobus foraminibus.

[2.21] Deinde experimentator accipiet lignum minutum, sicut
acum, et applicet eam in exteriori parte superioris foraminis quod est
in hora instrumenti, et preservet se quod acus transeat per medium
foraminis. Deinde aspiciat supra vas, et preservet positionem quam
255   prius mensuravit. Tunc videbit umbram acus in medio lucis. Deinde
incurret acum, attrahendo ipsam donec extremitas eius sit in medio
foraminis, et intueatur lumen quod est intra aquam et quod est in superficie aque. Tunc inveniet umbram extremitatis acus in medio lucis
que est intra aquam et in medio lucis que est in superficie aque.

260   [2.22] Deinde mutet positionem acus, et ponat extremitatem eius
etiam apud medium foraminis, et intueatur umbram. Tunc inveniet
umbram extremitatis acus apud medium lucis. Deinde elevet acum,
et inveniet lucem redeuntem ad suum statum intra aquam et in superficie aque. Deinde applicet acum in latere foraminis, et ponat eam
265   cordam in foramine non diametrum, et intueatur lumen quod est intra aquam et in superficie aque. Tunc inveniet in utroque illorum
umbram que est corda. Deinde elevet acum. Tunc inveniet lumen
rediens ad suum locum, et si mutaverit situm acus in lateribus foraminis, inveniet umbram semper in latere luminis.

270   [2.23] Declarabitur ergo ex hac experientia quod punctus qui est
in medio lucis que est intra aquam, que est circumferentia medii circuli, non exivit lux ad illum nisi ex puncto qui est medium lucis que
est in superficie aque, et quod punctus qui est medium lucis que est
in superficie aque non exibit lux ad ipsum nisi ex puncto quod est
275   centrum foraminis superioris, et transit per punctum quod est centrum foraminis inferioris, scilicet foraminis quod est in horis, nam si

249 loco *om. L3 /* interiori: interiore *R*   250 ex: a *FP1*; de *C1*   251 accipiet: accipiat *L3ER /*
sicut: sive *ER*   252 in: ex *L3 /* exteriori: exteriore *R*; interiori *P1L3 /* est *om. O*   253 preservet:
observet *R /* se *om. R /* quod: ut *R /* acus *corr. ex* arcus *C1 /* transeat: pertranseat *FP1 / post* per *scr. et
del.* foramen *S*   254 preservet: servet *R*   255 prius mensuravit *transp. ER /* mensuravit:
mensuraverat *SC1O /* videbit: videbunt *S /* acus *corr. ex* arcus *C1*   256 incurret: incurvet *FP1R /*
acum *om. P1*   259 in¹ *om. E*   261 apud: quod *S*   262 deinde . . . acum *om. L3 /* elevet:
levet *R*   264 deinde . . . aque (266) *mg. a. m. L3*   267 elevet: levet *R /* acum *corr. ex* arcum *L3*
268 suum locum *transp. R / post* mutaverit *scr. et del.* suum *C1 /* lateribus *corr. ex* partibus *O*
270 experientia *corr. ex* experiment *L3 / post* quod *add.* ad *R /* punctus qui: punctum quod *R*
271 intra: inter *L3 / post* intra *scr. et del.* a *P1 /* que²: et *E /* que est² *om. R /* est² *om. E / post* est²
add. in *SR /* circumferentia: circumferentiam *E*   272 ad illum *om. R /* qui: quod *C1R /* qui . . .
puncto (274) *mg. C1* (que [272]: quod; qui [273] *corr. ex* quod; que [273]: quod)*/* medium . . . est¹
(273) *mg. F /* que: qui *F*   273 in . . . est³ *om. E /* superficie . . . in (274) *mg. O / post* quod *add.* ad
R / punctus qui: punctum quod *R / post* est² *add.* inter *L3 /* medium: medius *O*   274 superficie:
medio *SP1L3OE /* exibit: exivit *R /* ad ipsum *om. R*   275 *post* foraminis *scr. et del.* foram *F*;
scr. et del. et *P1 /* superioris . . . foraminis¹ (276) *rep. S /* transit: transivit *R /* punctum . . . est *om. R*
276 inferioris . . . foraminis² *om. FP1 /* scilicet foraminis *om. R / post* horis *add.* aliis *FP1C1R*

non transisset per centrum foraminis inferioris, non manifestaretur
medium lucis que est in superficie aque, cum acus esset in medio
foraminis inferioris, sed non manifestaretur de luce que est in super-
280 ficie aque nisi locus alius a medio eius.

[2.24] Lux ergo que pervenit ad punctum quod est centrum lucis
que est in superficie aque et lux que extenditur in aere non extenditur
nisi secundum lineas rectas. Lux ergo que transit per centra duorum
foraminum extenditur secundum rectitudinem linee transeuntis per
285 centra duorum foraminum. Hec autem lux est illa que pervenit ad
medium lucis que est in superficie aque. Punctus ergo qui est in medio lucis que est in superficie aque est in linea recta transeunte per
centra duorum foraminum, et hec linea est in superficie medii circuli
de tribus circulis signatis in interiori parte hore instrumenti, et est illi
290 diameter, quia hec linea est equidistans diametro circuli qui est in superficie lamine. Cum ergo punctus qui est in medio lucis que est in
superficie aque fuerit super hanc lineam, tunc iste punctus est in superficie circuli medii predicti. Punctus autem qui est in medio lucis
que est intra aquam est in circumferentia medii circuli; ergo hec duo
295 puncta sunt in superficie medii circuli.

[2.25] Si ergo lux que est in superficie aque latuerit et non fuerit
bene manifesta, tunc experimentator mittet illam regulam in aquam,
et applicet horam eius in superficie lamine, et ponat superficiem in
qua signata est linea sequentem superficiem aque, et moveat illam
300 donec superficies eius fiat cum superficie aque. Cum ergo superficies
regule fuerit cum superficie aque, et regula fuerit erecta super horam
eius, tunc linea que est in superficie ipsius erit in superficie circuli
medii que transit per centra duorum foraminum. Hac autem positione preservata, apparebit lux que est in superficie aque super super-
5 ficiem regule, et inveniet medium lucis super lineam que est in medio
regule. Et si acus fuerit posita super medium superioris foraminis,
tunc linea que est in medio regule obumbrabitur, et si extremitas acus
fuerit posita super centrum foraminis, apparebit umbra extremitatis

277 non¹ *om. S/*non² *om. P1*   278 *post* in² *add.* foraminibus *C1*   279 inferioris *om. S/*sed *om.*
*L3/*manifestaretur: manifestatur *P1*   280 medio: centro *R*   281 *post* ergo *scr. et del.* pro *E/*ad:
a *O*   285 lux est *transp. FP1/*est *inter. a. m. L3*   286 punctus: punctum *R/*punctus ergo *transp.*
*F/*ergo *mg. F/*qui: quod *R*   288 centra: centrum *FP1O*   289 interiori: interiore *R/*illi: illud
O; illis *S;* illius *R*   291 punctus qui: punctus quod *R*   292 iste: ille *FP1;* illud *R; corr. ex* est
*S/*punctus: punctum *R/post* punctus *inter.* qui *a. m. E*   293 punctus: punctum *R/*qui: quod *R*
296 latuerit: latuit *O/*et *inter. a. m. F*   297 *post* illam *add.* minorem *R*   298 *post* applicet
add. illam *L3; scr. et del.* in *C1*   299 qua *corr. ex* aqua *O/*illam: eam *ER*   300 fiat cum: sit in
*FP1/*cum² . . . aque (1) *mg. a. m. E*   1 regula fuerit *transp. ER* (regula *inter. a. m. E*)   3 que:
qui *R/*transit *corr. ex* transsit *L3/*centra: centrum *O*   4 lux *inter. a. m. E/*lux que *transp. deinde*
*corr. L3/*est *om. P1L3*   5 regule *corr. ex* aque *C1/*inveniet *corr. ex* veniet *C1*   6 fuerit: sit *R; om.*
*E/*fuerit posita *transp. L3C1O/*medium superioris *transp. deinde corr. C1*   8 umbra extremitatis
corr. ex existentis *a. m. L3/*extremitatis: extremitas *E*

acus in medio lucis que est super regulam. Et si acus fuerit ablata,
10 redibit lux sicut erat.

[2.26] Cum hac ergo regula apparebit lux que est in superficie
aque apparitione manifesta, et manifestabitur quod est super lineam
transeuntem per centra duorum foraminum. Et iam posueramus superficiem aque apud centrum lamine. Cum ergo superficies regule
15 cum superficie aque fuerit, erit superficies regule transiens per centrum lamine, et tunc erit remotio centri lucis a centro lamine equalis
medietati latitudinis regule, que est equalis perpendiculari cadenti a
centro foraminis super superficiem lamine. Et sic erit centrum lucis
que est in superficie regule centrum circuli medii.

20 [2.27] Deinde oportet experimentatorem auferre regulam subtilem, et mittere eam iterum in aquam, et applicare superficiem latitudinis eius cum superficie lamine, et ponere angulum eius acutum
apud centrum lucis que est intra aquam, scilicet angulus qui est in
superficie eius superiori. Deinde moveat regulam donec acuitas eius
25 inferior, que est in superiori lamine, transeat per centrum lamine, et
sic acuitas eius superior transibit per centrum circuli medii. Punctus
ergo ex linea superiori regule qui est in superficie aque est centrum
circuli medii; est ergo centrum lucis que est in superficie aque, et erit
longitudo eius diameter ex diametris medii circuli.

30 [2.28] Hac autem ratione preservata, accipiat experimentator
acum longam, et mittet eam in aquam, et ponat capud suum in puncto ultimitatis regule, et intueatur lucem que est intra aquam. Tunc
inveniet umbram acus secantem lucem, et inveniet umbram capitis
acus apud cornu regule que est apud medium lucis. Deinde mutet
35 positionem acus, et capud eius sit loco eius ex fine regule. Tunc mutabitur situs umbre ex luce que est intra aquam, et erit umbra capitis
acus inseperabilis a medio lucis. Deinde auferat acum, et tunc redibit
lux ad suum locum. Deinde mittat acum in aquam iterum, et ponat
capud eius in alio puncto finis regule, et intueatur umbram. Tunc

9 super *inter. a. m.* E   12 apparitione: aparitione *S*/super: supra *FP1R*   13 posueramus:
proservamus *S*   15 cum . . . fuerit: fuerit cum superficie aque *R*/erit: transibit *R*/transiens
om. R   16 *post* lucis *add.* que est in superficie regule *C1*   17 latitudinis: latitudini *S*; *om.* P1
18 super *om.* F/lucis *corr. ex* locus *C1*   19 que *corr. ex* qui *C1*   20 auferre: afferre *SL3C1E*
21 superficiem *inter. a. m.* E   23 angulus: angulum *C1R*/*post* angulus *scr. et del.* in *F*   24 superiori: superiore *R*/eius inferior (25) *transp.* F   25 que: qui *C1*/que . . . lamine[1] *om.* R
26 acuitas *corr. ex* adcuitas *L3*/punctus: punctum *R*   27 ergo ex linea *corr. ex* ex linea ergo *L3*/
superiori: superiore *R*/qui: quod *R*; *corr. ex* que *C1*   28 et *inter.* O   29 diameter: diametri *R*
30 autem ratione *transp. SC1O* (autem: ante *S*; ratione: positione *C1O*)   31 mittet: mittit *FP1*;
mittat *R*/puncto: punctum *ER*   32 ultimitatis *corr. ex* ultimietatis *L3*   33 *post* umbram[1] *add.*
capitis *S*/*post* acus *scr. et del.* apud cornu regule *S*   34 que: quod *R*/medium lucis *transp. FP1*
35 eius[1] *inter. a. m. L3*/sit: si *O*/*post* sit *add.* in *FP1L3R* (*inter. a. m. L3*)   36 est *om.* E/capitis *inter.
a. m. C1*   37 inseperabilis: inseparabilis *FP1R*/tunc *om.* R   38 suum locum *transp.* R/mittat
corr. ex mittet *L3*   39 alio: aliquo *L3*/tunc: donec *R*; *corr. ex* donec *L3E* (*a. m. L3*)

40   inveniet secantem lucem que est intra aquam, et inveniet umbram ca-
      pitis acus in medio lucis. Deinde mutet positionem acus super mul-
      titudinem punctorum ex acuitate regule, et inveniet umbram capitis
      eius semper in medio lucis.
          [2.29] Declarabitur ergo ex hac experientia declaratione manifes-
45   ta quod lux que est in puncto mediante lucem que est intra aquam,
      que est super circumferentiam medii circuli, est perveniens ad illum
      punctum a puncto quod est medium lucis que est in superficie aque.
      Et declarabitur cum hoc quod hec lux extenditur super lineam rec-
      tam que est finis regule, nam experientia eius per extremitatem acus
50   ex diversis locis in fine regule ostendit illam transeuntem per omne
      punctum finis regule. Hac ergo via experimentabitur transitus lucis
      per corpus aque, ex qua declarabitur quod extensio lucis per corpus
      aque est secundum verticationes rectarum linearum.
          [2.30] Deinde oportebit experimentatorem quod ponat super
55   centrum lucis signum fixum cum scalpsione. Deinde quando fue-
      rit experimentator intuens punctum quod est in medio lucis que est
      intra aquam, inveniet ipsum non equidistans duabus extremitatibus
      diametri lamine, scilicet extra duas lineas perpendiculares super ex-
      tremitatem diametri lamine qui est intra aquam. Et inveniet decli-
60   nationem eius ab ista linea ad partem in qua est sol, et inveniet inter
      punctum quod est centrum medii lucis et punctum quod est com-
      munis differentia linee perpendiculari super extremitatem diametri
      lamine et puncto medio quod est extremitas diametri medii circuli
      transeuntis per centrum foraminum, inveniet dico distantiam sensi-
65   bilem.
          [2.31] Hoc declarato, oportet mittere regulam subtilem in aquam,
      et applicare eam cum superficie lamine, et ponere terminum regule
      super centrum lamine, et movere regulam quousque acuitas eius sit
      perpendicularis super superficiem aque quoad sensum. Tunc igitur
70   inveniet centrum lucis que est intra aquam inter acuitatem regule
      et lineam perpendicularem super diametrum lamine. Declarabitur

40 secantem: secante L3 / intra: infra FP1; inter O   43 semper: super S   45 in om. S /
lucem: luce SE / intra corr. ex inter F   46 est perveniens: pervenit R / illum: illud P1L3ER
47 a corr. ex ad P1 / post est¹ scr. et del. in C1   48 post et rep. et del. et F / extenditur corr. ex
ostenditur P1   49 eius: cuius C1O / per: secundum O / post extremitatem scr. et del. eius F
51 post regule scr. et del. hec F / post experimentabitur scr. et del. transl F   52 qua: quo R / extensio: extensum O   54 quod om. R / ponat: ponere R   55 scalpsione: sculpsione C1E;
sculptione R / fuerit om. R   56 intuens: intuebitur R   57 post inveniet scr. et del. e L3
58 scilicet: sed FP1C1R / lineas perpendiculares transp. FP1 / post perpendiculares add. que sunt
L3C1ER (inter. a. m. L3)   59 qui: que ER   60 et inter. O / inter om. S   61 quod est
centrum: centrum quod est O / medii: medie R   64 foraminum: foraminis R   66 mittere
mg. F   70 est om. O / inter: intra E / inter . . . regule: interiacentem regulam FP1

ergo ex hoc quod hec reflexio est ad partem perpendicularis exeuntis
a loco reflexionis perpendicularis super superficiem aque. Cum ergo
certus fuerit experimentator de hoc, oportebit eum signare apud ex-
75 tremitatem regule que est super circumferentiam medii circuli que
est extremitas perpendicularis exeuntis a centro medii circuli perpendicularis super superficiem aque signum fixum, ut primum quod signatum est apud centrum lucis.

[2.32] Et iam declaratum est quod lux que pervenit ad punctum
80 quod est centrum lucis que est intra aquam est lux extensa secundum
rectitudinem linee continuantis duo centra foraminum, et hec linea
pervenit ad centrum medii circuli equidistantis superficiei lamine, et
est illi diameter. Si hec linea fuerit extensa in ymaginatione secundum rectitudinem intra aquam donec perveniat ad horam lamine,
85 tunc igitur erit equidistans diametro lamine, et perveniet ad lineam
perpendicularem in interiori parte hore lamine. Et cum centrum lucis que est intra aquam non est super perpendicularem lineam hore
lamine, tunc lux que extenditur a medio lucis que est in superficie
aque ad medium lucis que est intra aquam non extenditur secundum
90 rectitudinem linee transeuntis per centra duorum foraminum, sed
refertur.

[2.33] Declaratum est autem quoniam hec lux extenditur recte a
medio lucis que est in superficie aque ad medium lucis que est intra
aquam. Ergo reflexio huius lucis est apud superficiem aque. Et iam
95 declaratum est quoniam hec lux transit per centra duorum foraminum et in medio lucis que est in superficie aque quod est centrum
circuli medii equidistantis superficiei lamine et medio lucis que est
intra aquam quod est in circumferentia medii circuli, ex quo patet
quod lumen perveniens ad centrum lucis que est intra aquam, dum
100 extenditur in aere et postquam reflectitur intra aquam, est in eadem

72 ergo *om.* L3/reflexio: refractio R/est ad partem: ad partem est L3/ad . . . perpendicularis *corr. ex* perpendicularis ad partem O/exeuntis: extremitatis S; *corr. ex* existentis *a. m.*
L3; *om.* FP1   73 reflexionis: refractionis R/*post* perpendicularis *add.* exeuntis F; *add.*
existentis P1   74 certus *inter.* O/certus . . . experimentator: experimentator fuerit certus O   75 est *inter. a. m.* E/circumferentiam . . . est (76) *om.* P1   76 extremitas:
extremitatis L3C1E/*post* perpendicularis *add.* exeuntes perpendicularis S   78 est *corr.*
ex sit O   80 est[1] *mg.* F/lux *inter. a. m.* E   81 rectitudinem: latitudinem L3/linee:
line S/continuantis *corr. ex* continentis F   83 illi: illius R   85 tunc . . . lamine *rep.*
P1/ad *corr. ex* a O   86 interiori: interiore R   87 que: qui P1/intra *inter. a. m.* L3/
super *om.* S/*post* perpendicularem *scr. et del.* lu C1/lineam: illam C1; *om.* L3   88 lamine
corr. ex lami F   89 *post* extenditur *scr. et del.* nisi E   91 refertur: reflectitur P1C1;
refringitur R   92 autem *inter. a. m.* E/quoniam: quod R/*post* quoniam *scr. et del.* hec P1
94 reflexio: refractio R   95 quoniam: quod ER/centra: centrum L3O   96 et *inter.*
a. m. E/in medio: per medium R   97 circuli medii *transp.* L3/*post* et *add.* per R; *add.* in
P1L3 (*inter. a. m.* L3)/medio: medium R   98 quod . . . aquam (99) *mg.* O   99 *post*
est *scr. et del.* lucis S   100 reflectitur: refringitur R

superficie equali, scilicet in superficie circuli medii trium circulorum
qui sunt in interiori parte hore instrumenti.

    [2.34] Et hec reflexio invenitur quando linea transiens per centrum
foraminum fuerit declinis super superficiem aque, non perpendicu-
105 laris, et numquam erit hec linea perpendicularis super superficiem
aque in hora transitus lucis solis nisi quando sol fuerit in verticatione
capitis. Et hoc erit in aliquibus locis et non in omnibus, in quibusdam temporibus, non in omnibus, neque sol transit per verticationem capitis habitantium in pluribus locis habitationis, et in istis locis
110 distinguetur hec experimentatio in omni tempore; illi autem super
quorum cenit transit sol, si voluerint hoc experiri, cavebunt tempus
in quo sol transit per capita eorum.

    [2.35] Item accipiat experimentator frusta vitri clari quorum fig-
ure sunt cubice, et sit longitudo uniuscuiusque eorum dupla diametri
115 foraminis quod est in hora instrumenti. Et adequentur superficies
eorum vehementer per confricationem quousque superficies eius sint
equales et equidistantes et latera eius sint recta. Deinde poliantur.
Hoc autem completo, signetur in medio lamine linea recta transiens
per centrum eius, et sit perpendicularis super diametrum eius super
120 cuius extrema sunt due linee perpendiculares in interiori parte instrumenti, et transeat in utramque partem. Et signetur hec linea ferro
ut descendat in corpus lamine, et remaneat ibi.

    [2.36] Deinde ponat unum vitrorum cubicorum super superficiem
lamine, et applicet unum latus suorum laterum cum hac perpendicu-
125 lari, et ponat medium lateris vitri vere super centrum lamine, et ponat
corpus vitri ex parte foraminum. Est ergo diametrum lamine super

101 equali: circuli L3   102 interiori: interiore R/hore inter. O   103 et om. FP1/hec
reflexio: refractio hec R   104 declinis: declivis R/non . . . hora (106) mg. a. m. E/non . .
. aque (106) mg. a. m. L3 (non: in)/post non add. quia si esset perpendicularis tunc non fieret
reflexio ut dictum est S   106 sol fuerit transp. ER/verticatione: vertitione F; vertice R
107 post omnibus add. et L3OER (mg. a. m. E)/in² . . . neque (108) mg. a. m. E/quibusdam corr.
ex quibus O   108 ante temporibus scr. et del. corpor S/post temporibus add. et L3C1E/sol
transit transp. R/verticationem: verticem R   109 habitantium corr. ex habitantibus S; corr.
ex habitium a. m. L3/post pluribus scr. et del. enim L3/locis inter. a. m. E/post et add. in quibus
transit R   110 hec: hoc L3/super: sub L3   111 cenit: cenith FP1C1O; zenith R/post
cenit scr. et del. quorum C1/transit: transibit E/transit sol transp. C1/si inter. O/voluerint:
voluerit SFP1L3O/cavebunt: cavebit FP1   112 in quo om. S   113 accipiat: accipiet FP1/
frusta: frustra SC1O   114 sunt: sint P1R/ alter in sint F/cubice: cubite P1; corr. ex gibbose a.
m. L3; om. E/post cubice inter. vel O/add. gibbose SOE; scr. et del. gibbose C1/dupla inter. a.
m. L3   115 post instrumenti scr. et del. ad S   116 superficies eius om. R/sint: sunt FP1
117 latera: latere S/eius om. R/sint: sunt SFP1/sint recta transp. SL3E   118 post lamine
rep. et del. lamine P1/linea inter. a. m. L3   119 eius¹ corr. ex visus a. m. C1   120 due
linee transp. R/linee om. C1/in om. P1/interiori: interiorhori O; interiore R/post parte add.
hore R   121 et¹: item FP1   122 ibi inter. E/post ibi add. hoc E   123 unum mg.
a. m. L3/vitrorum om. FP1/cubicorum: cubitorum P1L3OE   126 vitri corr. ex vite F/est:
super O; transibit R/diametrum: diameter ER

cuius extrema sunt due linee perpendiculares transiens per medium
superficiei vitri superposite lamine. Hac positione preservata, applicetur vitrum applicatione fixa per inglutum tali modo quod possit
130   evelli.

[2.37] Deinde accipiatur secundum vitrum, et ponatur ultra primum, scilicet ex parte foraminum, et applicetur aliqua superficierum
eius superficiei primi vitri. Hoc preservato, applicetur secundum vitrum lamine applicatione fixa. Deinde accipiatur tertium vitrum, et
135   applicetur secundo vitro, et adequetur superficies eius cum duabus
superficiebus laterum secundi vitri, et applicetur lamine. Et sic fiat
de pluribus vitris quousque perveniant intra ad horam perpendicularium super superficiem instrumenti, aut prope.

[2.38] Cum ergo intra fuerint applicata superficiei lamine secun-
140   dum positionem predictam, erit diameter lamine super cuius extremitates sunt due linee perpendiculares in extremitate instrumenti
transiens per mediam superficiem vitrorum superpositorum lamine.
Altitudo autem istorum vitrorum in latitudine est dupla diametri
foraminis, sed diameter foraminis est equalis perpendiculari exeun-
145   ti a centro foraminis super superficiem lamine et super diametrum
eius. Ergo unaqueque perpendicularium exeuntium a centris superficierum vitrorum, scilicet superficierum perpendicularium super superficiem lamine secantium diametrum opposite duobus foraminibus, est equalis perpendiculari exeunti a centro foraminis super su-
150   perficiem lamine et super diametrum lamine. Et erunt perpendiculares exeuntes a centris superficierum vitrorum ad superficiem lamine
cadentes super diametrum lamine super cuius extremitates est perpendicularis egrediens a centro foraminis. Linea ergo que transit per
centra duorum foraminum, si extendatur in ymaginatione secundum
155   rectitudinem, transibit per centra superficierum vitrorum, scilicet
superficierum perpendicularium super superficiem lamine opposite
duobus foraminibus.

127 sunt due *transp.* C1/*post* perpendiculares *scr. et del.* transeuntes P1/transiens *om.* R/per *corr.
ex* ad *a. m.* L3   128 superficiei: superficiem E/superposite: supposite S/positione: portione P1;
corr. ex portione F   129 inglutum: inglutinum E; glutinum R/quod: ut R   131 secundum:
alterum ER   132 foraminum *corr. ex* foraminis P1   135 adequetur: adequatur L3; sequetur FP1/eius *om.* L3   137 perveniant: pervenient L3/perveniant intra *transp.* FP1/intra: vitra
ER/*post* intra *add. et* L3/perpendicularium: perpendicularem O   139 intra: vitra ER (*mg. a. m.*
E)/superficiei: superficie O   140 erit: transibit R   142 transiens *om.* R/superpositorum:
supperpositorum OE   143 latitudine: altitudine L3C1   144 sed: si L3/equalis perpendiculari *transp.* C1/exeunti *corr. ex* exeunte E   146 superficierum *corr ex* super *a. m.* L3   147 superficierum: superficies E; *corr. ex* superficies F; *corr. ex* superficie *a. m.* L3   148 opposite: oppositam R   150 erunt: cadent R   151 centris: centros O/superficierum *corr. ex* superficie
a. m. L3   152 cadentes: cadentis OE; *om.* R   153 *ante* egrediens *scr. et del.* a F/que transit:
transiens R/per *corr. ex* a *a. m.* L3   155 *post* scilicet *scr. et del.* super L3   156 superficierum:
superficies SE; *corr. ex* superficie *a. m.* L3/*post* superficiem *scr. et del.* perpendicularem L3

[2.39] Deinde experimentator accipiat regulam subtilem predictam, et erigat illam super horam ipsius in superficie lamine, et ponat
160 faciem eius in qua signata est linea ex parte primi vitri quod est super
centrum lamine. Et ponat regulam prope vitrum, et ponat finem longitudinis regule secantem diametrum lamine perpendiculariter. Hoc
ergo preservato, applicet regulam lamine applicatione fixa ita quod
possit separari. Hac autem positione preservata in regula, tunc linea
165 que est in superficie regule erit in superficie medii circuli ex tribus
circulis signatis in interiori parte hore instrumenti, et erit linea recta
transiens per centra duorum foraminum et per media superficierum
vitrorum secans lineam que est in regula.

[2.40] Hoc toto completo, ponatur instrumentum in vas predic-
170 tum. Sit autem vas vacuum aqua, et ponat vas in sole, et moveat instrumentum quousque lux solis transeat per duo foramina, et erit lux
apud secundum foramen equalis, scilicet quod sit super omnia foramina, et si excesserit super foramen, erit vitrum continens foramen.
Tunc igitur intueatur experimentator superficiem regule oppositam
175 vitro, et inveniet lucem exeuntem a duobus foraminibus super superficiem regule, et inveniet illud quod circumdat lucem ex superficie
regule obumbratum umbra hore instrumenti, et inveniet centrum
lucis super lineam que est in superficie regule.

[2.41] Hoc ergo declarato, accipiat festucam subtilem, ut acum, et
180 ponat illam super superius foramen, et ponat extremitatem perpendicularis super centrum foraminis, et intueatur lucem que est super
regulam. Tunc inveniet umbram extremitatis festuce super centrum
lucis, et inveniet illam super lineam que est in superficie regule. Tunc
igitur accipiet experimentator pennam intinctam incausto, et signet
185 super extremitatem umbre que est in medio lucis que est super regu-

158 accipiat: accipiet C1/regulam: regula S/subtilem predictam transp. O   159 illam: eam ER
160 faciem: superficiem FP1   161 prope vitrum mg. a. m. L3/post ponat² add. super F   162 regule corr. ex regulam L3/lamine: regule SFL3OE; corr. ex regule a. m. C1   163 ergo: autem
R/quod: ut R   164 separari: evelli R   166 interiori: interiore R/parte hore transp. FP1O/
erit: transibit R/linea inter. F   167 transiens om. R/duorum: duo S   168 post in add. hac
FP1; scr. et del. medio S   169 in inter. F   170 ponat: ponatur ER/vas inter. a. m. E/moveat:
moveatur R   171 erit: sit R   172 secundum om. FP1/secundum foramen: secundam
formam S/scilicet . . . foramen² (173) om. R   173 et: sed E/erit rep. F   174 intueatur corr.
ex intuet F/superficiem: superficies S; corr. ex superficiei C1/regule om. L3   175 vitro: centro
P1; alter. in centro F   176 illud: illum F   177 obumbratum: obumbram S; obumbrant F;
obumbratarum O; obumbratam C1; corr. ex obumbratam a. m. L3; om. P1/post obumbratum add.
causarum S/umbra: umbram SFP1L3/hore om. P1/post centrum add. visus vel C1   178 lucis: visus SR; alter. in visus a. m. E; corr. ex visus a. m. L3   179 ergo: autem L3ER/festucam:
festinam S/ut: vel FP1ER   180 ponat²: ponant FP1/perpendicularis: perpendiculariter R;
corr. ex perpendicularem C1   182 umbram corr. ex mimbram L3   183 inveniet: invenient
FP1/post illam scr. et del. que C1/lineam: illam S   184 accipiet: accipiat P1L3R   185 post
super² scr. et del. umb P1

lam punctum. Ergo erit iste punctus super lineam que est in superficie regule.

[2.42] Deinde auferat acum a superiori foramine, et ponat ipsam
super inferius foramen, scilicet quod est in hora, et ponat extremita-
190   tem acus super centrum foraminis, et intueatur lucem que est super
regulam. Tunc inveniet umbram extremitatis acus super punctum
quod est in superficie regule. Deinde auferat acum, et redibit lux ad
suum locum. Declarabitur ergo ex hac experimentatione quod lux
que est super punctum quod est in superficie regule est lux que tran-
195   sit per centra duorum foraminum.

[2.43] Deinde accipiat experimentator calamum tinctum incausto,
et signet punctum in vero medio superficiei vitri ex parte regule. Si
vero non comprehendit medium vitri quoad sensum, signet in ipso
duos diametros secantes se, et locus sectionis est medium superficiei
200   vitri. Hoc autem facto, intueatur lucem que est super regulam, et
inveniet umbram puncti que est in medio vitri super punctum quod
est superficiei regule. Declarabitur ergo ex hoc quod lux que transit
per duo centra duorum foraminum transit per punctum quod est in
medio vitri.

205   [2.44] Hoc ergo declarato, oportet experimentatorem vitrum evellere et componere instrumentum secundo, et moveat ipsum quousque lux transeat per duo foramina. Deinde intueatur superficiem
regule que est centrum lucis et lucem pervenientem ad centrum lucis
que est in superficie regule, et est lux que transit per centra duorum
210   foraminum. Declarabitur ergo ex hoc quod lux que transit per centra
duorum foraminum transit etiam per punctum quod est in medio
superficiei secundi vitri, et situs lucis transeuntis per centra duorum
foraminum de superficiebus vitrorum in prima experimentatione, et
cum hec lux transit per punctum qui est in medio vitri secundi, tunc
215   lux que transit per centra duorum foraminum in prima experimentatione transit etiam per punctum quod est in medio vitri secundi.

186 ergo erit *transp.* FP1C1O / erit *inter. a. m.* L3 / erit iste *transp.* L3 / iste punctum: istud punctum R /
que *om.* P1   188 auferat *corr. ex* auferatur P1 / superiori: superiore R   189 inferius: superius
F / foramen: foramine O; *om.* FP1 / et *inter.* O   190 acus: lucis L3   192 lux: umbra SFP1L3OER;
corr. ex umbra *mg. a. m.* C1   193 *post* hac *scr. et del.* figura E   194 est¹ *om.* O   196 accipiat:
accipiet FP1 / tinctum: tinctam S   197 signet: signetur L3E / vero medio *transp.* S / vitri *inter. a. m.* E
198 comprehendit: comprehendat R / *post* medium *rep. et del.* medium F   199 duos: duas R / sectionis: secationis S / superficiei: superficie FP1   202 *post* est *add.* in medio P1 / *post* ergo *add.* quod
O / ex hoc *mg.* F   203 duo *om.* P1 / *post* foraminum *scr. et del.* et L3 / transit: transibit ER / per² *inter.*
a. m. L3   205 ergo: autem FP1ER / oportet experimentatorum *corr. ex* operimentatorum F / *post*
vitrum *add.* primum R   206 *post* et¹ *add.* signare in superficie secundi vitri punctum medium ut
prius et R   207 deinde intueatur *transp.* C1 / *post* intueatur *add.* et inveniet R / superficiem: superficies L3 / superficiem . . . et (208) *om.* R   208 lucis² *om.* FP1   210 declarabitur . . . foraminum
(211) *om.* E / lux *mg.* F   211 est *mg. a. m.* L3 / *post* medio *scr. et del.* secund P1   212 *post* et *add.*
situs eius est R   214 *post* hec *add.* hoc quando R / qui: quod R; que FP1; *corr. ex* que L3

[2.45] Deinde oportet experimentatorem evellere secundum vitrum et experiri tertium, et sic de ceteris usque ad ultimum. Patebit ergo experimentatione hac quod lux que transit per centra duorum foraminum perveniens ad superficiem regule transit etiam per centra superficierum vitrorum omnium positorum super superficiem lamine. Manifestum est ergo quoniam sunt in rectitudine linee recte transeuntis per centra duorum foraminum, et lux que transit per centra duorum foraminum in experimentatione omnium vitrorum extenditur in rectitudine linee continuantis centra duorum foraminum.

[2.46] Manifestum est ergo quoniam lux que extenditur per lineam rectam transeuntem per centra duorum foraminum transit etiam per centra superficierum vitrorum, ex quibus patet quod lux transit in corpus vitri in quo extenditur, postquam transit secundum lineas rectas, et quod lux que transit per centra duorum foraminum extenditur etiam in corpus vitri secundum rectitudinem linee per quam extendebatur in aere antequam pertransiret vitrum. Et illa linea per quam extenditur lux in aere est perpendicularis super superficiem vitri oppositam foramini, nam linea que transit per centra duorum foraminum est equidistans diametro lamine perpendicularis super primam superficiem superficierum vitrorum, quia est perpendicularis super differentiam communem inter superficiem vitri et superficiem lamine.

[2.47] Item accipiat experimentator medietatem spere vitree munde clare aut cristalline cuius semidiameter sit minor distantia que est inter hora et centrum lamine, et inveniat centrum basis eius super quod signet lineam subtilem cum incausto. Postea separet ex hac linea ex parte centri basis, quod est centrum spere, lineam equalem diametro foraminis quod est in hora instrumenti. Erit ergo hec linea equalis linee que est inter centrum foraminis quod est in hora instru-

---

217 oportet experimentatorem *transp.* L3 / evellere *inter. a. m.* L3   218 tertium: centrum L3
220 etiam *om.* P1R / etiam per centra *om.* E   221 *post* vitrorum *add.* et P1; *scr. et del.* et F /
positorum: positarum S / super *rep.* P1 / superficiem: superficies C1   222 *post* lamine *add.*
et per centra superficierum vitrorum positorum super superficiem lamine SFL3O / quoniam:
quod R / sunt: sit L3ER / recte *ŏm.* ER   223 transeuntis: transeuntes F; *corr. ex* transeuntes
P1 / per¹ . . . transit *om.* S / centra² *inter. a. m.* E   224 duorum: ?? E / in *inter.* O   226 quoniam: quod R / extenditur: transit R   227 rectam . . . foraminum *ŏm.* FP1 / per centra *inter. a.
m.* E   228 etiam per centra: per centra etiam FP1 / *post* centra *scr. et del.* duorum P1 / quibus:
quo R   229 postquam: per quam C1 / transit: pertransit FP1   231 vitri *corr. ex* vistri F
233 est *inter.* F   234 foramina: foramini R / nam . . . lamine (238) *mg. a. m.* L3   235 *post*
duorum *add.* oppositorum F / *post* foraminum *add.* oppositorum P1 / *post* lamine *add.* que est
R / perpendicularis: perpendiculari C1   236 quia: que FP1   240 munde clare *transp.*
deinde corr. E / aut: ut R / distantia: differentia S / que est (241) *om.* R   241 hora: tabulam R;
cilia SFP1C1OE; *corr. ex* cilia L3 / inveniat: inveniet L3   242 *post* lineam *add.* eius L3 / separet: seperet S   244 erit . . . instrumenti (245) *mg.* O   245 foraminis *om.* S

menti et superficiem lamine que est perpendicularis super superficiem lamine. Deinde statuamus super extremitatem linee separate a diametro lineam perpendicularem, et extrahamus illam in utramque partem.

250 　　　[2.48] Deinde secemus vitrum super hanc lineam in confrictorio vel tornatorio donec locus sectionis fiat superficies equalis et perpendicularis super superficiem basis semicirculi, et mensuremus angulum qui est inter duas superficies per angulum rectum factum ex cupro donec verificetur superficies ista. Et tunc differentia communis huic

255 superficiei et superficiei basis spere erit linea recta, et linea copulans centrum spere cum hac linea erit perpendicularis super superficiem factam. Postea sumatur in medio huius linee que est communis differentia particula parva que erit signum medii eius.

　　　[2.49] Hoc completo, poliatur vitrum vehementissime, et pona-

260 tur super superficiem lamine et gibbositas eius ex parte foraminum, et sit pars facta in vitro super superficiem lamine. Et superponatur linea recta que est differentia communis duabus superficiebus equalibus que sunt in vitro super lineam scilicet signatam in lamina secantem diametrum perpendiculariter, et ponatur medium linee super

265 centrum lamine. Hac ergo positione preservata, applicetur vitrum lamine applicatione fixa.

　　　[2.50] Deinde ponamus regulam subtilem super superficiem instrumenti, sicut ponebamus in experimentatione vitrorum cubicorum, et ponat superficiem regule in qua est linea ex parte vitri et

270 prope illum. Deinde ponat instrumentum in predictum vas, et ponat vas in sole vacuum aqua, et moveat instrumentum donec lux solis transeat per duo foramina, et sit situs lucis de secundo foramine situs mediocris. Et intueatur regulam, et inveniet lucem transeuntem per duo foramina super superficiem regule. Deinde applicet stilum cum

275 superiori foramina, et ponat extremitatem stili super centrum foraminis, et intueatur lucem que est in regula. Tunc inveniet umbram ex-

246 et . . . lamine *om. R* / superficiem[1]: superficie *C1; corr. ex* superficiei *L3* / super *inter. a. m.* *L3* / super . . . lamine (247) *corr. ex* lamine super superficiem *L3*　　　247 *post* lamine *rep. et del.* que (246) . . . lamine *F* / statuamus: statuimus *C1* / linee *inter. a. m. E*　　　250 confrictorio: confictorio *S*　　　251 vel: et *SC1O* / *post* vel *add. in ER; scr. et del.* tora *P1* / *post* fiat *add.* super *P1E* / superficies *corr. ex* superficie *L3*　　　255 superficiei[1] *corr. ex* superficies *S* / superficiei[2]: superficie *S* 256 hac *inter. O* / superficiem *corr. ex* superficiei *L3*　　　257 huius: huis *S* / est *om. L3*　　　258 erit: est *L3ER*　　　260 super *om. P1* / *post* eius *add.* sit *R*　　　261 pars facta: perfecta *F* / in vitro *om. C1* / superponatur: supperponatur *E*　　　262 est *om. S* / differentia communis *transp. L3ER*　　　263 in[1] *inter. O* / lamina *corr. ex* lamine *L3* / secantem: secante *SC1O*　　　264 medium linee *transp. L3* 265 *post* centrum *scr. et del.* linee *C1*　　　268 cubicorum: cubitorum *L3*　　　269 ponat: ponamus *R* / *post* linea *add.* recta latitudinis sit *R*　　　270 illum: ille *S*; illud *P1R; alter. in* illud *F* / ponat[1,2]: ponatur *R* / vas *inter. a. m. L3* / et . . . vas (271) *om. P1*　　　271 vas *inter. O* / vas in sole: in sole vas *O* / *post* vacuum *add.* sine *C1ER* / moveat: moveatur *R; corr. ex* movet *F*　　　273 *post* intueatur *add.* experimentator *R*　　　274 cum *om. R*　　　275 foramina: foramini *R* / stili *corr. ex* celi *F*

tremitatis stili apud centrum lucis. Deinde auferat stilum, et redibit
lux ad suum locum.

[2.51] Postea applicet stilum ad secundum foramen, et ponat ex-
280 tremitatem eius apud centrum secundum, et intueatur lucem que est
in regula. Tunc inveniet umbram extremitatis stili apud centrum lucis. Postea ponat extremitatem stili apud centrum basis vitri quod est
centrum spere, et intueatur lucem que est super regulam, et inveniet
umbram extremitatis regule super centrum lucis. Deinde ponat sti-
285 lum in medio lucis que est super convexum vitri oppositi foramini
secundo quod est prope illum, et intueatur lucem que est super regulam, et inveniet umbram extremitatis stili apud centrum lucis, ex quo
patet quod lux que transit per centra duorum foraminum transit
etiam per centrum basis vitri et per medium superficiei lucis que est
290 in convexo vitri.

[2.52] Manifestum est igitur quod lux que transit in corpus vitri extenditur secundum rectitudinem linee transeuntis per centra duorum
foraminum. Hec autem linea est diameter spere vitree, nam perpendicularis exiens a centro basis vitri ad laminam est equalis diametro
295 foraminis. Diameter autem foraminis est equalis perpendiculari exeunti a centro foraminis ad superficiem lamine. Ergo perpendicularis
exiens a centro basis vitri super superficiem lamine est equalis perpendiculari exeunti a centro foraminis ad superficiem lamine, et hee
due perpendiculares cadunt super diametrum lamine.

300 [2.53] Linea ergo que transit per centra duorum foraminum, si
fuerit extensa in rectitudine, perveniet ad centrum spere vitree. Erit
ergo diameter huius spere; est ergo perpendicularis super superficiem huius spere. Experimentatione autem cubicorum vitrorum patuit quod lux que extenditur in corpus vitri est in rectitudine linee per
5 quam extendebatur in aere, et linea per quam extendebatur in aere
erat illic perpendicularis super superficiem vitri.

[2.54] Et oportet experimentatorem auferre regulam subtilem applicatam ad superficiem lamine, et ponat instrumentum secundo, et
moveat ipsum quousque lux transeat per duo foramina, et intueatur
10 horam instrumenti que est intra vas. Et inveniet lucem super horam

279 postea: de *alter. in* deinde *a. m. C1* / *post* stilum *scr. et del.* post *P1*   280 *post* centrum *add.* lucis *L3*
282 *post* vitri *add.* quod est centrum basis vitri *P1*   283 super *inter. E; mg. F*   284 regule *inter*
O; stili *R*   285 super *inter. a. m. C1*   286 illum: illud *R*   289 centrum: centra *E*   291 est
inter. F; om. L3   293 vitree *corr. ex* vitre *S*   296 *post* lamine *add.* est equalis *deinde del.* equalis
C1 / ergo perpendicularis *transp. C1* / ergo . . . lamine (297) *mg. a. m. E*   297 exiens *om. R* / *post* centro
add. foraminis *SC1OER; scr. et del.* foraminis *L3* / est *inter. O*   300 si *om. S*   1 in: et *C1*   2 est
ergo *transp. C1*   3 cubicorum: cubitorum *P1L3* / cubicorum vitrorum *transp. R*   5 *post* quam[1]
add. lux *S* / et . . . aere[2] *mg. FL3* (*a. m. L3*) / per *om. FP1* / quam[2]: que *F*   6 illic: illi *FP1*; illuc *L3*
7 et: item *FP1* / applicatam *corr. ex* applicabilem *a. m. L3*; *alter. in* applicabilem *a. m. E*   8 ad: ab *L3* /
ponat: componat *R*   10 *post* instrumenti *scr. et del.* in puncto *E* / que . . . instrumenti (11) *mg. O*

instrumenti, et inveniet centrum lucis in puncto que est differentia
communis inter circumferentiam circuli medii et lineam perpendicularem in hora instrumenti quod est extremitas diametri circuli medii
transeuntis per centra duorum foraminum. Et lux que extenditur per
hanc lineam erit differentia communis perveniens ad centrum spere
vitree. Centrum ergo lucis que est in hora instrumenti, et centrum
spere vitree, et centra duorum foraminum sunt in eadem linea recta,
ex quo patet quoniam lux que transit in corpus vitri perveniens ad
centrum eius, cum extrahitur in aere, extenditur in rectitudine linee
per quam extendebatur in corpus vitri.

[2.55] Hec autem linea est perpendicularis super superficiem basis
vitri que est equidistans diametro lamine qui est perpendicularis super superficiem basis vitri, quia est perpendicularis super lineam rectam que est differentia communis duabus superficiebus vitri equalibus, quarum altera est superposita superficiei lamine et reliqua erecta
super superficiem lamine. Linea ergo transiens per centra duorum
foraminum et per centrum spere vitree est perpendicularis super superficiem vitri; est ergo perpendicularis super superficiem aeris qui
tangit hanc superficiem. Et si experimentator effuderit aquam in vas,
remanente vitro in sua positione, et posuerit aquam supra centrum
vitri, et inspexerit lucem que est in hora instrumenti, inveniet centrum lucis super extremitatem diametri medii circuli.

[2.56] Et si evulserit vitrum et posuerit illud in lamina econtra
huic ordinationi, scilicet quod superficies equalis sit ex parte foraminum et convexitas vitri sit ex parte interioris vasis, et superposuerit
lineam rectam que est in vitro que est differentia communis duabus
suis superficiebus equalibus super lineam rectam que est in lamina
secantem perpendiculariter diametrum lamine, et posuerit medium
huius linee, scilicet que est in vitro, super centrum lamine, et inspexerit lucem sicut fecit in prima positione, inveniet lucem cadentem

11 que: qui *FP1*; quod *R*   14 duorum *om. P1* / duorum foraminum *transp. F*   15 communis: communi *O*   16 *post* et *add.* ad *FP1*   17 spere . . . centrum (19) *mg. F* (recta [17]
inter.) / centra: centrum *SR* / in *om. O*   18 patet *corr. ex* petet *O* / quoniam: quod *R* / vitri *corr.*
ex vitrei *L3*   19 *post* centrum *add.* spere *L3ER* / extrahitur: extrahatur *O*; extenditur *C1* / aere:
aera *E*; aerem *R* / rectitudine linee *transp. C1*   20 corpus: corpore *R* / vitri *corr. ex* vitrei *L3*
22 *post* vitri *scr. et del.* quia est *O* / que: quia *O* / qui: que *L3ER*   23 superficiem . . . super
om. P1; *rep. S* / est *om. L3E* / super: supra *SL3C1E*   24 communis *om. P1*   25 superposita:
supperposita *L3E* / erecta *corr. ex* rerecta *C1*   28 est ergo *transp. C1* / super *om. P1* / qui: que *E*
29 et: item *FP1* / effuderit: infunderit *E*; infuderit *R*   30 *post* remanente *add.* in *L3* / *post* centrum *scr. et del.* lucis *C1*   32 diametri . . . circuli: circuli medii diametri *L3O*; medii circuli diametri *E* / medii circuli *transp. F*   33 et: item *FP1* / illud: illum *FL3C1E* / econtra: econtrario *R*
34 quod: ut *R* / foraminum . . . parte (35) *om. S*   35 et[1] *inter. a. m. L3* / *post* parte *add.* foramini
utriusque *P1* / interioris: interiore *R*; *om. P1* / superposuerit: supperposuerit *L3E*   36 est[2]
om. P1   37 est *om. O*   38 *post* et *scr. et del.* inspexerit lucem *S*   39 scilicet *om. L3* / est
om. FP1 / super: supra *FP1*   40 *post* inveniet *scr. et del.* centrum lucis super punctum *C1*

super horam instrumenti, et inveniet centrum lucis super punctum quod est differentia communis circumferentie medii circuli et linee stanti in hora instrumenti, ex quibus declarabitur quod lux que tran- sit per centra duorum foraminum transit etiam in corpus vitri secun-
45 dum rectitudinem linee per quam extendebatur in aere, et postquam egreditur corpus vitri, extenditur etiam in aere secundum rectitudi- nem linee per quam extendebatur in vitro.

[2.57] Et linea que transit per centra duorum foraminum est in hac positione etiam perpendicularis super superficiem vitri opposi-
50 tam foramini, scilicet superficiem que est basis semispere, et hec linea est etiam perpendicularis super superficiem convexam, nam in hac positione etiam est diameter spere. Est ergo perpendicularis super superficiem spere; est ergo perpendicularis super superficiem aeris continentis superficiem spere. Et si experimentator infuderit aquam
55 in vas, et relinquerit vitrum in sua positione, et posuerit aquam infra centrum vitri, et aspexerit lucem que est in hora instrumenti, inveniet centrum lucis in extremitate diametri medii circuli.

[2.58] Ex hiis igitur experimentationibus que fiunt per cubicum et spericum vitrum patet quod, si lux occurrerit corpori diaffono di-
60 verse diaffonitatis a corpore in quo est, et linea per quam extenditur fuerit perpendicularis super superficiem secundi corporis, tunc lux extenditur in secundo corpore in rectitudine linee per quam extende- batur in corpore primo, nec differt si secundum corpus fuerit gros- sius primo aut subtilius.

65 [2.59] Item oportet experimentatorem evellere vitrum, et revertere illud ad laminam, et ponere medium linee recte que est in ea super centrum lamine, et ponere superficiem equalem ex parte duorum foraminum et lineam que est in vitro que est differentia communis duabus suis superficiebus obliquam super diametrum lamine qua-
70 libet obliquatione, et ponere obliquationem diametri lamine super hanc lineam ad illam partem ad quam declinabat apud experimen- tationem aque. Necesse est igitur ut perpendicularis que egreditur

41 super[1]: supra L3/inveniet corr. ex invenit L3   42 quod: que SC1/est om. S/circumferentie om. R/medii circuli transp. L3   43 post lux add. solis ER   45 quam alter. in quod P1/postquam corr. ex post C1   46 in … rectitudinem corr. ex secundum rectitudinem in aere L3   48 et linea: lineaque ER (alter. ex linea a. m. E)/in: per O   49 super corr. ex superfi S   50 que est om. O/semispere: hemisperii R   51 etiam om. O   52 post spere scr. et del. est C1/est[2] … est (53) mg. C1/super corr. ex superfic P1   53 spere … superficiem[2] om. L3ER/est … superficiem[2] om. P1   54 continentis: contingentis R   55 relinquerit: relinquit FP1; reliquerit R/posuerit: exposuerit E/infra: intra C1   56 centrum: centra E   58 que … vitrum (59) mg. a. m. E/cu- bicum: cubitum P1   59 si inter. O/occurrerit: occurret FP1   60 quo inter. O   61 secundi corporis transp. deinde corr. P1   63 grossius corr. ex grossus F   65 revertere corr. ex reverte C1; referre R   66 medium corr. ex medie O/ea: eo R/super: supra FP1   69 obliquam: obli- qua SL3   70 ponere corr. ex opponere C1/super corr. ex per a. m. L3   71 ad quam: in qua P1 72 est igitur transp. C1/ut inter. E/que inter. a. m. L3/egreditur: regreditur O

a centro vitri que est super superficiem vitri perpendiculariter que extenditur in corpus vitri obliqua sit a linea transeunte per centra
75 duorum foraminum ad partem in qua sunt duo foramina. Et applicet experimentator vitrum secundum hunc situm applicatione fixa, et ponat instrumentum in vas et vas in sole, et moveat instrumentum donec lux transeat per duo foramina, et intueatur lucem que est intra vas.

80 [2.60] Tunc inveniet illam in interiori hore instrumenti, et inveniet centrum lucis in circumferentia medii circuli sed extra punctum qui est differentia communis circumferentie circuli medii et linee stanti in hora instrumenti, et declinatio eius erit ad partem in qua est sol. Erit ergo ad partem perpendicularis exeuntis a loco reflexionis, et hec lux
85 extenditur in aere in rectitudine linee transeuntis per centra duorum foraminum, et hec linea in hoc situ pervenit ad centrum spere vitree, et est obliqua super superficiem vitri equalem.

[2.61] Huius autem lucis terminatio extensionis in vitro est a centro vitri; extenditur ergo in corpus vitri secundum lineam rectam
90 exeuntem a centro spere; ergo est illi diameter. Hec igitur lux extenditur in corpus vitri secundum verticationem diametri alicuius eius; cum ergo pervenerit ad superficiem eius spericam, erit perpendicularis super illam, et cum extrahetur in aere, erit perpendicularis super aerem continentem superficiem spericam.

95 [2.62] Non ergo reflectitur in aere, nec extenditur recte. Ergo reflectitur, sed non in corpus vitri, nec in convexo eius, neque in primo aere, neque in secundo. Ergo reflectitur apud centrum vitri, et hec lux est obliqua super superficiem vitri equalem in qua est centrum vitri, ex quibus patet quod, cum lux extenditur in aere, et transit in vitrum,
100 et fuerit obliqua super superficiem vitri, reflectitur et non transibit recte. Et reflexio eius erit ad partem in qua est perpendicularis exiens a loco reflexionis, et corpus vitri grossius est corpore aeris.

73 perpendiculariter: perpendicularis *R* / que extenditur (74) *om. S*   74 corpus: corpore *R* / sit *om. O*   80 interiori: interior *O*; interiore *R* / hore: hora *R*; *alter. in* hora *E*   81 qui: quod *R*; *corr. ex* quod *P1*   82 circumferentie . . . medii: circuli medii circumferentie *E* / circuli medii *transp. L3* / stanti: stantis *P1*   84 reflexionis: refractionis *R*   86 in . . . pervenit: pervenit in hoc situ *FP1* / pervenit: perveniet *L3ER* (*alter. ex* pervenit *a. m. L3*)   87 est: erit *R* / vitri *om. ER*   88 est *om. O*   89 corpus: corpore *R*   90 *post* ergo *scr. et del.* in corpus vitri *F* / est *om. S* / est illi *transp. ER* / illi: illius *R* / diameter *corr. ex* diametro *L3*   91 corpus: corpore *R*   92 pervenerit: pervenit *FP1* / superficiem . . . spericam: spericam superficiem *ER* / erit . . . spericam (94) *mg. a. m. L3*   93 super illam *transp. deinde corr. F* / cum extrahetur *transp. deinde corr. S* / extrahetur: extrahitur *FP1L3* / aere: aerem *R*   94 continentem: contingentem *R*   95 reflectitur[1,2]: refringitur *R* / nec: neque *R*   96 corpus: corpore *R* / nec: neque *R*   97 reflectitur: refringitur *R* / apud: ad *FP1* / centrum vitri: centri *L3* / et hec *inter. a. m. E*   98 super *inter. O* / *post* superficiem *add.* duas *O* / vitri[1] *om. L3C1ER*   99 et *inter. O*   100 reflectitur: refringetur *R*   101 reflexio: refractio *R* / eius erit *transp. C1*   102 reflexionis: refractionis *R* / et corpus *om. S*

[2.63] Manifestum est igitur ex hac experimentatione et prima de reflexione lucis ab aere ad aquam, luce existente obliqua super super-
105 ficiem aque, quoniam cum lux fuerit extensa in corpore subtiliori et occurrerit grossius corpus, et fuerit obliqua super grossius corpus, re-flectetur ab ipso, et erit reflexio eius ad partem in qua est linea exiens a loco reflexionis que est perpendicularis super superficiem corporis grossioris.
110 [2.64] Item oportet experimentatorem evellere vitrum et ponere ipsum econtra, scilicet quod superficies convexa sit ex parte forami-num, et ponat medium differentie communis que est in vitro super centrum lamine, et ponat differentiam communem obliquam super diametrum lamine, et applicet vitrum applicatione fixa. Et extrahat
115 a centro lamine lineam in superficie lamine perpendicularem super differentiam communem que est in vitro. Erit ergo hec linea perpen-dicularis super superficiem vitri, nam superficies vitri equalis est per-pendicularis super superficiem lamine.
[2.65] Deinde experimentator ponat instrumentum in vase, vase
120 existente sine aqua, et moveat instrumentum quousque lux transeat per duo foramina, et intueatur lucem que est intra vas. Tunc in-veniet illam in interiori hore instrumenti, et inveniet centrum lucis in circumferentia medii circuli et extra punctum qui est differentia com-munis circumferentie medii circuli et linee perpendiculari in hora in-
125 strumenti, qui est extremitas diametri medii circuli. Et inveniet dec-linationem eius ad contrariam partem illi in qua est perpendicularis.
[2.66] Hec autem lux extenditur in vitro secundum rectitudinem linee transeuntis per centra duorum foraminum, quia hec linea est diameter vitri in hac etiam positione, quia transit per centrum vitri.
130 In hac igitur positione reflexio lucis est etiam apud centrum vitri, et hec lux est obliqua super superficiem vitri equalem et super superfi-ciem aeris contingentem vitrum, ex quibus patet quod, cum lux ex-

103 de *inter.* O; *om.* E   104 reflexione: refractione R/aquam *corr. ex* quam O/existente *corr. ex* extente O   105 quoniam: quod R/lux fuerit *transp.* FP1/subtiliori: subtiliore R   106 *post* occurrerit *add.* illi R/et . . . corpus² *om.* ER/corpus² *om.* P1/reflectetur: refringetur R   107 reflexio: refractio R   108 reflexionis: refractionis R/super *inter.* O   111 econtra: contra S; econtrario R/quod: ut R   112 super . . . communem (113) *om.* FP1   113 obli-quam: obliqua P1; *alter. ex* obliqua *in* obliquum F   115 superficie: superficiem E/lamine² *om.* C1R/super . . . perpendicularis (116) *mg.* C1   116 ergo *om.* ER/hec linea *transp.* L3   117 super *om.* S/equalis est *transp.* E   118 superficiem *om.* FP1/lamine: laminam FP1   119 in . . . instrumentum (120) *om.* S/vase² *om.* ER   121 intueatur *corr. ex* intuetur L3/lu-cem *inter. a. m.* L3   122 interiori: interiore R/*post* lucis *add.* que O   123 medii circuli *transp.* L3/qui: que P1L3E; quod R   125 qui: que L3OE; quod R/*post* qui *add.* punctum R/diametri . . . circuli: medii circuli diametri FP1/inveniet: inveniat SO   126 eius *inter.* O   128 linee transeuntis *transp. deinde corr.* C1/centra . . . foraminum: duo foraminum centra C1   129 *post* positione *add.* qua S/quia: qua S   130 positione reflexio *transp. deinde corr.* E/reflexio: refractio R/est etiam *transp.* L3ER; *rep.* P1/apud: capud O   131 lux est *transp.* O/equalem . . . superficiem² *om.* S/et *inter.* C1/super² *om.* L3ER

tenditur in vitro, et egreditur ad aerem, et fuerit obliqua super superficiem aeris, reflectetur, et reflexio eius erit in superficie circuli medii
135   et ad partem contrariam illi in qua est linea exiens a loco reflexionis
que est perpendicularis super superficiem aeris.

[2.67] Et si experimentator effuderit aquam in vas, existente vitro
in sua positione, et posuerit aquam super centrum vitri, et aspexerit
lucem que est intra vas, inveniet lucem in interiori parte hore instru-
140   menti. Et inveniet centrum lucis in circumferentia medii circuli, et
inveniet illud extra extremitatem diametri medii circuli obliquum ad
partem contrariam illi super quam cadit perpendicularis. Et inveniet
distantiam centri lucis ab extremitate diametri medii circuli minorem
distantia centri lucis ab hoc puncto in experientia egressus lucis a
145   vitro ad aerem, quia aer est subtilior aqua, aqua autem est subtilior
vitro.

[2.68] Ex hac ergo experimentatione et predicta patet quoniam,
quando lux extenditur in corpore grossiori, et occurrerit corpori subtiliori, et fuerit obliqua super superficiem corporis subtilioris, reflec-
150   tetur et non transibit recte, et reflexio eius erit ad partem contrariam
illi in qua est perpendicularis exiens a loco reflexionis que est perpendicularis super superficiem corporis subtilioris. Et tantum declinabit
a perpendiculari quanto corpus erit subtilius.

[2.69] Item oportet experimentatorem evellere vitrum et ponere
155   etiam ipsum in superficie lamine, et superponat lineam rectam que
est in eo super lineam rectam que est in lamina, et ponat superficiem
eius convexam ex parte duorum foraminum et lineam rectam que est
in vitro extra centrum lamine. Et coniungat vitrum bene, et ponat regulam subtilem super superficiem lamine, et erigat eam super horam
160   eius, et ponat superficiem eius in qua signatur linea ex parte vitri,
et terminus eius secet diametrum lamine perpendiculariter, et appli-

133 ad: in *S*   134 reflectetur: refringetur *R*; *corr. ex* reflectitur *E* / reflexio: refractio *R* / medii
om. SFO; scr. et del. P1; inter. a. m. L3   135 reflexionis: refractionis *R*   137 et: item *FP1* /
effuderit: infuderit *R*   138 super: supra *SL3C1* (*alter. ex* super *S*)   139 *post* lucem
inter. que est *a. m. E* / in *om. P1* / interiori: interiore *R* / hore instrumenti *transp. C1*   140 lucis
om. P1 / in: et *S*   141 illud: illum *FP1* / extra *rep. S* / *post* extremitatem *add.* medii *L3* / medii
inter. a. m. L3 / medii circuli *transp. L3* / obliquum *corr. ex* oblique *a. m. E* / *post* ad *scr. et del.*
mediam *C1*   142 et *om. P1*   143 ab . . . lucis[1] (144) *mg. a. m. E* / diametri . . . circuli:
medii circuli diametri *S*   144 distantia *corr. ex* distantiam *C1* / in *inter. O* / egressus: egrus *S*
145 vitro: centro *ER* / aqua[1] *inter. a. m. L3* / est[2] *om. P1*   147 ergo: autem *L3ER* / quoniam:
quod *R*   148 quando *om. SE* / *post* lux *add.* quomodo *S* / grossiori: grossiore *R* / occurrerit: occurrit *S*; occurret *O*; occurrat *L3E*   149 obliqua *om. F* / reflectetur: refringetur *R*
150 reflexio: refractio *R* / eius *om. FP1* / ad: in *O*   151 reflexionis: refractionis *R*   152 superficiem: superficie *O* / tantum: tanto *R* / *post* tantum *add.* magis *P1R*   153 *post* corpus
add. eius *S*   155 etiam *om. FP1O* / superponat: supponat *P1* / rectam *om. L3* / que . . . rectam
(156) *mg. C1*   156 *post* in[2] *scr. et del.* eo *F* / lamina *corr. ex* lamine *F*   159 horam *corr. ex*
oram *O*   161 lamine *corr. ex* linee *P1* / applicet: applicetur *R*

cet regulam hoc modo. Sic ergo linea que transit per centra duorum foraminum non transit per centrum spere sed per alium punctum superficiei vitri equalis, et erit obliqua super spericam superficiem.

165 [2.70] Deinde oportet experimentatorem ponere instrumentum in vase et vas in sole, et moveat instrumentum quousque lux transeat per duo foramina, et intueatur superficiem regule. Tunc inveniet lucem super superficiem regule, et centrum eius super lineam que est in superficie regule, et centrum lucis extra rectitudinem linee que 170 transit per centra duorum foraminum. Et inveniet declinationem eius ad partem in qua est centrum vitri, et inveniet lineam que transit per centra duorum foraminum perpendicularem super superficiem vitri equalem, est enim equidistans diametro, et diameter lamine est perpendicularis super superficiem vitri equalem. Et si lux transiret 175 per duo centra foraminum et extenderetur secundum rectitudinem ad superficiem equalem, tunc extenderetur in rectitudine in aere. Sed cum centrum lucis que est in regula non est in rectitudine huius linee, ergo lux non extenditur in rectitudine ipsius ad superficiem equalem. Et lux in corpore vitri extenditur recte. Ergo lux que extenditur in 180 corpore vitri non est in rectitudine linee que transit per centra duorum foraminum.

[2.71] Ergo est reflexa, sed non in aere neque in corpore vitri; ergo reflectitur apud spericam superficiem vitri, et est obliqua super superficiem spericam, quia linea que transit per duo centra foraminum 185 non transit per centrum vitri, et hec lux, cum egreditur a superficie vitri equali, reflectitur. Sed cum regula subtilis fuerit valde propinqua superficiei vitri, tunc declinatio centri lucis que est in regula a rectitudine linee que extenditur in corpore vitri non latebit in tantum quod possit occultare reflexionem lucis in corpore vitri aut partem 190 eius. Et hec reflexio erit ad partem in qua est centrum vitri; ergo est ad perpendicularem exeuntem a loco reflexionis perpendiculariter super superficiem vitri spericam, quia linea exiens a centro vitri ad

162 regulam *om. R*/sic: si *O* 163 alium: aliud *SR* 164 spericam superficiem *transp. FP1*
165 *post* instrumentum *scr. et del.* instr *P1* 166 moveat: movet *E*/lux transeat *transp. C1*
167 per duo foramina *corr. ex* per centra duorum foraminum *a. m. E* 168 *post* eius *add.* et *S*
169 *post* regule *rep. et del.* et (168) . . . regule *F* 173 est[1] . . . equalem (174) *om. S* 174 vitri
equalem *transp. E*/transiret: transisset *R* 175 duo centra: centra duorum *R*/extenderetur . . .
tunc (176) *om. P1* 177 est[2]: sit *R*/huius linee *transp. O* 179 vitri *inter. a. m. L3*/*post* vitri *add.*
non *SO*; *scr. et del.* non *L3* 180 non *alter. in* que *O*/que *om. O* 182 est *om. FP1L3*/reflexa:
refracta *R* 183 reflectitur: refringitur *R*/apud: super *S*/et . . . spericam (184) *om. R* 184 *post*
spericam *inter.* vitri *a. m. E*/quia: et *ER*/duo centra: centra duorum *R* 185 centrum *corr.*
ex centra *S* 186 reflectitur: refringitur *R* 188 in[1]: a *L3* 189 quod: ut *R*/reflexionem:
refractionem *R*/lucis *inter. O* 190 reflexio: refractio *R*/est[1] *om. L3* 191 reflexionis:
refractionis *R*/perpendiculariter: perpendicularem *R* 192 *post* spericam *add.* nam linea
que egreditur in ymaginatione a centro vitri ad punctum reflexionis perpendiculariter super
superficiem vitri spericam *SC1O* (*mg. O*; vitri spericam *transp. S*)/linea: linee *C1*

punctum reflexionis est perpendicularis exiens a loco reflexionis super superficiem spericam.

195   [2.72] Deinde oportet experimentatorem evellere vitrum et ponere econtra huic compositioni, scilicet quod ponat superficiem vitri equalem ex parte duorum foraminum, et ponat differentiam communem duabus superficiebus equalibus vitri super lineam secantem diametrum lamine perpendiculariter, et ponat medium differentie
200   communis extra centrum lamine. Vitro autem coniuncto hoc modo, linea que transit per centra duorum foraminum non transit per centrum vitri, sed perveniet ad punctum de superficie eius equali in qua est centrum eius extra punctum centri, et erit perpendicularis super superficiem equalem, sicut predictum est. Et cum linea que transit
205   per centra duorum foraminum extensa fuerit recte in ymaginatione, perveniet ad punctum quod est extremitas diametri circuli medii.

[2.73] Et cum experimentator posuerit vitrum hoc modo, ponet instrumentum in vase et vas in sole, et moveat instrumentum donec lux transeat per duo foramina, et intueatur horam instrumenti, et
210   inveniet lucem in interiori parte hore instrumenti. Et inveniet centrum lucis in circumferentia circuli medii et extra punctum quod est extremitas diametri circuli medii et declinans ad partem in qua est centrum spere vitree. Et linea que egreditur a centro huius spere in ymaginatione ad locum reflexionis est perpendicularis super super-
215   ficiem huius spere; est ergo perpendicularis super superficiem aeris qui continet superficiem spere. Hec ergo reflexio est ad partem contrariam illi in qua est perpendicularis exiens a loco reflexionis super superficiem aeris continentis superficiem que extenditur in corpore aeris.

220   [2.74] Lux autem que transit per centra duorum foraminum transit in corpus vitri recte, quia est perpendicularis super superficiem vitri equalem oppositam duobus foraminibus, et perveniet ad con-

193 reflexionis[1,2]: refractionis *R*/*post* reflexionis[1] *scr. et del.* super superficiem *L3*/est *om. O*/ exiens *inter. O*   196 econtra: econtrario *R*/compositioni: positioni *SP1C1R*; *corr. ex* positioni *E*/quod: ut *R*   199 perpendiculariter . . . lamine (200) *mg. a. m. L3*   201 *post* foraminum *scr. et del.* extensa fuerit *C1*   202 ad: in *S*/de *inter. a. m. L3*/superficie *corr. ex* superficiebus *C1*/eius . . . punctum (203) *inter. a. m. L3*/equali: equalis *S*; *corr. ex* equalis *C1*   203 extra: super *S*/*post* punctum *rep. et del.* punctum *F*   204 equalem *mg. a. m. E*/predictum: supradictum *ER*   205 centra: centrum *F*/recte: recta *L3*; *corr. ex* recta *F*/ ymaginatione *corr. ex* ymagine *a. m. E*   207 ponet *corr. ex* pone *a. m. C1*   210 inveniet[1]: intueatur *S*/interiori: interiore *R*/hore *inter. a. m. L3*/instrumenti *inter. a. m. E*   211 in *om. S*/circuli medii *transp. FP1*   212 medii *mg. C1*/et *om. L3*/ad *mg. F*   213 in . . . spere (215) *inter. a. m. L3*   214 ymaginatione *corr. ex* ymagine *O*/ad locum: a loco *L3*/reflexionis: refractionis *R*   215 huius . . . superficiem *om. S*   216 qui: que *FP1*/continet: contingit *R*/reflexio: refractio *R*   217 illi *mg. a. m. L3*/reflexionis: refractionis *R*   218 continentis: contingentis *R*/*post* superficiem *add.* spere *R*/que: qua *S*/que . . . aeris (219) *om. R*   221 recte: recto *P1*/recte . . . vitri (222) *rep. S*   222 convexum: convexitatem *L3ER*

vexum spere vitree. Et cum pervenit ad illam superficiem, non erit perpendicularis super illam, cum non sit diameter in spera, et omnis
225 perpendicularis super superficiem spere est diameter illius aut secundum rectitudinem diametri illius. Sed lux que extenditur in corpore vitri hoc modo non est perpendicularis super superficiem aeris continentis convexum vitri, et hec lux invenitur reflexa. Ergo reflectitur apud convexum spere.

230 [2.75] Et si experimentator effuderit aquam intra vas, vitro remanente in suo situ, et posuerit aquam infra centrum lamine, et aspexerit lucem que est in hora instrumenti, inveniet etiam lucem reflexam et ad partem in qua est centrum vitri. Erit ergo ad partem contrariam illi in qua est perpendicularis exiens a loco reflexionis que extenditur
235 in corpore vitri a corpore aeris perpendicularis super concavitatem aeris continentis convexum vitri.

[2.76] Ex omnibus ergo hiis experimentationibus patet quod lux solis transit in omne corpus diaffonum secundum verticationes linearum rectarum, et si occurrerit corpori diaffono diverse diaffonitatis diaf-
240 fonitati corporis in quo est, et linee per quas extenditur in primo corpore fuerint declinantes super superficiem secundi corporis, tunc lux reflectetur in secundo corpore in verticatione linearum rectarum aliarum a primis per quas extendebatur in primo corpore. Et si linee recte per quas extendebatur in primo corpore fuerint perpendiculares
245 super superficiem secundi corporis, tunc lux extendetur in rectitudine eius, et non reflectetur.

[2.77] Et cum lux obliqua exiverit a corpore subtiliori ad grossius, reflectetur ad partem perpendicularis exeuntis a loco reflexionis perpendicularis super superficiem secundi corporis. Et cum lux obli-

---

223 pervenit: pervenerit *R* 224 *post* super *scr. et del.* superficiem *C1* 225 superficiem spere *transp. R* 227 *post* superficiem *scr. et del.* co *P1* / continentis: contingentis *R* 228 reflexa: refracta *R* / reflectitur: refringitur *R* 230 et: item *FP1* / effuderit: effunderit *E*; infuderit *R* / remanente: permanente *L3* 231 suo situ *transp. FP1* 232 etiam *om. FP1ER* / *post* lucem[2] *scr. et del.* convexam *P1* / reflexam: refractam *R* / *post* reflexam *rep. et del.* reflexam *F* 233 et *om. R* / est *om. E* / erit *om. ER* 234 reflexionis: refractionis *R* 235 in: a *ER* / in . . . aeris *mg. C1* / in . . . vitri (236) *mg. a. m. E* / vitri *om. SFP1L3*; *scr. et del.* aeris *O* / vitri a corpore *om. C1* / a: in *ER* 236 continentis: contingentis *R* / *post* convexum *scr. et del.* vite *P1* 238 solis transit *transp. C1* / verticationes: verticationis *FP1* 239 rectarum *corr. ex* rectu *F* / si: cum *FP1R* / *post* si *scr. et del.* o *L3* / occurrerit: occurrit *ER* / diaffonitatis *corr. ex* diaffonitas *S* / *post* diaffonitatis *add.* a *ER* / diaffonitati: diaffonitate *ER* 240 corporis: corpori *SL3O*; *corr. ex* corpori *a. m. C1* / et *om. L3ER* / linee: lineeque *R*; *alter. in* suntque *a. m. E* / in . . . fuerint (241): fuerint in primo corpore *O* 241 secundi *inter. a. m. L3* 242 reflectetur: reflectitur *E*; refringitur *R* / secundo corpore *transp. ER* 243 et . . . corpore (244) *mg. a. m. E* 244 fuerint *om. P1* 245 *post* corporis *scr. et del.* tunc lux extenditur super superficiem secundi corporis *E* / extendetur: extenditur *ER* 246 eius *om. O* / reflectetur: refringitur *R* 247 et: item *FP1* / et . . . reflectetur (248) *mg. C1* / *post* obliqua *add.* fuerit et *ER*; *scr. et del.* o *F* / exiverit: exierit *R* / subtiliori: subtiliore *R* 248 reflectetur: refringetur *R* / reflexionis: refractionis *R* 249 et *om. ER* / et . . . corporis (252) *om. O* / *post* cum *add.* vero *R* / lux *om. FP1*

250   qua fuerit extensa a grossiori ad subtilius, reflectetur ad partem con-
      trariam perpendicularis exeuntis a loco reflexionis super superficiem
      secundi corporis. Cum ergo lux solis transit per omnia diaffona cor-
      pora secundum lineas rectas, ergo omnes luces extendentur in omni-
      bus corporibus diaffonis, quia declaratum est in primo tractatu huius
255   libri quod proprium est lucis semper extendi secundum lineas rectas,
      sive lux fuerit essentialis sive accidentalis, sive fortis sive debilis.
         [2.78] Preterea potest experimentator experiri luces accidentales
      in illo predicto instrumento et illis viis predictis in aliqua domo in
      quam intret lux diei per aliquod foramen alicuius quantitatis, si clau-
260   serit laminam, et posuerit instrumentum in oppositione foraminis, et
      aspexerit lucem que est intra aquam et ultra vitrum in hora instru-
      menti, et processerit per vias preostensas et in experimentatione lucis
      solis. Cum ergo experimentator expertus fuerit lucem accidentalem
      hiis predictis viis, inveniet lucem accidentalem transeuntem per cor-
265   pus aque et per corpus vitri, et inveniet extensionem eius in vitro
      secundum verticationes linearum rectarum, et reflexionem si fuerit
      obliqua super superficiem corporis secundi, et rectam si fuerit per-
      pendicularis super superficiem corporis secundi. In primo autem
      tractatu declaratum est quod lux omnis, sive essentialis aut acciden-
270   talis, fortis aut debilis, semper extenditur a quolibet puncto cuiuslibet
      corporis secundum lineam rectam.
         [2.79] Ex istis ergo omnibus que declaravimus experientia et ra-
      tione patet quod omnis lux in omni corpore lucido essentialiter aut
      accidentaliter, fortiter aut debiliter extenditur a quolibet puncto illius
275   per corpus diaffonum contingens illud corpus per omnem lineam rec-
      tam per quam poterit extendi, sive illud corpus contingens sit aer, aut
      aqua, aut lapis diaffonus. Et si luces extense per corpus contingens
      lucem que est principium eius occurrerint corpori diverse diaffonitatis

---

250 a *inter.* C1/grossiori: grossiore R/reflectetur: reflectitur E; refringetur R   251 reflexionis: refractionis R/super *om.* P1   252 ergo: vero E/solis *om.* ER/transit: transeat R/omnia . . . corpora *corr. ex* diaffona corpora omnia P1/corpora *om.* R   254 *post* tractatu *scr. et del.* quod P1/huius: huis S   255 *post* libri *scr. et del.* et E/est *om.* FP1/est lucis *transp.* L3ER/semper extendi *transp.* ER   257 potest: postea S; docet P1   258 *post* et add. in C1/*post* predictis add. si R   259 quam: qua L3O/si *correxi ex* et; *om.* R   260 laminam: iamiam P1L3C1E; ianuam R/oppositione: appositione FP1   261 aspexerit: inspexerit ER/*post* vitrum *add.* et SC1O; *scr. et del.* que est L3   262 preostensas *corr. ex* ostensas E/et² *om.* ER   264 hiis *om.* F/hiis . . . accidentalem *om.* P1/*post* viis *add.* et E   265 per *om.* C1   266 reflexionem: re-flexio SFP1L3OE; refractam R/si fuerit *rep.* S   267 corporis secundi *transp.* R/et rectam: est recta E/et . . . secundi (268) *mg. a. m.* E   268 corporis *om.* P1/*post* secundi *rep. et del.* et (267) . . . secundi F   269 sive *om.* FP1/aut: sive ER   270 *ante* fortis *add.* si FP1; *add.* vel E; *add.* sive R/aut: sive R/semper extenditur *transp.* L3   273 omni *om.* L3ER   275 contingens: contingentem O   276 sit *corr. ex* fuerit P1   277 diaffonus: diaffonis L3   278 lucem *mg. a. m.* L3/principium *corr. ex* prins L3/occurrerint: occurrerit SL3E; occurreret F; occurret P1; concurret O

ad diaffonitatem corporis in quo existit, si fuerint in lineis perpen-
280 dicularibus super superficiem secundi corporis, extenduntur recte in
secundo corpore. Et si fuerint in obliquis lineis super superficiem
secundi corporis, reflectentur in secundo corpore, cum in secundo
corpore extendentur in verticatione rectarum linearum aliarum a
primis.

285 [2.80] Et si lux fuerit reflexa, tunc linea per quam extendebatur
lux in primo corpore et linea per quam reflectebatur in secundo erunt
in eadem equali superficie, et quod reflexio eius, cum egressa fuerit a
corpore subtiliori ad grossius, erit ad partem perpendicularis exeuntis a loco reflexionis super superficiem grossioris corporis. Et cum
290 egressa fuerit a grossiori corpore ad subtilius, tunc reflexio eius erit
ad partem contrariam illi in qua est perpendicularis exiens a loco
reflexionis super superficiem subtilioris corporis.

[2.81] Quare autem reflectitur lux quando occurrit corpori diaffono diverse diaffonitatis hoc est quia transitus lucis per corpora diaf-
295 fona fit per motum velocissimum, ut iam declaravimus in tractatu secundo. Luces ergo que extenduntur per corpora diaffona extenduntur
motu veloci qui non patet sensui propter suam velocitatem. Preterea
motus eorum in subtilibus corporibus, scilicet in illis que valde sunt
diaffona, velocior est motu eorum in eis que sunt grossiora. in eis sci-
300 licet que minus sunt diaffona. Omne enim corpus diaffonum, cum
lux transit in ipsum, resistit luci aliquantulum secundum quod habet
de grossitie, nam in omni corpore naturali necesse est quod sit aliqua
grossities, nam parve diaffonitatis non habet finem in ymaginatione,
que est ymaginatio lucide diaffonitatis, et est quod omnia corpora
5 naturalia perveniunt ad finem quem non possunt transire. Corpora

279 ad diaffonitatem: a diaffanitate R/existit: extat P1; alter. ex existuat in existat F   280 extenduntur: extenditur FP1L3E; extendentur R   281 et . . . corpore (282) om. S/fuerint: fuerit FP1O   282 reflectentur: reflectuntur FP1; reflectetur E; refringentur R/cum: tamen FC1E;
tum R/in² om. FP1   283 extendentur: extenditur FP1; extendetur L3; alter. ex extenditur
in extendetur E/rectarum linearum transp. L3ER/aliarum inter. O   285 reflexa: refracta R
286 reflectebatur: refringebatur R   287 superficie: superficiei C1/quod om. R/reflexio: refractio R   288 subtiliori: subtiliore R/erit om. S   289 reflexionis: refractionis R/super
alter. in ad a. m. E   290 grossiori: grossiore R/reflexio: refractio R/erit inter. a. m. E   291 in
inter. a. m. E   292 reflexionis: refractionis R   293 reflectitur: reflectatur E; refringatur
R/occurrit: occurrerit S/corpori diaffono transp. C1   294 post diaffonitatis add. causa R/hoc:
hec R   295 fit: scilicet L3/ut corr. ex et a. m. E/iam om. R   296 post extenduntur² scr. et
del. per corpora E   297 non inter. E/preterea: propterea S   298 eorum: earum R/in¹
inter. a. m. F/subtilibus: subtilioribus P1/post subtilibus scr. et del. cor P1/sunt diaffona (299)
transp. FP1   299 velocior: velociter S/eorum: earum R/eis¹: iis R/eis que sunt diaffona. O/in² om.
C1ER; scr. et del. L3/eis²: illis R   1 in: per SO   2 quod: ut R   3 post nam add. corpus
C1ER/parve alter. in parva a. m. L3; alter. in parvitas a. m. O/diaffonitatis corr. ex quantitatis a. m.
E; alter. in diaffonitas a. m. L3/ymaginatione corr. ex ymagine a. m. L3   4 est¹: etiam S/post
est¹ add. et FP1/ymaginatio: ymago C1/et: etiam S/est² inter. L3E (a. m. L3)/est quod om. R
5 perveniunt: pervenient L3/quem: quoniam L3/corpora ergo (6) transp. L3

ergo naturalia diaffona non possunt evadere aliquam grossitiem. Luces ergo, cum transeunt per corpora diaffona, transeunt secundum diaffonitatem que est in eis, et sic impediunt lucem secundum grossitiem que est in eis.

[2.82] Cum ergo lux transiverit per corpus diaffonum et occurrerit alii corpori diaffono grossiori primo, tunc corpus grossius resistet luci vehementius quam primum resistabat, et omne motum, cum movetur ad aliquam partem essentialiter aut accidentaliter, si occurrerit resistenti, necesse est ut motus eius transmutetur. Et si resistentia fuerit fortis, tunc motus ille reflectetur ad contrariam partem. Si vero debilis, non revertetur ad contrariam partem, nec poterit per illam procedere per quam inceperat, sed motus eius mutabitur.

[2.83] Omnium autem motorum naturalium que recte moventur per aliquod corpus passibile transitus super perpendicularem que est in superficie corporis in quo est transitus erit facilior. Et hoc videtur in corporibus naturalibus, si enim aliquis acceperit tabulam subtilem, et paxillaverit illam super aliquod foramen amplum, et steterit in oppositione tabule, et acceperit pilam ferream, et eicerit eam super tabulam fortiter, et preservaverit quod motus pile sit super lineam perpendicularem super superficiem tabule, tunc tabula cedet pile aut frangetur, si tabula fuerit subtilis et vis qua spera movetur fuerit fortis. Et si steterit in parte obliqua ab oppositione tabule et in illa eadem distantia in qua prius erat, et eicerit pilam super tabulam illam eadem vi qua prius eicerat, tunc spera labetur de tabula, si tabula non fuerit valde subtilis, nec movebitur ad illam partem ad quam primo movebatur, sed declinabit ad aliquam partem.

[2.84] Et similiter si acceperit ensem, et posuerit coram se lignum, et percusserit eum ense ita quod ensis sit perpendicularis su-

6 ergo: grossa *O*/luces *inter. a. m. L3*   7 ergo: que *L3*/per ... transeunt[2] *om. S*/diaffona *om. P1*/secundum *corr. ex* per *E*   8 est in eis: in eis est *E*   10 transiverit: transivit *S*/occurrerit: occurreret *L3*; occurrit *R*   11 corpori: corporii *S*/diaffono *om. R*/primo *inter. O*/primo tunc *transp. deinde corr. L3*/corpus grossius *transp. C1*/resistet: resistit *FP1R*; *corr. ex* resistit *E*   12 resistabat: restabat *S*; resistebat *R*/omne: omnem *O*/cum *corr. ex* quod *C1*   13 occurrerit: occurrit *L3*   14 trans-mutetur: transmittetur *S*   15 reflectetur: refringetur *R*/si ... partem (16) *om. P1*; *mg. a. m. E*   16 non *inter. a. m. L3*/revertetur: reflectecetur *L3*; reflectetur *E*; refringetur *R*   17 eius *inter. a. m. E*   18 naturalium: naturaliter *FP1R*/post naturalium *add.* et hec *S*/que *inter. a. m. E*   19 passibile: passibilem *S*; *corr. ex* possibile *C1*   20 est *om. SO*; *mg. a. m. L3*/hoc: hec *S*   21 aliquis: quis *E*/acceperit *corr. ex* acceperat *a. m. E*   22 paxillaverit: paxillam *FP1*; *corr. ex* passillaverit *O*/illam *om. FP1*   23 *post* et[1] *add.* sic *P1*/post acceperit *scr. et del.* palam *S*/eicerit: eiecerit *P1R*/eam: illam *C1*   24 *post* et *scr. et del.* quod *E*/preservaverit: observaverit *R*/quod: ut *R*/lineam *om. R*   26 frangetur: frangitur *F*/fuerit subtilis *transp. L3ER*   27 parte obliqua *transp. C1*/ab *om. FP1*   28 eicerit: eiecerit *R*/illam: illa *SL3C1O*/eadem: eandem *FP1ER*   29 vi: in *FP1OER*/qua: quam *FP1ER*/post prius *scr. et del.* erat *S*/eicerat: eicerat *S*; eiecerat *R*/tabula non *mg. F*   30 valde *om. P1*/partem *om. F*/movebatur *corr. ex* intuebatur *E*   31 *post* partem *add.* aliam *ER*   32 et[1]: item *FP1*/post si *inter.* quic *a. m. C1*/post acceperit *add.* quis *FP1*/se *inter. a. m. L3*   33 percusserit: percussit *S*/eum: eam *L3O*; cum *P1C1R*/quod: ut *R*

per superficiem ligni, tunc lignum secabitur magis. Et si ensis fuerit
35 obliquus et percusserit lignum oblique, tunc lignum non secabitur
omnino, sed forte secabitur in parte, aut forte ensis errabit deviando,
et quanto magis ensis fuerit obliquus, tanto minus aget in lignum. Et
alia multa sunt similia, ex quibus patet quod motus super perpendicularem est facilior et fortior et quod de obliquis motibus ille qui
40 vicinior est perpendiculari est facilior remotiori.

[2.85] Lux ergo, si occurrerit corpori diaffono grossiori illo corpore in quo existit, tunc impedietur ab illo ita quod non transibit in
partem in qua movebatur, sed quia non fortiter resistit, non redibit
in partem ad quam movebatur. Si ergo motus lucis transiverit super
45 perpendicularem, transibit recte propter fortitudinem motus super
perpendicularem, et si motus eius fuerit super lineam obliquam, tunc
non poterit transire propter debilitatem motus. Accidet ergo quod
declinetur ad partem in quam facilius movebitur quam in parte in
quam movebatur. Sed facilior motuum est super perpendicularem,
50 et quod vicinius est perpendiculari est facilius remotiori.

[2.86] Et motus in corpore in quod transit, si fuerit obliquus super superficiem illius corporis, componitur ex motu in parte perpendicularis transeuntis in corpus in quo est motus et ex motu in parte
linee que est perpendicularis super perpendicularem que transit in
55 ipsum. Cum ergo lux fuerit mota in corpore diaffono grosso super
lineam obliquam, tunc transitus eius in illo corpore diaffono erit per
motum compositum ex duobus predictis motibus. Et quia grossities
corporis resistit ei a verticatione quam intendebat, et resistentia eius
non est valde fortis, ex quo sequeretur quod declinaret in partem ad
60 quam facilius transiret, et motus super perpendicularem est facilimus
motuum, necesse est ut lux que extenditur super lineam obliquam
moveatur super perpendicularem exeuntem a puncto in quo lux occurrit superficiei corporis diaffoni grossi.

[2.87] Et quia motus eius est compositus duobus motibus, quo-
65 rum alter est super lineam perpendicularem super superficiem cor-

34 si *inter. O* / ensis *om. ER*   35 lignum oblique *transp. R* / lignum² *om. SP1* / secabitur: sepa-
rabitur *FP1*   37 ensis fuerit *transp. L3R* / aget: agit *P1*   39 facilior et fortior: fortior et facilior
ER / et fortior *om. L3*   40 est¹ *om. FP1* / facilior *om. P1* / remotiori: remotiore *R*; remotior *S*; *corr. ex*
remotior *O*   41 occurrerit: occurit *L3C1OR* / corpori diaffono *transp. C1*   42 impedietur *alter.*
in impeditur *E* / illo: eo *ER*   43 qua: quam *R* / sed: si *C1*   44 in: ad *P1* / si ergo *transp. FP1* /
post transiverit *scr. et del.* tunc impedietur ab illo *C1*   46 tunc non poterit (47): non poterit tunc *O*
47 accidet: accidit *C1* / quod: ut *R*   48 *post* partem *add.* motus *L3C1ER* / quam² *inter. O* / parte:
partem *ER*   49 sed *corr. ex* hec *L3* / motuum: motus *FP1*   50 facilius *corr. ex* facilis *S* / remotiori:
remotiore *R*   51 *post* motus *add.* etiam *P1*   56 *post* obliquam *scr. et del.* erit *C1* / tunc transitus
transp. deinde corr. C1   57 duobus: duabus *S*   58 a verticatione: ad verticationem *ER*   59 in:
ad *P1R*   61 *post* est *add.* ergo *R* / ut: quod *E* / obliquam: obliquum *S*   64 est *corr. ex* ex *P1*; *om.*
C1 / *post* compositus *add.* ex *FP1C1R* / quorum: quarum *O*   65 alter: altera *O* / lineam *om. C1*

poris grossi et reliquus super lineam perpendicularem super perpendicularem hanc, et motus compositus qui est in ipso non omnino demittitur sed solummodo impeditur, necesse est ut lux declinet ad partem faciliorem parte ad quam prius movebatur, remanente in ipso
70 motu composito. Sed pars facilior parte ad quam movebatur, remanente motu in ipso, est illa pars que est vicinior perpendiculari, unde lux que extenditur in corpore diaffono, si occurrerit corpori diaffono grossiori corpore in quo existit, reflectetur per lineam propinquiorem perpendiculari exeunti a puncto in quo occurrit corpori grossiori que
75 extenditur in corpore grossiore quam linea per quam movebatur.

[2.88] Hec igitur est causa reflexionis splendorum in corporibus diaffonis que sunt grossiora corporibus diaffonis in quibus existunt, et ideo reflexio proprie inventa in lucibus obliquis. Cum ergo lux extenditur in corpore diaffono, et occurrerit corpori diaffono diverse
80 diaffonitatis corporis in quo existit et grossioris, et fuerit obliqua super superficiem corporis diaffoni cui occurrit, reflectetur ad partem perpendicularis super superficiem corporis diaffoni extense in corpore grossiore.

[2.89] Causa autem que facit reflexionem lucis a corpore grossiori
85 ad corpus subtilius ad partem contrariam parti perpendicularis est quia, cum lux mota fuerit in corpore diaffono, repellet eam aliqua repulsione, et corpus grossius repellet eam maiori repulsione, sicut si lapis, cum movetur in aere, movetur facilius et velocius quam si movetur in aqua eo quod aqua repellit ipsum maiori repulsione quam
90 aer. Cum ergo lux exiverit a corpore grossiori ad subtilius, tunc motus eius erit velocius, et cum lux fuerit obliqua super duas superficies corporis diaffoni que est differentia communis ambobus corporibus, tunc motus eius erit super lineam existentem inter perpendicularem exeuntem a principio motus eius et inter perpendicularem super li-

66 *ante* grossi *scr. et del.* et F / *post* super² *scr. et del.* superficiem corporis grossi C1   67 in *inter. a. m.* L3   68 demittitur: dimittitur R; *corr. ex* dimittetur E / est *inter.* O   70 quam *om.* L3   71 vicinior: vicinor R   72 occurrerit: occurrit R   73 grossiori *corr. ex* grossori C1 / reflectetur: reflectitur L3; refringetur R   74 exeunti: exeunte L3 / occurrit *corr. ex* occurrerit S; *corr. ex* concurrit *a. m.* E   75 *post* grossiore *add.* per aliam lineam R / *post* quam *add.* sit R; *scr. et del.* per P1   76 est causa *transp.* R / reflexionis splendorum: refractionis splendoris R   77 diaffonis *inter. a. m.* E / existunt: existit O   78 reflexio: refractio R / *post* reflexio *inter.* est *a. m.* L3 / proprie: propria O / *post* proprie *add.* est ER / proprie inventa *transp.* L3 / lux extenditur (79) *transp.* C1   80 *post* diaffonitatis *add.* a R / corporis: corpore R / grossioris: grossiori R / et² *inter. a. m.* L3 / fuerit obliqua *transp.* L3   81 cui: cum L3 / reflectetur: refringetur R   82 extense *corr. ex* extensio P1   84 reflexionem: refractionem R / grossiori: grossiore R   86 cum lux *transp.* C1   87 et . . . repulsione² *mg.* C1 / maiori: maiore R   88 si¹ *om.* P1R / et *inter. a. m.* L3   89 movetur: moveretur L3C1ER / aqua² *om.* P1 / ipsum: ipsam F / maiori: maiore R   90 lux *om.* P1 / exiverit: exierit ER / grossiori: grossiorie S; grossiore R / ad: in ER; *inter. a. m.* L3 / *post* ad *add.* corpus C1   91 velocius: velocior R / duas superficies *transp.* C1   92 que: quod R / communis *inter.* O

95 neam perpendicularem exeuntem etiam a principio motus. Resistentia
ergo corporis grossioris erit a parte ad quam exit secunda perpendicularis. Cum ergo lux exiverit a corpore grossiori et pervenerit ad corpus subtilius, tunc resistentia corporis subtilioris luci que est in parte
ad quam exit secunda perpendicularis erit minor prima resistentia, et
100 sic motus lucis ad partem a qua resistebatur maior, et sic est de luce
in corpore subtiliore ad contrariam partem parti perpendicularis.

CAPITULUM TERTIUM
De qualitate reflexionis lucis in corporibus diaffonis

[3.1] In predicto capitulo declaratum est quod omnis lux que re-
105 flectitur a corpore diaffona ad aliud corpus diaffonum semper erit in
una superficie equali. Lux ergo que reflectitur ab aere ad aquam est
semper in eadem superficie equali; linea ergo recta que est per quam
extenditur lux in aere et linea recta per quam reflectitur in aqua semper erunt in eadem superficie equali. Hec autem superficies apud
110 inspectionem instrumenti predicti est medius circulus illis tribus predictis signatis in interiori parte hore instrumenti.
[3.2] Sed superficies interioris lamine est equidistans superficiei
dorsi, cui superponitur superficies regule quadrate. Ergo superficies
circuli medii est equidistans superficiei regule quadrate. Et superfi-
115 cies regule quadrate que est superposita dorso lamine est perpendicularis super alteram superficiem secantem superficiem superposi-
tam dorso lamine, et hec superficies regule superponitur superficiei
duarum differentiarum sibi applicatarum in duabus extremis regule.
Sed superficies duarum differentiarum superponitur hore instru-
120 menti.

---

95 *post* exeuntem *scr. et del.* a E / a . . . motus *corr. ex* motus a principio L3 / *post* motus *add.* eius FP1
96 corporis grossioris *inter. a. m.* E / *post* corporis *scr. et del.* eius P1 / *post* erit *scr. et del.* a corpore p
C1 / a: ex S / exit: erit SO   97 exiverit: exivit S   98 *post* subtilioris *add.* facta R   99 exit
mg. a. m. E / exit secunda *transp.* R   100 sic: fit R / resistebatur: resistabatur S   101 subtiliore: subtiliori SP1 / contrariam partem *transp.* ER / parti perpendicularis *transp.* E; *transp. deinde*
corr. L3   102 capitulum . . . diaffonis (103) *om.* S; de qualitate refractionis lucis in corporibus
diaffonis capitulum tertium R   103 corporibus *corr. ex* operibus O   104 reflectitur: refringitur R   106 lux . . . equali (107) *om.* L3ER / aquam: quam FP1; *corr. ex* quam O   107 que
est *om.* R   108 reflectitur: refringitur R   109 erunt *om.* S   110 *post* circulus *inter.* de
a. m. C1 / illis: ille R; *corr. ex* ille *a. m.* E / *post* illis *add.* ex R / predictis *om.* L3R   111 interiori:
interiore R / *post* instrumenti *add.* et ille circulus est equidistans superficiei interioris lamine R
112 interioris: interior FP1   113 cui: circuli S; et O / superponitur: supperponitur P1 / quadrate
corr. ex quarte O / *post* superficies² *rep. et del.* superficies O   114 et . . . quadrate (115) *inter. a. m.*
L3; *mg. a. m.* E   115 superposita: supposita S / *post* lamine *scr. et del.* que C1   116 super:
supra E / superficiem² *om.* SFP1O   117 dorso lamine *om.* L3ER   118 sibi . . . differentiarum
(119) *mg. a. m.* E / applicatarum: applicarum F; *corr. ex* applicata O / extremis: extremitatibus ER
119 superponitur: supperponitur FP1

[3.3] Ergo superficies medii circuli est perpendicularis super superficiem transeuntem super horam instrumenti, et superficies transiens per horam instrumenti est equidistans orizonti apud experimentationem. Superficies ergo circuli medii est perpendicularis super superficiem orizontis. Cum ergo declaratum sit quod lux que est in aere et reflectitur in aqua est apud experimentationem in circumferentia medii circuli, manifestum est quoniam lux que extenditur in aere et reflectitur in aqua est semper in eadem superficie equali super superficiem orizontis.

[3.4] Et etiam ymaginemur lineam a centro medii circuli ad centrum mundi. Sic ergo hec linea erit perpendicularis super superficiem aque, quia est diameter mundi. Sed hec linea est in superficie medii circuli; ergo est in superficie reflexionis. Ergo superficies reflexionis est perpendicularis super superficiem aque. Et iam declaratum est quod, cum lux reflectitur ex aere ad aquam, erit inter primam lineam per quam extenditur in aere, que est inter diametrum medii circuli, et inter perpendicularem exeuntem a centro medii circuli super superficiem aque. Et declaratum est etiam quoniam lux que est in puncto quod est centrum lucis que est intra aquam non pervenit ad ipsum nisi ex luce que extenditur a centro circuli medii. Lux ergo que reflectitur ex aere ad aquam reflectitur in superficie perpendiculari super superficiem aque, et reflexio eius erit ad partem perpendicularis exeuntis a loco reflexionis super superficiem aque, et non perveniet ad perpendicularem.

[3.5] Reflexio autem lucis ab aere ad vitrum hoc modo fit, declaratum enim est in experimentatione vitri quod, cum linea que transit per centra duorum foraminum, cum fuerit obliqua super superficiem vitri equalem et transiverit per centrum vitri, et superficies vitri equalis fuerit ex parte foraminum, tunc linea reflectetur apud cen-

122 super: per *S*/*post* et *add.* hec *ER*/transiens *om. S*   123 equidistans: equidistanti *P1*
124 ergo *inter. a. m. L3*/circuli medii *transp. R*   126 et *inter. C1E* (*a. m. E*)/reflectitur: refringitur *R*/aqua: aere *P1*/*post* aqua *add.* et reflectitur *P1*/est *inter. a. m. E*/in . . . circuli (127): circuli medii in circumferentia *L3*   127 medii circuli *transp. C1*/est *om. ER*/quoniam: quod *R*   128 reflectitur: refringitur *R*/*post* equali *scr. et del.* tunc superficies *P1*/super *inter. O*   130 etiam ymaginemur *corr. ex* in ymaginationem *P1*   131 ergo *mg. a. m. L3*/ hec linea *transp. ER*/*post* linea *scr. et del.* est in superficie medii circuli *L3*   132 quia *corr. ex* que *a. m. E*   133 reflexionis^{1,2}: refractionis *R*/ergo^{2} . . . reflexionis^{2} *om. S*   134 est^{2} *om. S*   135 reflectitur: refringitur *R*   137 centro medii *transp. L3*   138 *post* et *add.* iam *L3ER*/etiam *om. O*/quoniam: quod *R*   139 intra: inter *P1*   140 circuli medii *transp. R*/reflectitur: refringitur *R*   141 aquam: quam *S*/reflectitur: refringitur *R*/*post* reflectitur *scr. et del.* ex aere *P1*/perpendiculari: perpendicularis *L3*/super *om. F*   142 et . . . aque (143) *om. S*/reflexio: reflexionis *P1*; refractio *R*/eius *om. P1*   143 reflexionis: refractionis *R*/super *inter. F*/perveniet: pervenit *L3E*; *corr. ex* pervenit *F*   145 reflexio: refractio *R*/ab aere *om. L3E*   146 enim est *transp. ER*   147 cum *om. R*   148 et^{2} . . . vitri *inter. O*/ superficies: superficiem *S*; *corr. ex* superficiem *L3*/superficies vitri *om. P1*   149 linea *om. R*/reflectetur: refringetur *R*

150 trum vitri, et reflexio eius erit in superficie circuli medii ad partem in
qua est perpendicularis exiens a centro vitri super superficiem vitri
equalem.

[3.6] Et declaratum est etiam quod, cum linea que transit per centra duorum foraminum fuerit obliqua super superficiem vitri speri-
155 cam, et superficies sperica fuerit ex parte foraminum, tunc lux reflectetur in corpore vitri et apud superficiem vitri spericam. Et erit
reflexio eius in superficie medii circuli et ad partem perpendicularis
exeuntis a loco reflexionis super superficiem vitri spericam. Et superficies vitri equalis in qua est centrum circuli vitrei est perpendicularis
160 super superficiem lamine; est ergo perpendicularis super superficiem
medii circuli.

[3.7] Superficies ergo medii circuli est perpendicularis super superficiem vitri equalem, et superficies circuli medii transit etiam per
centrum spere vitree; in omnibus experimentationibus vitri ergo est
165 perpendicularis super superficiem vitri spericam etiam. Lux ergo
que extenditur in aere et reflectitur in corpore vitri, apud extensionem eius in aere et postquam reflectitur in vitro semper est in superficie perpendiculari super superficiem vitri, et semper reflexio eius erit
ad partem perpendicularis exeuntis a loco reflexionis super superfi-
170 ciem vitri, sive superficies vitri fuerit equalis, sive sperica.

[3.8] Item declaratum est etiam quod linea que transit per duo
centra foraminum, cum fuerit perpendicularis super superficiem vitri, et extensa fuerit in corpus vitri secundum rectitudinem, et superficies sperica fuerit ex parte foraminum, et fuerit hec linea—scilicet
175 que transit per centra duorum foraminum—declinans super superficiem vitri equalem, et transiverit per centrum vitri et reflexa in corpore aeris contingentis superficiem vitri equalem et apud centrum
vitri, tunc reflexio eius erit in superficie circuli medii et ad contrariam
partem illi in qua est perpendicularis exiens a centro vitri super su-
180 perficiem vitri equalem.

150 reflexio: refractio R/post eius add. apud centrum P1/ad corr. ex a O   153 que: qua S
154 fuerit corr. ex fuerint F   155 reflectetur: refringetur R   156 post apud add. super F
157 reflexio: refractio R   158 reflexionis: rationis S; refractionis R/post vitri scr. et del.
foramini C1   159 circuli vitrei transp. L3E/vitrei: vitri SO; corr. ex vitri L3C1E/post vitrei add.
medii P1   160 lamine . . . superficiem² mg. a. m. E   161 medii circuli transp. P1/medii . . .
superficiem (162) om. S   163 post et scr. et del. super E/circuli medii transp. C1   165 etiam:
et E; om. R   166 reflectitur: flectitur S; reflectetur L3O; refringitur R   167 post postquam
add. iterum R/reflectitur: refringitur R   168 super mg. F/reflexio: refractio R/eius erit transp.
deinde corr. C1/erit inter. a. m. E   169 reflexionis: refractionis R   170 post vitri¹ rep. et del. et
(168) . . . partem (169) C1   171 declaratum: declaratio FP1/etiam inter. O/que om. FP1/duo
centra (172): centra duorum O   174 ex . . . foraminum: foraminum ex parte O/foraminum
corr. ex foramine L3   175 declinans corr. ex declinansus C1   176 equalem inter. O/reflexa:
refracta R/post refracta add. fuerit R   177 vitri¹ om. FP1/equalem: equale S   178 reflexio:
refractio R/erit: eius S/post circuli scr. et del. et C1

[3.9] Et declaratum est etiam quod linea que transit per centra duorum foraminum, cum fuerit perpendicularis super superficiem vitri
equalem, et si fuerit extensa in corpore vitri secundum rectitudinem,
et superficies equalis fuerit ex parte foraminum, et hec linea—scilicet
185   que transit per centra duorum foraminum—fuerit obliqua super superficiem vitri spericam et non transiens per centrum eius, et fuerit
reflexa apud superficiem spericam in corpore aeris continentis superficiem spericam, tunc reflexio eius erit in superficie medii circuli et ad
partem contrariam illi in qua est perpendicularis exiens a loco reflexio-
190   nis super superficiem reflexionis. Et in hiis duobus sitibus superficies etiam medii circuli est perpendicularis super superficiem vitri
equalem et spericam. Lux ergo que extenditur in corpore vitri et reflectitur in aere, dum extenditur in vitro et reflectitur in aere, semper
est in superficie perpendiculari super superficiem aeris, et semper
195   reflexio eius erit ad partem contrariam illi in qua est perpendicularis
exiens a loco reflexionis super superficiem aeris.

[3.10] Ex omnibus ergo istis predeclaratis patet quod omnis lux
reflexa a corpore diaffono ad aliud corpus semper reflectitur in superficie perpendiculari super superficiem secundi corporis, et si secun-
200   dum corpus fuerit grossius primo, tunc reflexio eius erit ad partem
perpendicularis exeuntis a loco reflexionis super superficiem secundi
corporis, et non pervenit ad perpendicularem. Et si secundum corpus fuerit subtilius primo, reflexio erit ad partem contrariam illi in
qua est perpendicularis exiens a loco reflexionis super superficiem
205   secundi corporis secundum diversitatem figurarum superficierum
corporum diaffonorum.

[3.11] Et ex hiis etiam patet quod, cum lux reflectitur a corpore diaffono ad secundum corpus diaffonum et de secundo ad tertium, reflectetur etiam in superficie tertii si diaffonitas tertii differt a diaffonitate

181 centra duorum *transp. SFC1E* 184 et^2 ... foraminum (185) *mg. a. m. E* 185 transit
per *inter. a. m. L3* / fuerit obliqua *om. P1* 187 reflexa: refracta *R* / *post* superficiem *add.* vitri
R / in ... spericam (188) *om. P1* / continentis: contingentis *R* 188 reflexio: refractio *R* / et
mg. C1 189 contrariam *inter. a. m. L3* / qua *inter. O* / reflexionis: refractionis *R* 190 super
... reflexionis *om. P1C1*; *inter. a. m. E* / reflexionis: secundi corporis *R* / reflexionis ... superficiem
(194) *mg. a. m. L3* / duobus *mg. a. m. C1* 192 reflectitur: refringitur *R* 193 dum ... aere^2
om. L3 / reflectitur: refringitur *R* 194 est *rep. P1* / super *inter. O* / et ... aeris (196) *om. L3*
195 reflexio: refractio *R* / eius *om. ER* / partem *om. C1* 196 *post* a *scr. et del.* vobis *F* / reflexionis: refractionis *R* 198 reflexa: refracta *R* / aliud *corr. ex* aliquid *C1* / reflectitur: refringitur *R*
199 superficiem *om. S* 200 reflexio: refractio *R* 201 *post* loco *scr. et del.* foramini *P1* /
reflexionis: refractionis *R*; *corr. ex* rationis *S* 203 fuerit *corr. ex* fuerint *C1* / reflexio: refractio
R / *post* partem *scr. et del.* perpen *P1* 204 est *om. FP1* / reflexionis: refractionis *R* 205 *post*
superficierum *scr. et del.* corporum *F*; *rep. et del.* superficierum *O* 207 etiam *om. O* / lux *corr.*
ex lex *C1* / reflectitur: reflectatur *FP1*; refringitur *R* 208 secundum corpus *transp. deinde corr.*
C1 / tertium: centrum *L3* / reflectetur: refringetur *R* 209 *post* tertii^2 *scr. et del.* diaffonitatis
C1 / differt *mg. a. m. C1* / diaffonitate: diaffonitatem *O*

210 secundi. Si vero tertium fuerit grossius secundo, tunc reflexio lucis
erit ad partem perpendicularis exeuntis a loco reflexionis super superficiem tertii. Si autem tertium fuerit subtilius secundo, tunc reflexio
lucis erit ad partem contrariam illi in qua est perpendicularis; similiter si lux reflexa fuerit ad quartum corpus, et ad quintum, aut ad
215 plura.

[3.12] Hoc autem quod declaravimus in hoc capitulo est qualiter
omnes luces reflectuntur in corporibus diaffonis diverse diaffonitatis. Quare autem fit reflexio in superficie perpendiculari super superficiem corporis diaffoni hoc est quia linea per quam extenditur lux
220 in primo corpore diaffono reflectitur ad partem perpendicularis in
hac superficie, scilicet in qua est perpendicularis et prima linea, pars
enim perpendicularis est in hac superficie. Ideo reflexio fit in superficie perpendiculari super superficiem corporis diaffoni.

[3.13] Quantitates autem angulorum reflexionis differunt secun-
225 dum quantitates angulorum quos continet prima linea per quam
extenditur lux in primo corpore et perpendicularis exiens a loco reflexionis super superficiem secundi corporis secundum diaffonitatem
secundi corporis, nam quanto magis crescit angulus quem continet
prima linea et perpendicularis, crescit angulus reflexionis, et quan-
230 to decreverit ille angulus, decrescit angulus reflexionis. Sed anguli
reflexionum non observant eandem proportionem ad angulos quos
continet prima linea cum perpendiculari; sed differunt hee proportiones in eodem corpore diaffono. Cum ergo prima linea per quam
lux extenditur in primo corpore continuerit cum perpendiculari duos
235 angulos inequales in duobus temporibus diversis aut in duobus locis

210 reflexio: refractio R; corr. ex ratio a. m. S   211 partem . . . exeuntis corr. ex perpendicularis exeuntis partem C1/perpendicularis . . . partem (213) om. S/reflexionis: refractionis R   212 tertium: tertius O; corr. ex tertius a. m. S/fuerit: fuit C1/reflexio: refractio R
213 erit: exit L3   214 reflexa: refracta R   216 quod inter. a. m. L3; om. R/post declaravimus add. quidem R/capitulo corr. ex speculo a. m. E/est om. FP1ER/qualiter: equaliter SFP1
217 reflectuntur: refringantur R   218 fit reflexio: fiat refractio R/perpendiculari corr. ex
perpendicularis C1   219 diaffoni om. P1/hoc: hec R/quia: quod L3   220 primo corpore transp. FP1/corpore diaffono transp. R/reflectitur: refringitur R/post perpendicularis
add. reflectitur O   221 superficie: susuperficie S   222 est . . . perpendiculari (223) mg.
a. m. L3/superficie¹ alter. in linea a. m. E/ideo: cum L3/reflexio: refractio R; corr. ex ratio a. m.
S/superficie²: corpore SO; corr. ex corpore a. m. E   224 post autem add. scilicet FP1O/reflex-
ionis: refractionis R; scr. et del. S   225 quos: quas SO/continet: continent SR   226 post
extenditur scr. et del. lumen P1/post primo add. in S/reflexionis: refractionis R   227 post
corporis add. et SC1O   228 crescit: cressit S/angulus corr. ex anus L3/quem . . . linea (229)
rep. L3/continet: continent SL3C1R; corr. ex continent O   229 post perpendicularis add.
tanto R/angulus: anus L3/reflexionis: refractionis R   230 ante decreverit add. magis R/
decreverit: decressit S; decrescit R/ille inter. a. m. E/angulus¹ om. C1/post angulus¹ add. quem
continent perpendicularis et prima linea tanto ER (tanto om. E)/decrescit: decrescet SP1O;
alter. ex decrescit in decrescet F/reflexionis: refractionis R; scr. et del. S   231 reflexionum:
refractionum R/eandem: eadem S   233 post eodem scr. et del. tem S   234 lux inter. O
235 temporibus diversis transp. FP1R

diversis, tunc proportio anguli reflexionis que est ab angulo minori ad angulum minorem minor erit proportione anguli reflexionis anguli maioris ad angulum maiorem.

[3.14] Cum ergo experimentator voluerit experiri istos angulos,
240 dividat a circulo medio qui est in circumferentia instrumenti ex parte centri foraminis quod est in circumferentia instrumenti arcum decem partium ex illis partibus quibus medius circulus dividitur in 360. Deinde extrahamus a loco differentie lineam rectam perpendicularem super superficiem lamine, et copulemus extremitatem eius que
245 est in lamina cum centro lamine per lineam rectam, et protrahamus ipsam in aliam partem.

[3.15] Deinde dividamus in circumferentia medii circuli etiam arcum sequentem primum, cuius quantitas sit 90, et signemus in extremitate huius arcus signum. Linea ergo que exit a centro medii
250 circuli ad hoc signum erit perpendicularis super lineam exeuntem a centro medii circuli ad primum signum quod est in circumferentia medii circuli. Et erit arcus residuus qui est inter secundum signum et extremitatem diametri medii circuli qui transit per centra duorum foraminum 80 partium. Signemus igitur in extremitate huius dia-
255 metri etiam signum.

[3.16] Deinde ponamus instrumentum in vase, et preservemus ut circumferentia vasis sit equidistans orizonti, et incipiamus experiri ab hora ortus solis. Et infundamus in vas aquam claram quousque perveniat ad centrum lamine, et moveamus instrumentum donec prima
260 linea signata in superficie lamine sit contingens superficiem aque. In hoc ergo statu linea que transit per centrum circuli medii equidistans est prime linee signate in superficie lamine cuius extremitas pervenit ad primum signum signatum in circumferentia circuli medii. Tanget etiam superficiem aque, locus enim harum duarum linearum non dif-

236 proportio: proportior *O*/reflexionis: refractionis *R*/minor: minore *R*   237 minorem *corr. ex* maiorem *O*/minor . . . maiorem (238) *om. SFP1L3O*/reflexionis: refractionis *R*/reflexionis anguli (238) *inter. a. m. E*   238 maioris: minoris *C1*   239 istos: illos *R*/angulos *mg. a. m. C1*/*post* angulos *add.* iam dictos *mg. a. m. C1*   242 *post* quibus *scr. et del.* m *S*/in: scilicet *L3; om. ER*/360: 36 *SFO*   244 copulemus *corr. ex* copilemus *C1*/extremitatem: extremitates *SFL3OE*   245 est: sunt *SFL3OE*/*post* in *scr. et del.* linea *C1*/protrahamus: pertrahamus *S*   246 aliam *corr. ex* illam *S*/ *post* partem *add.* linee *L3*   247 *post* deinde *add.* et *L3*/circumferentia: circumferentiam *S*/etiam *corr. ex* per *a. m. L3*   248 sequentem *corr. ex* secantem *E*/primum *om. S*/quantitas: qualitas *P1*; *corr. ex* qualitas *F*/*post* 90 *add.* partium *R*   250 erit perpendicularis *transp. C1*   251 *post* ad *scr. et del.* hoc signum erit perpendicularis *S*   252 arcus *corr. ex* acus *O*/secundum *om. R*/secundum signum *transp. C1*/signum *mg. a. m. L3*   253 diametri medii: diamedii *S*/qui: que *R* 254 80: 8 *S*/80 partium: centrum *O*/*post* partium *add.* kr *S*/igitur *om. SR*   256 in *inter. a. m. L3*/ preservemus: observemus *R*   257 *post* experiri *add.* et *C1*   258 hora *om. S*/ortus: hortus *S* 259 moveamus: movemus *C1*; *corr. ex* movemus *O*   260 linea *inter. a. m. E*/*post* linea *scr. et del.* figura que distat per 10 gradus a foramine *E*/contingens: contingat *R*/superficiem: superficiei *L3* 261 statu: situ *ER*   262 est *om. O*   263 circuli medii *transp. C1R*/*ante* tanget *add.* et *R*/tanget: tangit *C1*   264 *post* etiam *scr. et del.* in *L3*/harum *alter. in* horarum *E*

265 fert in respectu superficiei aque quoad sensum. Et hec linea continet
cum linea exeunte a centro medii circuli ad secundum signum quod
est in circumferentia medii circuli perpendicularem super superficiem aque angulum rectum, et diameter circuli medii qui transit per
centra duorum foraminum continet cum hac perpendiculari exeunti
270 a centro circuli medii super superficiem aque angulum cuius quantitas erit 80 partes, hunc enim angulum cordat arcus medii circuli qui
est inter secundum et tertium signum. Arcus autem qui est inter centrum foraminis et primum signum, qui est decem partium, cordat
angulum declinationis.

275   [3.17] Deinde oportet experimentatorem considerare solem et
mutare instrumentum donec lux transeat per duo foramina, et tunc
aspiciat lucem que est in hora instrumenti que est intra aquam, et signet super centrum lucis signum. Hoc ergo signum erit in circumferentia medii circuli. Deinde auferat instrumentum, et aspiciat tertium
280 signum, quod est inter extremitatem medii circuli et inter secundum
signum quod est extremitas perpendicularis exeuntis a centro medii
circuli super superficiem aque. Ex hac ergo experimentatione patebit quod angulus reflexionis est ille quem cordat arcus qui est inter
centrum lucis et tertium signum quod est extremitas linee transeun-
285 tis per centra duorum foraminum per quam extendebatur lux. Et ex
numero partium huius arcus patebit quantitas anguli reflexionis et
quantitas proportionis anguli reflexionis ad 80 partes que sunt angulus quem continet linea per quam extendebatur lux cum perpendiculari exeunti a puncto reflexionis super superficiem aque.

290   [3.18] Deinde oportet experimentatorem delere signum et lineam
signatam in lamina et distinguere inter circumferentiam medii circuli
ex parte centri foraminis quod est in hora instrumenti arcum cuius
quantitas sit viginti partes. Et signet in extremitate eius signum, et
extrahat ab hoc signo perpendicularem super superficiem lamine, et
295 extrahat ab eius extremitate lineam ad centrum lamine. Et protrahamus illam in utramque partem, et dividamus etiam arcum sequentem

266 secundum *scr. et del. E; om. R*   267 perpendicularem: perpendiculare *C1*; perpendiculari
ER / super *om. FP1*   268 angulum rectum *om. SFP1L3O; inter. a. m. E* / circuli medii *transp. ER* /
qui: que *R*   269 exeunti: exeunte *R*   270 circuli medii *transp. R*   271 80: 8 *S*; 60 *O* / partes: partium *R*   273 et: ad *O* / post est *scr. et del.* inter *P1*   275 et mutare (276) *mg. a. m. E*
278 signum[1] *om. FP1*   279 aspiciat: respiciat *C1*   280 signum *om. L3E* / est *mg. a. m. L3* / inter
extremitatem: in extremitate *O*   281 est *om. S; inter. a. m. L3*   282 experimentatione *corr. ex* ex-
temp *O*   283 reflexionis: refractionis *R* / quem: que *S* / qui *inter. O*   285 et *om. SE*   286 post
numero *scr. et del.* ?? *O* / reflexionis: refractionis *R* / et . . . reflexionis (287) *om. S*   287 reflexionis:
refractionis *R* / sunt: est *R*   288 continet *inter. a. m. L3; om. E* / post lux *inter.* ?? *O*   289 exeunti:
exeunte *R* / reflexionis: refractionis *R*   290 delere: dese *S*   291 lamina: laminam *FO* / inter: in
C1 / circumferentiam *alter. in* circumferentia *C1*   292 centri *inter. a. m. E* / arcum *om. P1* / post arcum
add. de E; scr. et del. ce *L3*   293 signet: signe *S*   295 eius *om. O*   296 etiam *om. L3ER*

illum cuius quantitas erat viginti in 90, et signemus in ipso signum.
Et sit arcus qui est inter secundum signum et extremitatem linee transeuntis per centra duorum foraminum 70 partes, et signemus in ex-
300 ·   tremitate huius linee signum.

[3.19] Deinde ponamus instrumentum in vas et revolvamus illud quousque linea signata in lamina tangat superficem aque. Linea
ergo que exit a centro circuli medii ad secundum signum erit perpendicularis super superficiem aque, ut predictum est, et linea que tran-
5   sit per centra duorum foraminum continet cum hac perpendiculari
angulum 70 partium. Deinde experimentator consideret solem, et
moveat instrumentum quousque lux pertranseat per duo foramina,
et signemus super centrum lucis signum. Et auferat instrumentum,
et inspiciat signa que sunt in circumferentia medii circuli, ex qua ex-
10   perimentatione habebit quantitatem anguli reflexionis et proportionem eius ad angulum quem continet linea per quam extenditur lux
cum perpendiculari exeunte a loco reflexionis que est in hoc statu 70
partes.

[3.20] Deinde experimentator auferat instrumentum et deleat sig-
15   na et lineam que est in lamina, et dividat arcum ex parte foraminis
cuius quantitas sit 30 partes. Et procedat ut in primis ablationibus,
et sic habebit quantitatem anguli reflexionis et proportionem eius ad
angulum quem continet linea per quam extendebatur lux cum perpendiculari exeunte a loco reflexionis, qui est in hoc situ 60 partes.
20   Deinde dividamus arcum cuius quantitas sit 40 partes; deinde arcum
cuius quantitas sit 50 partes; deinde 60; deinde 70; deinde 80; et consideret unumquemque istorum arcuum, et sic habebit quantitates angulorum reflexionis et angulorum declinationis quos cordant primi
arcus distincti ex parte centri foraminis, et habebit proportionem an-
25   gulorum reflexionis ad angulos quos continent prime linee per quas

297 erat: erit *S*; *corr. ex* erit *O*; *om. L3ER*/viginti: inginta *F*/*post* in¹ *add.* partes *R*/90: 9S *S*
298 et¹ . . . signum *om. P1*/sit: erit *R*/secundum signum *transp. L3ER*/*post* et² *add.* etiam *C1*/
transeuntis: pertranseuntis *P1*   299 70: 7S *S*/partes: partium *R*   1 illud: illum *FP1*; istud *O*
2 lamina *corr. ex* lamine *F*   3 *post* erit *scr. et del.* perp *P1*   4 super *inter. F*; *om. P1*/superficiem:
superficiei *P1*   5 hac perpendiculari *transp. S*   6 70: 7S *S*; 7 partes *P1*/*post* deinde *scr. et del.*
p *C1*   7 moveat: removet *FP1*/pertranseat: transeat *ER*   8 signemus: signet *R*   9 inspiciat:
aspiciat *C1ER*/signa *corr. ex* signam *C1*/in circumferentia *om. S*   10 reflexionis: refractionis *R*
12 reflexionis: refractionis *R*/70: 7S *S*   13 partes: partium *R*   14 deinde *om. P1*   16 *post*
cuius *scr. et del.* arcus *C1*/30: 20 *E*/partes: partium *R*   17 anguli *corr. ex* antiqui *L3*/reflexionis:
refractionis *R*   18 quem: quoniam *S*; *mg. F*/linea *mg. a. m. C1*/extendebatur: extenditur *L3*
19 reflexionis: refractionis *R*/partes: partium *R*   20 arcum¹: ipsum *L3*/*post* sit *scr. et del.* 50 *F*/40:
4S *S*/40 partes *transp. P1*/40 . . . 50 (21) *mg. F*/partes: partium *R*/*post* deinde² *add.* dividamus *P1*
21 50: 5S *S*/partes: partium *R*; *om. FP1C1O*/60 *corr. ex* 70 *P1*/70: 7S *S*/80 *alter. in* 90 *O*/consideret:
considerat *L3*   22 *ante* unumquemque *add.* in *S*/unumquemque: unamquamque *L3O*   23 reflexionis: refractionis *R*/et *inter. C1*/et . . . declinationis *mg. F*/angulorum *om. C1*   24 proportionem: proportiones *R*   25 reflexionis: refractionis *R*/linee *corr. ex* figure *E*

extendebatur lux cum perpendiculari que est in superficie aque, qui crescunt per decem. Et si experimentator voluerit quod anguli crescant per quinque, bene poterit facere, et si voluerit per minus quam quinque, bene poterit facere predicto ordine.

30 [3.21] Et cum experimentator voluerit experiri per vitrum, dividat arcus, et signet predicta signa, et superponat vitrum predictum superficiei lamine, et superponat differentiam eius communem linee signate in lamina. Et ponet superficiem vitri equalem ex parte foraminum, et applicet vitrum bene. Et ponat instrumentum in vase,
35 et moveat ipsum quousque lux transeat per duo foramina, et signet super centrum lucis signum. Et auferat instrumentum, et intueatur arcus. Et deinde deleat signa, et dividat alios arcus, et signet alia signa, et inspiciat arcus prout aspexerit per aquam, et sic habebit quantitates reflexionum in transitu lucis de aere ad vitrum.

40 [3.22] Et si voluerit experiri reflexionem lucis de vitro ad aerem et ad aquam, applicet vitrum econtra primi situs, scilicet quod ponat convexum eius ex parte duorum foraminum, et ponat medium communis differentie que est in vitro super centrum lamine. Tunc ergo lux que transit per centra duorum foraminum pervenit recte ad cen-
45 trum vitri et reflectitur apud illum de vitro ad aerem. Deinde dividat arcus successive, et mutet positionem vitri, et sic habebit angulos reflexionis particulares et proportionem eorum ad angulos quos continet prima linea per quam extenditur lux cum linea perpendiculari super superficiem contingentem superficiem vitri.

50 [3.23] Et cum experimentator expertus fuerit hos duos predictos situs, videbit quoniam quantitates angulorum reflexionis de aere ad vitrum et de vitro ad aerem semper erunt equales, cum angulus quem continet linea per quam extenditur lux ad locum reflexionis cum linea perpendiculari, cum reflectatur de aere ad vitrum, equalis sit angulo

26 que: qui F / in superficie: super superficiem SC1   27 crescunt: crescit E / post voluerit add. experiri deinde scr. et del. quantitates angulorum reflexionis qui sunt apud convexum vitri L3 / quod: ut R   28 et ... facere (29) mg. a. m. E   30 et: item FP1 / experiri corr. ex expere L3 / dividat corr. ex dividet a. m. E   31 arcus: arcum FP1 / predicta signa transp. S / superponat: supperponat P1L3; superponet C1 / post predictum add. in P1   32 superficiei: superficie FP1 / et superponat inter. O   33 ponet: ponat SL3R   34 bene: linee C1   35 post moveat scr. et del. instrumentum P1   37 alia: talia C1   38 inspiciat: in S / quantitates corr. ex quantitatem C1   39 reflexionum: angulorum refractionis R / ad: in S; corr. ex a O   40 reflexionem: refractiones R   41 aquam corr. ex quam O / post aquam add. et FP1 / econtra: econverso FP1; econtrario R / scilicet om. S / quod: ut R   45 reflectitur: refringitur R; corr. ex reflectetur O / illum: illud SC1R   46 reflexionis: refractionum R; alter. in reflexionum a. m. L3   47 proportionem: proportiones R   49 superficiem corr. ex superficies C1   51 quoniam: quod R / reflexionis: refractionis R   52 angulus: angulo O; corr. ex anus L3   53 linea[1]: lineam S / locum inter. C1 / reflexionis: refractionis R / cum ... reflexionis (55) mg. C1   54 reflectatur: refringitur R / equalis sit transp. FP1 / sit inter. O

55   quem continet linea per quam extenditur lux a loco reflexionis cum
     perpendiculari cum reflectitur a vitro.

     [3.24] Et si quis voluerit experiri quantitates angulorum reflexio-
     nis qui sunt apud convexum vitri, dividat de circumferentia medii
     circuli ex parte centri foraminis quod est in hora instrumenti arcum
60   cuius quantitas sit decem partium, et extrahat ab extremitate eius
     perpendicularem super superficiem lamine in superficie hore instru-
     menti, sicut prius fecerat. Deinde dividat ex hac linea incipiens a
     centro lamine lineam equalem semidiametro vitri, et ab extremitate
     huius linee extrahat perpendicularem super diametrum lamine super
65   cuius extremitates sunt due linee perpendiculares in hora instrumen-
     ti, et protrahat hanc perpendicularem in utramque partem. Deinde
     superponat vitrum super superficiem lamine, et superponat differen-
     tiam eius communem predicte perpendiculari, et ponat medium dif-
     ferentie communis super punctum a quo extracta fuerit perpendicu-
70   laris.

     [3.25] Et sic erit centrum vitri in superficie medii circuli, et linea
     que transit per centra duorum foraminum erit perpendicularis su-
     per superficiem vitri equalem, est enim equidistans diametro lamine,
     qui est perpendicularis super differentiam communem que est in vi-
75   tro. Et centrum circuli medii erit in convexo vitri, nam linea que exit
     a centro circuli medii ad centrum lamine est equalis linee exeunti a
     centro vitri ad medium differentie communis, et utraque istarum li-
     nearum est perpendicularis super superficiem lamine. Ergo due linee
     sunt equales et equidistantes, et linea que copulat centrum vitri cum
80   centro medii circuli est equalis linee que copulat centrum lamine et
     medium differentie communis que est in vitro. Hec autem linea posi-
     ta fuit equalis semidiametro vitri; ergo linea equidistans ei est equalis
     semidiametro vitri. Centrum ergo medii circuli est in convexo vitri;
     linea ergo que transit per centra duorum foraminum, que transit per
85   centrum medii circuli, tenet cum linea exeunti a centro vitri angulum
     equalem angulo qui est apud centrum lamine.

55 linea *corr. ex* lineam *L3* / a loco: ad locum *P1* / reflexionis: refractionis *R*   56 *post* vitro *add.*
ad aerem *R*   57 reflexionis: refractionis *R*; *scr. et del. S*   59 quod . . . instrumenti *om.*
P1 / hora: ora *R* / instrumenti: foraminis *SFC1OE*; *corr. ex* foraminis *L3*   60 extrahat *corr. ex*
extrehat *C1*   62 *post* fecerat *add.* et protrahamus ad la lineam ad centrum lamine *deinde del.* ad
la *P1* / ex: ab *S*   63 vitri *inter. a. m. E* / et *inter. F*; *om. P1*   64 extrahat *corr. ex* extrahant *L3*
67 differentiam eius (68) *transp. S*   68 predicte: dicte *FP1*   69 fuerit: fuit *O*   73 vitri
equalem *transp. FP1*   74 qui: que *ER* / *post* super *add.* illam superficiem et *R*   75 exit:
erit *S*   76 circuli medii *transp. C1*   77 istarum linearum *transp. E*   78 *ante* est *scr. et del.*
est *L3*   79 *post* equales *scr. et del.* sunt *P1* / vitri . . . centrum (80) *mg. C1*   80 medii circuli
transp. ER / circuli *inter. E*   81 posita . . . equalis[1] (82): equalis posita fuit *R*   82 fuit *corr.*
ex fit *O* / ergo . . . vitri[1] (83) *mg. a. m. L3*; *scr. et del. E*; *om. R* / equalis[2]: equidistans *L3*   83 vitri[1]
corr. ex vitro *F*   85 centrum . . . circuli: medii circuli centrum *C1R* / tenet *inter. a. m. L3*; *om.*
O / exeunti: exeunte *R*   86 equalem: equale *S*

[3.26] Extendantur ergo due linee in ymaginatione recte in utramque partem, scilicet diameter vitri predictus et linea que transit per centra duorum foraminum. Perveniet ergo ad circumferentiam
90 medii circuli, sunt enim ambe in superficie medii circuli. Ergo due linee divident a circumferentia medii circuli ex utraque parte arcum cuius quantitas est decem partium, et extremitates linee que transit per centra duorum foraminum sunt note, altera enim earum est centrum foraminis, et altera punctus oppositus centro foraminis, et altera
95 duarum extremitatum linee que transit per centrum vitri est extremitas arcus quam separaverat a circumferentia medii circuli, qui distat a centro foraminis decem partibus. Reliqua ergo extremitas linee que transit per centrum vitri distat a linea que transit per centra duorum foraminum decem partibus in parte opposita primo signo. Signemus
100 igitur extremitatem huius diametri et extremitatem linee que transit per centra duorum foraminum, quamvis locus iste sit notus, quia est super lineam perpendicularem in hora instrumenti.

[3.27] Et intueatur experimentator signum, et inveniet illud remotius ab extremitate linee que transit per centrum vitri plus quam
105 sit extremitas linee que transit per centra duorum foraminum. Hec ergo reflexio est ad partem contrariam perpendiculari a loco reflexionis, quia perpendicularis exiens a loco reflexionis est linea que transit per centrum vitri. Et arcus circumferentie medii circuli que est inter centrum lucis et extremitatem linee que transit per centra duorum
110 foraminum est quantitas anguli reflexionis, angulus enim reflexionis est apud centrum circuli medii, lux enim extenditur super lineam transeuntem per centra duorum foraminum recte donec perveniat ad convexum vitri et spericum. Angulus ergo reflexionis erit apud centrum circuli medii, qui est super convexum vitri, et arcus qui est inter
115 centrum lucis et extremitatem linee que transit per centra duorum foraminum est ille qui cordat angulum reflexionis qui est decem partium.

87 extendantur *corr. ex* extendebatur *E* / in[2] *inter. O*   88 *post* utramque *scr. et del.* lineam *C1* / vitri predictus *transp. ER* (predictus: predicta *R*)   89 perveniet: pervenient *R* / ergo *inter. O* / circumferentiam *corr. ex* circumferentium *F*   90 medii circuli[1] *transp. P1* / sunt . . . circuli[2] *mg. a. m. E; om. S* / medii circuli[2] *transp. L3* / medii[2] . . . circuli (91) *mg. C1* / circuli[2] *om. P1*   91 divident: dividunt *C1O*; dividuntur *S* / *post* circumferentia *add.* circuli *P1* / arcum: arcuum *P1*   93 earum: illarum *S*; harum *L3O*   94 punctus: punctum *R* / oppositus: oppositi *L3*; opposito *O*; oppositum *R*   96 quam: quem *R* / a *om. L3*   97 reliqua: aliqua *S*   98 per[1] . . . transit[2] *om. FP1*   99 *post* foraminum *add.* distat *P1* / in: a *P1*   100 *post* que *scr. et del.* transit *L3*   101 quamvis: quoniam *R* / sit: est *R* / quia *corr. ex* qui *a. m. E*   103 experimentator *corr. ex* experimentatorum *C1*   104 centrum . . . per (105) *om. R*   106 reflexio: refractio *R* / perpendiculari *om. SFL3C1OE* / reflexionis: refractionis *R*   107 reflexionis: refractionis *R*   108 que: qui *FP1R*; *alter. in* qui *L3* / est *om. SO*   110 reflexionis[1,2]: refractionis *R*   111 circuli medii *transp. ER*   113 reflexionis *inter. O*; refractionis *R* / erit *om. P1*   114 qui: quod *P1* / et . . . *inter. om. P1*   116 reflexionis: refractionis *R* / qui: que *E*

[3.28] Deinde oportet experimentatorem evellere vitrum et divi-
dere incipiens a centro foraminis arcum cuius quantitas sit viginti, et
120   procedat, ut prius. Et sic habebit quantitatem anguli reflexionis dif-
ferentem a quantitate anguli qui est viginti. Et sic dividat alios arcus
successive, et experiatur reflexiones eorum, sicut in primis, et habebit quantitates angulorum reflexionis qui sunt apud convexum vitri,
et hee eedem quantitates sunt quantitates angulorum reflexionis lu-
125   cis de aere ad vitrum, hoc enim declaratum est in predictis duabus
experimentationibus. Sed reflexio de aere ad vitrum est ad partem
perpendicularis, reflexio vero de vitro ad aerem est ad partem contrariam perpendiculari. Et si quis voluerit experiri vitrum et aquam
etiam a convexo vitri et a superficie eius equali, habebit quantitates
130   angulorum reflexionis de vitro ad aquam, aqua enim efficitur in loco
aeris.

[3.29] Et si quis voluerit experiri quantitates angulorum reflexionis apud concavum vitri, accipiat vitrum concavum concavitate
columpnali in quantitate semicolumpne. Et sit figura universi vitri
135   equidistantium superficierum, et longitudo eius sit maior diametro
vitri sperici uno grano ordei, et latitudo eius sit similiter. Et sit spissitudo eius sicut duplum diametri foraminis quod est in hora instrumenti, et concavitas eius sit in uno suorum laterum. Et vas concavitatis columpnalis sit in superficie vitri quadrati, et longitudo columpne
140   sit in spissitudine vitri. Et semidiameter basis columpne sit in quantitate semidiametri vitri sperici, et sint fines vitri linee recte veris-
sime. Hoc autem instrumentum sic bene potest fieri super formam,
ita quod forma fiat eadem doctrina predicta, et dissolvatur vitrum et
infundatur super formam predictam.

145   [3.30] Si ergo experimentator voluerit experiri reflexionem hoc instrumento, dividat de circumferentia medii circuli arcum cuius quantitas sit illa quam vult experiri, et extrahat ab extremitate arcus perpendicularem super superficiem lamine, ut predictum est. Et copu-

119 incipiens *om. R*/cuius: qui *R*/quantitas *om. R*/*post* viginti *add.* partium *R*   120 reflexionis: refractionis *R*   121 *post* viginti *add.* partium *R*/alios: duos *O*   122 experiatur:
experiantur *L3O*/reflexiones: refractiones *R*   123 reflexionis: refractionis *R*/qui: que *C1E*
124 hee *om. ER*/hee eedem: heedem *L3*/quantitates: quantitatis *S; om. ER*/reflexionis: refractionis *R*   125 *post* de *scr. et del.* centro *C1*/duabus *om. R*   126 reflexio: refractio *R*/*post* vitrum
scr. et del. et *F*   127 reflexio: refractio *R*/est *om. FP1*/contrariam *mg. a. m. L3; om. O*/contrariam
perpendiculari (128) *transp. L3*   128 perpendiculari: perpendicularis *L3OE*/si: cum *L3*   129 etiam: et *ER* (*inter. a. m. E*)   130 reflexionis: refractionis *R*/efficitur: ponitur *R*   132 reflexionis: refractionis *R*   133 *ante* apud *scr. et del.* lucis *C1*   134 columpnali: columpnari *R*   136 sit^2: sic *O; om. P1ER*   137 sicut: sit *OR*/duplum: dupla *R*   138 eius
om. R/suorum laterum *transp. C1*/et vas concavitatis *om. R*   139 columpnalis: columpnaris
R/sit: scilicet *R*/vitri: una *R*/quadrati: quadrata *R*; quadratum *SFP1C1L3O; corr. ex* quadratum *a. m. E*   140 spissitudine: longitudine *R*   142 super formam *transp. deinde*
corr. S   143 quod: ut *R*   145 reflexionem: refractionem *R*   147 extrahat *corr.*
ex trahat *E*/*post* arcus *rep. et del.* arcus *E*   148 copulet: copulat *L3; corr. ex* copulat *F*

let extremitatem perpendicularis cum centro lamine linea recta quam
150 protrahat in alteram partem, et dividat ex hac linea in altera parte,
scilicet in qua sunt duo foramina, lineam equalem semidiametro basis columpne. Et extrahat ab extremitate eius perpendicularem super
diametrum lamine, et protrahat illam in utramque partem. Deinde
superponat vitrum lamine, et ponat dorsum concavitatis ex parte
155 duorum foraminum. Et superponat duas superfluitates que super-
fluunt super diametrum columpne huic perpendiculari, et preservet
quod sint distantie duarum extremitatum diametri basis concavitatis a puncto a quo exivit perpendicularis distantie equales. Erit ergo
centrum basis concavitatis columpnalis super punctum a quo exivit
160 perpendicularis et super punctum cuius distantia a centro lamine est
in quantitate semidiametri basis concavitatis. Hoc situ preservato,
applicet vitrum fixa applicatione.

[3.31] Et erit superficies medii circuli secans foramen columpnale et equidistans basi eius, nam basis eius in hac dispositione est
165 in superficie lamine. Superficies ergo circuli medii facit in superficie columpnali concava semicirculum, et est diameter huius circuli
medii equidistans diametro basis concavitatis. Erit ergo linea que
egreditur a centro huius dimidii ad centrum basis concavitatis, que
est perpendicularis super superficiem lamine, equalis perpendicu-
170 lari exeunti a centro circuli medii perpendicularis super superficiem
lamine. Et perpendicularis que exit a centro medii circuli ad centrum
lamine est equalis semibasi columpne; ergo linea que exit a centro circuli medii ad centrum semicirculi qui sit in superficie columpne est
equalis semidiametro huius dimidii. Centrum ergo circuli medii est

---

149 recta *om. FP1*   150 protrahat: pertrahat *S*   152 ab extremitate *inter. O*   153 *post* lamine
rep. et del. lamine *F* / protrahat: pertrahat *S*   154 superponat: supponat *SC1*   156 columpne
... diametri (157) *mg. O* / et *om. E* / preservet: preservetque *E*; observetque *R*   157 quod: ut *R* /
diametri *corr. ex* diameter *C1*   158 exivit: exeunt *S; corr. ex* exeunt *L3C1* / perpendicularis:
perpendiculares *SL3O; corr. ex* perpendiculares *C1* / distantie ... perpendicularis (160) *om. L3*
159 columpnalis: columpnaris *R* / exivit: exeunt *E*   160 et *om. ER* / super: superque *ER* (que
inter. E); inter *SFP1C1O; corr. ex* inter *a. m. L3* / cuius *corr. ex* eius *a. m. L3*   161 preservato:
observato *R*   163 columpnale: columpne *ER*   164 equidistans *corr. ex* equidistisans
S / basi: basis *O; corr. ex* basis *L3*   166 *ante* columpnali *add.* circuli *mg. a. m. E* / columpnali:
columpnari *R* / semicirculum *corr. ex* circulum semi *E* / et est *transp. S* / diameter: diametrum
FP1 / circuli: semicirculi *R; inter. L3; om. SOE* / circuli medii (167) *transp. C1*   167 medii:
dimidii *SI3C1OE; om. R* / equidistans *corr. ex* equidistantis *P1C1* / equidistans ... dimidii (168)
mg. O   168 dimidii: semicirculi *R*   169 *post* lamine *scr. et del.* et perpendicularis que
exit a centro lamine *S* / *post* lamine *add.* quia lamina est *SO; scr. et del* quia lamina est *L3C1*
170 perpendicularis: perpendiculari *R*; perpendiculariter *FP1E; alter. in* perpendiculariter *C1*
171 *post* lamine *add.* et que est ab illa perpendicularis *E* / et *corr. ex* est *L3* / et ... circuli *om. C1E* /
medii circuli *transp. FP1R*   172 semibasi: semibasis *SC1O*; semidyametro basis *ER; alter. in*
semicirculus basis *a. m. L3* / ergo ... columpne (173) *mg. O*   173 sit: fit *ER*   174 dimidii:
semicirculi *R* / *post* ergo *scr. et del.* centrum *S* / circuli medii *transp. deinde corr. S*

175   in circumferentia semicirculi facti; est ergo in concavo columpne. Et
quia terminus vitri superponitur linee perpendiculari super punctum
lamine, erit diameter lamine perpendicularis super superficiem vitri
equalem, nam superficies vitri equales sunt perpendiculares super se
adinvicem. Erit ergo linea que transit per centra duorum foraminum
180   perpendicularis super superficiem vitri equalem que est in parte convexa vitri, quia est equidistans diametro lamine, et hec superficies
equalis vitri est ex parte foraminum.

[3.32] In hoc situ ergo lux que extenditur per lineam que transit
per centra duorum foraminum extenditur in corpore vitri recte donec
185   perveniat ad concavum vitri. Et tunc reflectitur apud concavum vitri,
cum non transeat per centrum circuli, qui est in concavo vitri, nec est
perpendicularis super concavum vitri; ergo reflectitur in concavo vitri. Et hec linea occurret concavo vitri in uno puncto; ergo differentia
communis huic linee et concavo vitri est centrum circuli medii. Ergo
190   lux que extenditur per lineam que transit per centra duorum foraminum reflectitur apud centrum circuli medii; ergo arcus qui est inter
centrum lucis et extremitatem linee que transit per centra duorum
foraminum cordat angulum reflexionis.

[3.33] Hac igitur via posset quis experiri quantitates angulorum
195   reflexionis qui fiunt in concavis vitri addendo in arcubus parum
parum. Et hec reflexio est a vitro concavo ad aerem, et erunt anguli
adquisiti hac reflexione idem illis qui fiunt ex aere ad vitrum in concavo vitri, declaratum est enim paulo ante quod angulus reflexionis a
vitro ad aerem et ab aere ad vitrum est idem, cum angulus quem con-
200   tinet prima linea per quam extenditur lux et perpendicularis exiens
a loco reflexionis sit idem angulus. Hac ergo via posset quis habere
quantitates angulorum reflexionis de aere ad aquam, et de aere ad
vitrum, et de vitro ad aerem, et de vitro ad aquam a superficie equali,
et concava, et convexa.

175 *post* circumferentia *scr. et del.* circuli medii *P1*   176 *post* superponitur *scr. et del.* vitro *O*/
perpendiculari: perpendicularis *E*   177 super *om. S*   178 super: inter *R*   179 adinvicem
om. R   181 *post* est *scr. et del.* perpendicularis *P1*   182 equalis vitri *transp. R*/est *om. S*/ex:
in *O*; *corr. ex* in *a. m. L3*   183 in . . . ergo: ergo in hoc situ *FP1*/situ ergo *transp. R*/per: super *R*
185 tunc *inter. O*/reflectitur: refringitur *R*   186 nec: neque *FR*   187 reflectitur: refringitur *R*
188 et . . . puncto *om. R*; *scr. et del. E*/occurret: concurret *S*; concurrit *E*/*post* vitri *scr. et del.* est
centrum intra *E*   189 huic: huius *L3O*/et *inter. O*; *om.* *C1*/concavo: concavitati *L3*/vitri *corr.*
ex eius *L3*/ergo . . . medii (191) *mg. O*   190 per[1]: super *R*/per[2] *corr. ex* super *E*/centra *corr. ex*
centrum *P1*   191 reflectitur: refringitur *R*/circuli medii *transp. R*   193 reflexionis: refractionis *R*   194 via *inter. O*   195 reflexionis: refractionis *R*/fiunt: sunt *C1*/concavis: concavo *R*
196 parum *inter. a. m. L3*; *om. SFP1R*/*post* hec *rep. et del.* hec *P1*/reflexio: refractio *R*/reflexio est
transp. S   197 reflexione: refractione *R*/*post* illis *scr. et del.* eis *L3*/qui: que *L3*/fiunt *corr. ex*
sunt *a. m. E*   198 est enim *transp. E*/enim: autem *R*/reflexionis: refractionis *R*   199 *post*
cum *scr. et del.* l *F*/angulus: angulo *ER*   200 reflexionis: refractionis *R*/sit . . . angulus *om. R*/
post hac *scr. et del.* rati *S*   202 reflexionis: refractionis *R*/ad aquam *corr. ex* aquam *P1*/aquam:
quam *S*; *corr. ex* quam *O*/et *inter. O*/et . . . aquam (203) *mg. C1*

205   [3.34] Hiis ergo angulis experimentatis et proportionibus eorum notis, experimentator inveniet [1] quoslibet duos angulos quorum utrumque continet prima linea per quam extenditur lux et perpendicularis exiens a loco reflexionis super superficiem corporis diaffoni, inveniet dico in eisdem corporibus diaffonis. Et erunt duo anguli

210 diversi, nam angulus reflexionis ab angulo maiori ex illis erit maior duobus angulis reflexionis ab angulo minori, et excessus anguli reflexionis super angulum reflexionis erit minor excessu anguli maioris quem continet prima linea cum perpendiculari super angulum minorem quem continet prima linea cum perpendiculari. [2] Et pro-

215 portio anguli reflexionis ab angulo maiori ad angulum maiorem erit maior proportione anguli reflexionis ab angulo minore ad angulum minorem. [3] Et illud quod restat post angulum reflexionis de angulo maiori est maius illo quod remanet post angulum reflexionis de angulo minore, [4] et remotio anguli reflexionis cum lux exiverit de

220 corpore subtiliori ad corpus grossius semper erit minor angulo quem continet linea per quam extenditur lux ad locum reflexionis cum perpendiculari exeunti a loco reflexionis. [5] Et si lux exiverit a corpore grossiori ad subtilius, tunc angulus reflexionis erit medietas coniuncti duorum angulorum. [6] Et si comparaveris angulos reflexionis

225 qui sunt inter aliquod istorum corporum diaffonorum et aliud corpus grossius illis ad angulos reflexionis qui sunt inter illud corpus idem diaffonum subtilius et aliud corpus grossius primo grosso, invenies proportiones maiores angulorum reflexionis ad angulos quos continet prima linea et perpendicularis, qui sunt inter corpus subtilius et

230 corpus grossius quod magis grossum est proportionibus angulorum

205 angulis *corr. ex* angulus *S*   206 quoslibet: quousque *FP1*; *om. R* / duos *inter. a. m. C1*
207 utrumque: vitrumque *S*   208 reflexionis: refractionis *R*   210 angulus reflexionis
transp. FP1 / reflexionis: refractionis *R* / maiori: maiore *R*   211 angulis: angulus *S* / reflexionis^{1,2}: refractionis *R* / ab . . . reflexionis (212) *om. S* / minori: minore *R* / *post* anguli *add.* maioris *C1*
212 reflexionis: refractionis *R*   213 *post* super *scr. et del.* lineam *C1* / angulum *om. FP1*   215 reflexionis: refractionis *R* / maiori: maiore *R*   216 maior: minor *FP1* / reflexionis: refractionis *R* / minore: minori *L3*   217 illud: illum *F*; *corr. ex* illum *O* / reflexionis: refractionis *R*
218 maiori: maiore *R* / maiori . . . angulo (219) *om. S* / reflexionis: refractionis *R*   219 *post* minore *add.* quantitas distantia *OL3 (deinde del. L3)* / reflexionis: refractionis *R*   220 subtiliori: subtiliore *R* / corpus *om. R*   221 reflexionis: refractionis *R*   222 exeunti: exeunte *R*; *corr. ex* exeuntis *L3* / *post* exeunti *scr. et del.* perpendiculari *C1*; *rep. et del.* grossius (220) . . . exeunti *O* / reflexionis: refractionis *R*   223 grossiori: grossiore *R* / reflexionis: refractionis *R* / coniuncti *om. R*   224 *post* angulorum *add.* coniunctorum *R* / reflexionis: refractionis *R*   225 *post* istorum *add.* duorum *C1* / diaffonorum *corr. ex* diaffonum *O* / aliud *corr. ex* alius *F*   226 reflexionis: refractionis *R* / corpus idem *transp. R* / idem . . . corpus (227) *om. FP1*   227 invenies *correxi ex* inveniet (invenies *R*)   228 reflexionis: refractionis *R*   229 et^1 *om. S* / *post* inter *add.* idem *SFP1* / *post* et^2 *scr. et del.* g *F*   230 corpus *om. R* / grossius *corr. ex* subtilius *C1* / quod . . . grossius (232) *om. S* / grossum: grossius *FP1L3O*

reflexionis ad angulos quos continet prima linea cum perpendiculari, qui sunt inter idem corpus subtilius et corpus grossius quod minus est grossum, scilicet quoniam, si fuerint duo anguli equales quorum utrumlibet continet prima linea per quam extenditur lux et perpen-
235 dicularis que exit a loco reflexionis, quorum alter est inter corpus subtilius et corpus grossius illo et alter inter illud idem corpus subtilius et corpus grossius primo grosso, tunc angulus reflexionis qui est in corpore grossiori erit maior angulo reflexionis qui est in corpore grossiori quod est minus grossum. [7] Et similiter, si reflexio fuerit a cor-
240 pore grossiori ad corpus subtilius quod magis est subtile, maior erit angulo reflexionis qui est ab illo corpore eodem grossiori ad corpus subtilius quod est minus subtile. Hec ergo sunt omnia que pertinent ad qualitates reflexionis lucis in corporibus diaffonis.

QUARTUM CAPITULUM
245 *Quoniam quicquid visus comprehendit ultra corpora diaffona que differunt in diaffonitate a corpore in quo est visus, cum fuerit obliquum a lineis perpendicularibus super superficiem corporis, est comprehensio secundum reflexionem*

[4.1] In predicto autem capitulo patuit quod lux transit de vitro
250 ad aerem, et de aere ad vitrum, et de aere ad aquam, et cum transit de vitro ad aerem et ad aquam, constat quod transit de aqua ad aerem, aqua enim est subtilior vitro, cum fuerit clara. Et cum transit de aere ad vitrum, transibit de aqua ad vitrum, cum aqua sit grossior aere. Preterea patuit quod omnes luces accidentales et essentiales, et fortes
255 et debiles transeunt per hec corpora diaffona hiis modis. Ergo omne

231 reflexionis: refractionis *R* / ad angulos *om. R* / cum *inter. a. m.* C1; *om.* O; et *L3ER* (*inter.* L3E) / perpendiculari: perpendicularis *C1ER; corr. ex* perpendicularis *L3*   233 est: erit *P1* / scilicet *om. FP1R*   235 reflexionis: refractionis *R*   236 corpus grossius *transp.* FP1 / illo . . . grossius (237) *rep. S* / inter *inter. a. m.* E   237 reflexionis: refractionis *R*   238 grossiori[1]: grossiore *R* / erit . . . grossiori[2] *mg. a. m.* E / reflexionis: refractionis *R* / grossiori[2]: grossiore *ER*   239 reflexio: refractio *R*   240 grossiori: grossiore *R* / corpus *om. R* / subtilius . . . corpus (241) *mg.* O / magis est *transp.* C1ER (magis: maius C1) / erit *om. S; inter. E; mg. a. m.* L3   241 reflexionis: refractionis *R* / post ab *scr. et del.* a *P1* / corpore eodem *transp. R* / corpore . . . grossiori: grossiori corpore eodem *E* / grossiori: grossiore *R*   242 subtilius *corr. ex* subtilissimum O / est *inter.* O   243 qualitates: quantitates *FP1*; qualitatem *L3R* / reflexionis: refractionis *R* / in: a *R*   244 quartum capitulum *om. R* / quartum . . . reflexionem (248) *om. S*   245 quoniam: quod *R* / quicquid *corr. ex* qui O / visus *om. R* / visus comprehendit *transp.* C1 / comprehendit: comprehenditur *R*   246 in[1]: a *P1* / fuerit obliquum (247) *transp.* FP1   247 superficiem: superficies C1   248 corporis: eorum *P1C1ER* / est comprehensio: comprehenditur *R* / reflexionem: refractionem *R* / post reflexionem *add.* capitulum quartum *R*   250 et[3] . . . aquam (251) *mg.* O / cum *inter. a. m.* L3   251 transit: transibit *R*   252 enim est *transp.* C1   253 transibit . . . vitrum[2] *mg.* P1   254 preterea *corr. ex* primo *a. m.* L3 / omnes luces *transp.* ER / et[2] *om.* SL3OER / post et[2] *scr. et del.* d F   255 modis ergo *transp. R*

corpus lucidum quacumque luce mittit lucem suam in omne corpus
diaffonum, et si occurrerit aliud corpus diaffonum, transibit in alio
corpore aut reflexe aut recte.

[4.2] Et in primo declaratum est quod a quolibet puncto cuius-
260 libet corporis lucidi oritur lux per quamcumque lineam rectam que
potest extendi ex illo puncto, ex quibus patet quod a quolibet puncto cuiuslibet corporis diaffoni contingentis aliquod corpus lucidum
quacumque luce oritur lux per omnem lineam rectam que poterit
extendi ex illo puncto, et transit in corpore diaffono tangenti illud
265 punctum. Et si occurrerit aliud corpus diaffonum diverse diaffonitatis a diaffonitate corporis tangentis illud, transibit etiam in ipsum
aut reflexe aut recte, sive primum corpus sit subtilius secundo, sive
secundum sit subtilius primo.

[4.3] Et in primo etiam declaratum est quod ab omni corpore colo-
270 rato lucido color oritur cum luce que est in ipso mixtus cum luce,
et quod visus, cum comprehenderit lucem, comprehendit formam
coloris mixtam sibi, ex quibus patet quod corpora colorata que sunt
in aqua et ultra corpora diaffona que differunt in diaffonitate a diaffonitate aeris, cum in eis fuerit lux essentialis aut accidentalis, fortis
275 aut debilis, tunc lux que est in eis oritur a quolibet puncto cum forma
coloris que est in illo puncto, et transit lux mixta cum colore in corpore aque et in omni corpore diaffono contingente ipsa, et extenditur
lux cum forma coloris in corpore aque et in omni corpore diaffono
per lineas rectas donec perveniat ad superficiem aque aut illius cor-
280 poris diaffoni.

[4.4] Et cum fuerit aer aut aliud corpus diaffonum tangens aquam,
tunc in illud corpus diaffonum transibit lux cum forma mixta sibi in
aere aut in alio corpore diaffono per lineas rectas, et hee linee secunde
in maiori parte secabunt primas lineas per quas extendebantur, et

285 quedam earum erunt in rectitudine primarum linearum. Et omnia
corpora que sunt in aqua et ultra diaffona corpora que differunt a diaffonitate aeris, cum fuerint in loco lucido, scilicet cum lux orta fuerit
super aquam in qua sunt, tunc lux perveniet ad ipsa, manifestum est
enim quod omnis lux transit in omne corpus diaffonum.

290 [4.5] Ergo omne corpus in aqua existens aut in alio corpore diaffono, cum super aquam illam aut super illud corpus diaffonum orta
fuerit lux, illud corpus erit lucidum, et a quolibet puncto ipsius orie-
tur forma lucis que est in ipso cum forma coloris, et extenditur in
universo illius aque aut illius corporis diaffoni per omnem lineam
295 rectam que poterit extendi ab ipso puncto donec perveniat lux cum
forma coloris qui est in illo puncto ad superficiem aque aut ad superficiem illius corporis diaffoni.

[4.6] Sed non potest extrahi ab eodem puncto alicuius superficiei
ad eandem superficiem linea perpendicularis nisi una. Ergo a quo-
300 libet puncto cuiuslibet corporis colorati lucidi existentis in corpore
diaffono oritur forma lucis cum forma coloris in universo corporis
diaffoni in quo existit secundum lineas rectas, et pervenit forma ad
universum oppositum de superficie corporis diaffoni. Et una illarum
linearum erit perpendicularis super superficiem corporis diaffoni et
5 super superficiem continuam cum superficie corporis diaffoni; reli-
que autem linee erunt oblique super superficiem corporis diaffoni.

[4.7] Sed in precedenti capitulo declaratum est quoniam lux, cum
extenditur in corpore diaffono et occurrerit alii corpori diaffono diverso a diaffonitate primi corporis, et linea per quam extensa est lux
10 in primo corpore fuerit perpendicularis super superficiem secundi
corporis, tunc lux extendetur in rectitudine eius in secundo corpore.
Et si linea per quam extenditur lux fuerit obliqua super superficiem
secundi corporis, tunc lux reflectitur. Et cuiuslibet puncti cuiuslibet

286 *post* que[1] *scr. et del.* differunt *S* / aqua: aliqua *P1*; *corr. ex* qua *a. m. L3* / diaffona corpora
transp. P1 / a *inter. a. m. L3; om. O*   288 super: per *L3* / qua: aqua *SL3OE; corr. ex* aqua
C1 / lux *om. E* / perveniet: perveniat *C1* / ipsa: ipsam *SO; corr. ex* ipsam *L3; corr. ex* ipsarum *F*
289 transit *corr. ex* transibit *a. m. E* / diaffonum . . . corpus (290) *om. R*   290 aqua *corr. ex*
aliqua *O*   291 *post* cum *add.* illa *SL3OE* (*deinde del. L3*) / super[2] *om. R* / illud corpus *transp.*
R / corpus *corr. ex* corpud *P1* / *post* diaffonum *add.* ceciderit *R* / orta . . . lucidum (292) *om. R*
292 a *om. P1* / ipsius: illius *S* / *post* ipsius *add.* corporis *R*   293 extenditur: extendetur *R*
295 perveniat: pervenit *FP1*   296 illo: ipso *S* / aut *om. P1*   298 alicuius: aliquo *FP1*;
corr. ex a cuius *O*   299 linea: linee *O*   300 lucidi *om. R*   1 *post* diaffono *scr. et del.*
si *C1* / in *inter. C1*   2 pervenit: perveniet *O*   3 et . . . diaffoni (4) *mg. a. m. L3* / una *om.*
C1 / illarum: earum *C1*   4 *post* linearum *add.* una earum *deinde del.* earum *C1* / corporis . .
. superficiem (5) *om. P1* / et: vel *R*   5 super *om. ER*   6 super *inter. E*   7 quoniam:
quod *R*   8 et . . . diaffono[2] *mg. a. m. E* / diverso: diverse *L3*   9 a *om. L3O* / primi *mg.*
F / primi corporis *transp. C1*   11 extendetur: extenditur *L3* / extendetur . . . lux (13) *om. S*
13 secundi corporis *transp. FP1* / reflectitur: refringetur *R*

corporis colorati et lucidi existentis in corpore diaffono forma lucis et
15 coloris extenditur in universo corpore diaffono, et pervenit opposite
ad superficiem corporis diaffoni.

[4.8] Et si fuerit aliud corpus diaffonum contingens illud corpus
diaffonum, et fuerit alterius diaffonitatis, tunc forma que pervenit ad
superficiem illius corporis diaffoni transit in corpus ipsum contin-
20 gens, et omnes erunt reflexe, preter quam forma que est in perpendiculari, extenditur enim secundum rectitudinem in corpore contingente. Et si forte perpendicularis ceciderit super punctum superficiei
continue cum superficie corporis qui non est in ipso corpore diaffono,
tunc illa forma delebitur, et tunc omnes forme que transeunt in cor-
25 pus contingens erunt reflexe.

[4.9] Ergo forme omnium visibilium que sunt in aqua, et in celo, et
in omnibus corporibus diaffonis contingentibus aerem que differunt
a diaffonitate aeris, extenduntur in universo aere opposito secundum
lineas rectas, et ille linee que fuerint ex istis lineis declinate per quas
30 extenduntur forme super superficiem aeris contingentis superficiem
corporis diaffoni, forma que extenditur per illas erit reflexa, et que
fuerint ex illis perpendiculares super superficiem aeris contingentis
superficiem corporis diaffoni, forma que extenditur per illas erit secundum rectitudinem ipsarum.

35 [4.10] Et cum iam declaratum sit quod a quolibet puncto cuiuslibet
corporis colorati et lucidi extenditur forma lucis et coloris in universo
corpore diaffono, et pervenit ad superficiem eius, et reflectitur a superficie eius, ergo forma que extenditur ab uno puncto ad superficiem corporis diaffoni erit continua coniuncta. Et cum forma fuerit
40 continua, et superficies corporis diaffoni fuerit continua coniuncta, et
forma fuerit reflexa in alio corpore diaffono, tunc reflectetur continua.
Et cum forma reflexa fuerit continua et occurrerit corpus densum,
tunc forma perveniet ad illud corpus diaffonum, et sic locus corporis
diaffoni per quem extenditur forma puncti quod est in primo corpore

14 existentis: existenti P1; corr. ex existentes F   15 opposite: oppositam FP1R/opposite ad
(16) transp. R   17 si om. P1/aliud: aliquod L3O/diaffonum: oppositum R; om. L3E/contin-
gens . . . corpus mg. a. m. E/contingens . . . diaffonum (18) om. P1/illud corpus om. R   18 tunc
om. FP1   19 illius corporis transp. C1   20 reflexe: refracte R   22 forte corr. ex forma L3
23 qui: quod R   24 omnes forme transp. C1   25 reflexe: refracte R   29 ille linee
transp. C1/declinate inter. a. m. L3E; om. SO   30 aeris: corporis P1/post contingentis scr. et del.
super C1   31 forma . . . reflexa: habebunt formas refractas R/extenditur: est L3/post et add.
si S/que² inter. O   32 fuerint: fuerit P1/post fuerint add. que L3O/perpendiculares: perpendicularis FP1/post contingentis add. super L3OE (deinde del. E)   33 forma . . . erit: habebunt
formas extensas R   35 a: ab O   36 et¹ om. L3O/coloris: color SO; corr. ex color a. m. E
37 eius inter. O/et² inter. a. m. E/reflectitur: reflectetur E; refringitur R   39 post continua add.
et R   40 fuerit: fuit L3   41 fuerit: fuit F; lucis P1/reflexa: refracta R/in: ab S/reflectetur:
reflectitur C1; refringetur R   42 reflexa: refracta R   43 diaffonum: densum P1

45 que reflectitur a superficie primi corporis ad illum locum, cum fuerit
lucidus coloratus, mittit formam lucis et coloris a quolibet puncto
ipsius per omnem lineam rectam que poterit extendi ex illo puncto.

[4.11] Accidit ergo ex hoc quod sint linee reflexe ad illum locum ex
lineis per quas extenditur forma illius loci. Et iam extendebatur forma
50 cuiuslibet puncti illius loci per unam illarum linearum reflexarum.
Forma ergo illius loci ex corpore denso colorato lucido erit in loco ex
superficie corporis diaffoni apud quem reflectitur forma unius puncti
extensi ad illum locum superficiei corporis diaffoni que reflectitur ad
eundum locum corporis densi, ex quo sequitur quod forma loci cor-
55 poris densi que extenditur ad illum locum corporis diaffoni reflectitur super easdem lineas extensas ab uno puncto ad illum locum corporis diaffoni.

[4.12] Et cum forma loci corporis diaffoni fuerit reflexa super illas
easdem lineas, tunc perveniet ad illud idem punctum, ex quo declara-
60 tur quod, si ymaginatus fueris piramidem extensam a quolibet puncto
aeris secundum lineas rectas, et piramis fuerit coniuncta continua, et
pervenerit illa piramis ad superficiem corporis diaffoni diverse diaffonitatis ab aere, et ymaginatus fueris omnem lineam rectam que possit extendi ex illa piramide reflecti apud superficiem corporis diaffoni
65 in loco quem exigit eius declinatio, et si aliqua fuerit perpendicularis,
extendetur recte. Tunc efficitur ex hoc corpus continuum reflexum in
corpore diaffono quod differt a diaffonitate aeris. Et cum hoc corpus
reflexum pervenerit ad corpus densum, tunc illud corpus densum, si
fuerit coloratum et lucidum, mittit formam lucis et coloris que sunt
70 in ipso in hoc corpore reflexo ymaginato per quamlibet lineam rectam que poterit extendi in hoc corpore reflexo a linea extensa in corpore piramidis a puncto qui est in aere, nam omne corpus coloratum

45 que *corr. ex* quod *L3* / reflectitur: refringitur *R* / a superficie: ad superficiem *S* / locum *inter. O*
46 mittit: mutet *S*; mittet *P1* / coloris *inter. O*   48 reflexe: refracte *R* / illum locum *transp. C1*
49 *post* loci *scr. et del.* per unam illarum linearum reflexarum *S* / iam *om. P1* / extendebatur: ostende-
batur *L3*   50 illius loci *transp. C1* / *post* illius *scr. et del.* per u *P1* / loci *inter. E* / reflexarum: refractarum *R*   51 loci *inter. O* / ex² *inter. F*   52 *post* quem *add.* locum *FP1* / reflectitur: refringitur
R; corr. ex reflectatur *F* / unius: illius *C1*   53 reflectitur: refringitur *R*   54 *post* densi *rep. et
del.* diaffoni (53) . . . densi *C1* / ex . . . densi (55) *mg. F*   55 reflectitur: refringitur *R*   56 super:
ad *R; alter. in* ad *a. m. E* / uno *corr. ex* illo *a. m. L3*   58 et . . . diaffoni *om. P1* / reflexa: refracta *R*
59 easdem lineas *om. C1* / perveniet: pervenit *FP1*; proveniet *L3* / illud: illum *F*   60 fueris: fuerit
SER (*mg. a. m. E*) / *post* fueris *add.* aliquis *ER*   61 secundum: super *P1*   62 pervenerit: pervenit *FP1* / illa piramis *om. P1*   63 fueris: fuerit *L3R; corr. ex* fuerit *E* / possit: poterit *L3*   64 ex:
ab *L3* / reflecti: refringi *R*   66 extendetur: extenditur *C1; corr. ex* extenditur *S* / reflexum: refractum *R*   67 *post* diaffonitate *scr. et del.* corporis *P1* / aeris *om. S*   68 reflexum: refractum *R; corr.
ex* refleum *S; om. FP1* / *post* reflexum *scr. et del.* in corpore diaffono *C1* / *post* densum *scr. et del.* si fuerit
coloratum *P1*   69 mittit: mutet *S* / *ante* formam *scr. et del.* foramen *S*   70 reflexo: refracto *R*
71 reflexo: refracto *R*   72 qui: quod *R* / coloratum lucidum (73) *transp. L3*

lucidum proprie mittit formam suam a quolibet puncto ipsius per omnem lineam rectam que poterit extendi ab illo puncto.

75   [4.13] Erit ergo forma puncti illius loci corporis densi extensa per quamlibet linearum reflexarum ad illum locum corporis densi. Perveniet ergo forma illius a corpore denso colorato lucido ad locum superficiei corporis diaffoni in quem reflectuntur ille linee, et cum pervenerit forma ad illum locum superficiei corporis diaffoni, neces-

80 sario reflectetur per easdem lineas extensas ad illum locum ab uno puncto qui est in aere. Et cum centrum visus fuerit in illo puncto qui est in aere, tunc forma que est forma loci colorati corporis densi quod est in corpore diaffono quod differt a diaffonitate aeris et est de numero illarum linearum per quas extenditur forma ad centrum visus,

85 tunc forma que extenditur per illam lineam pervenit ad centrum visus recte. Forme autem que extenduntur per omnes alias lineas que constituunt piramidem extensam a centro visus erunt reflexe, non recte.

    [4.14] Et in primo tractatu declaratum est quod aer recipit for-
90 mam visibilium et reddit illam omni corpori opposito et quod aer deferens formam, cum tetigerit visum, transibit forma que est in ipso corpus visus, et sic visus comprehendit visibilia que aer reddit visui. Ex omnibus ergo istis patet quod forma omnis corporis colorati lucidi existentis in corpore diaffono diverse diaffonitatis aeris extenditur
95 in corpore diaffono in quo existit, et reflectitur in aere, et extenditur in aere secundum lineas rectas, et quod quedam linearum rectarum per quas forma reflectitur in aere coniunguntur apud idem punctum aeris. Et cum centrum visus fuerit apud illum punctum, tunc visus comprehendet illud visum secundum reflexionem, et si aliquid ipsius
100 comprehenditur recte, non erit nisi unum punctum tantum. Hoc ergo

74 rectam *inter. a. m.* E / poterit: potest SC1 / puncto *inter.* O   75 loci *inter.* O   76 *post* linearum *add.* rectarum L3 / reflexarum: refractarum R / perveniet: pervenit FP1   77 forma illius *transp.* ER   78 reflectuntur: refringuntur R / *post* reflectuntur *scr. et del.* omnis C1 / ille linee *transp.* C1   79 pervenerit forma *transp.* C1 / superficiei . . . locum (80) *mg. a. m.* E   80 reflectetur: refringetur R / *post* locum *scr. et del.* superficiei corporis S   81 qui[1]: que S; quod R / et . . . aere (82) *om.* ER / illo *inter.* O   82 tunc forma: forma autem R / loci: luci S   83 et *om.* SO / *post* est[2] *add.* super lineam que est R   84 *post* forma *scr. et del.* que extenditur forma F   85 tunc *om.* R / *post* tunc *add.* illa C1 / *post* forma *add.* dico R   86 que *om.* E / extenduntur *corr. ex* extendetur P1   87 a *corr. ex* ad C1 / reflexe: refracte R   88 recte: directe ER   90 illam: eam ER   91 forma . . . comprehendit (92) *rep.* S / est *inter.* O / *post* ipso *add.* in R   92 corpus *mg.* F / *post* corpus *scr. et del.* est F / sic: si E   93 corporis: coloris E   94 *post* diaffonitatis *add.* a diaffonitate P1ER   95 in[1]: a FP1 / *post* diaffono *scr. et del.* diverse diaffonitatis O / *post* quo *scr. et del.* ex P1 / et[1] *inter.* C1 / reflectitur: refringitur R / et[2] . . . aere (96) *om.* C1   96 *post* aere *scr. et del.* et extenditur S / rectarum: istarum S   97 reflectitur: refringitur R / *post* reflectitur *rep. et del.* reflectitur O / coniunguntur: coniuntur S   98 et . . . punctum *mg.* F / illum: illud FP1R   99 comprehendet: comprehendit R; *corr. ex* comprehendit E / illud: illum FP1L3 (*alter. ex* illud F) / reflexionem: refractionem R   100 tantum *om.* P1

modo comprehendit visus res que sunt in aqua, et in celo, et omnia
visibilia que sunt ultra corpora diaffona que differunt a diaffonitate
aeris.

[4.15] Quoniam autem hoc verum sit sic poterit experimentari.
105 Accipiat ergo experimentator predictum instrumentum, et ponat ipsum in vase, et ponat vas in loco lucido quacumque luce ita quod lux
perveniat ad interius vasis. Et infundat intra vas aquam quousque
perveniat ad centrum lamine. Deinde diminuat foramina cum cera
ita quod non remaneat de foramine nisi modicum in medio eorum, et
110 mittat in duobus foraminibus unum calamum ita quod spatium quod
est inter duo foramina sit determinatum. Deinde moveat instrumentum donec diameter lamine super cuius extremitates sunt due linee
perpendiculares in hora instrumenti sit perpendicularis super superficiem aque. Deinde accipiat stilum subtilem album, et mittat eum
115 in vas, et eius extremitatem ponat in puncto circuli medii, qui est differentia communis circumferentie circuli medii et linee perpendiculari in hora instrumenti qui est extremitas diametri circuli medii qui
transit per centra duorum foraminum. Deinde ponat experimentator
alterum visum super superius foramen, et claudat reliquum, et in-
120 tueatur horam instrumenti quod est intra aquam, tunc enim videbit
extremitatem stili.

[4.16] Et declarabitur ergo ex hac experimentatione quod comprehensio eius ad extremitatem stili est secundum rectitudinem perpendicularis egredientis ab extremitate stili super superficiem aque, nam
125 linea que transit per centra duorum foraminum in qua est centrum
visus et extremitas stili, ex cuius verticatione comprehendit visus extremitatem stili, est perpendicularis super superficiem aque. In primo autem patuit quod nichil comprehendit visus nisi secundum rectitudinem linearum que extenduntur per centrum visus. Visus ergo
130 comprehendit extremitatem stili a verticatione linee que transit per
centra duorum foraminum, et hec linea extenditur ad extremitatem
stili recte, et est perpendicularis super superficiem aque.

101 comprehendit *alter. in* ostendit *a. m.* E   102 differunt *corr. ex* sunt *L3*   104 quoniam:
quod *R* / experimentari: experiri *L3*   105 accipiat: ac *C1* / ipsum *om.* ER   106 quod: ut *R*
107 interius . . . ad (108) *mg. a. m. L3* / infundat: infundatur *F* / intra: in *SR*   108 diminuat *corr.*
ex deminuat *F*   109 quod: ut *R* / post quod *scr. et del.* lux *S* / non *inter. F* / foramine: foraminibus *R*   110 mittat: mutat *S* / calamum *corr. ex* calameum *O* / quod: ut *R*   112 lamine
corr. ex linee *L3* / cuius: eius *SL3O* / linee *rep.* S   114 accipiat: accipiet *FP1*   115 et *inter.*
O / circuli medii *transp. R* / qui: que *P1*; quod *R*   116 circumferentie *om.* P1 / circuli: circulu
S / circuli medii *transp.* C1R   117 in *inter.* C1 / qui[1]: quod *R* / qui[2]: que *R*   119 super *om.* FP1
120 quod: que *L3ER* (*alter. ex* quod *L3*)   122 et *om. R* / ergo *om.* C1 / post experimentatione *scr.*
et del. et C1   123 eius *om.* S / est . . . stili (124) *mg. a. m. L3*   127 est perpendicularis: sunt
perpendiculares *R*   128 *post* autem *add.* libro *R* / nichil . . . visus: visus nichil comprehendit
R / nisi *om.* E   130 stili: facili *O*

[4.17] Deinde oportet experimentatorem declinare instrumentum donec linea que transit per centra duorum foraminum sit obliqua super superficiem aque. Et mittat stilum in aqua, et ponat extremitatem eius super primum punctum, scilicet super extremitatem diametri circuli medii qui transit per centra duorum foraminum, et ponat visum suum super superius foramen, et intueatur horam instrumenti que est intra aquam. Tunc enim non videbit extremitatem stili. Et deinde moveat stilum ad partem contrariam illi in qua est visus, et tunc non videbit extremitatem stili. Deinde moveat stilum ad partem in qua est visus, et moveat extremitatem stili per circumferentiam circuli medii suaviter et molliter, et intueatur horam instrumenti, tunc enim videbit extremitatem stili.

[4.18] Et tunc figat extremitatem stili in suo loco; deinde precipiat alii ut mittat in vas perpendicularem neque grossam nec gracilem, et ponat illam apud superficiem aque in oppositione secundi foraminis ut sit apud centrum circuli medii. Et intueatur experimentator interius vasis, et tunc non videbit extremitatem stili. Deinde precipiat auferre lignum, et tunc videbit extremitatem stili. Deinde figat extremitatem stili in suo loco, et elevet visum suum a foramina, et auferat instrumentum suum a vase, existente extremitate stili in suo loco, et intueatur locum in quo est extremitas stili. Tunc enim videbit inter ipsum et diametrum circuli medii distantiam sensibilem. Et si miserit regulam subtilem in aquam in hora experimentationis, et acumen eius fecerit transire per centrum lamine, et signaverit locum circuli medii qui est apud extremitatem regule signo, et abstulerit instrumentum, et aspexerit locum extremitatis stili, videbit locum extremitatis stili medium inter locum extremitatis regule et diametrum circuli medii.

[4.19] Deinde oportet eum auferre instrumentum, et infundere aquam in vas, et applicare vitrum lamine, et ponere superficiem vitri equalem ex parte foraminum, et ponere differentiam communem que est in ipso super lineam secantem diametrum lamine perpendicula-

165   riter. Sic ergo erit linea que transit per centra duorum foraminum
perpendicularis super superficiem vitri equalem et super superficiem
eius convexam. Deinde ponat instrumentum in aqua, et mittat stilum
in vas, et ponat extremitatem stili super extremitatem diametri circuli medii, et ponat visum suum super superius foramen, et intuea-
170   tur horam instrumenti. Tunc videbit extremitatem stili, et si moverit
extremitatem stili, et extraxerit illam a puncto quod est extremitas
diametri medii circuli, non videbit extremitatem stili, ex quo patet
quod extremitatem stili comprehendit recte, nam duo centra foraminum et extremitas diametri circuli medii sunt in eadem linea recta,
175   et experimentator non comprehendit extremitatem stili in hoc statu,
cum extremitas stili non fuerit super extremitatem diametri. Et si
evulserit vitrum, et posuerit ipsum econtra, scilicet quod ponat convexum vitri ex parte duorum foraminum et differentiam eius communem super primum locum, et expertus fuerit extremitatem stili,
180   videbit illam, cum fuerit in extremitate diametri circuli medii. Ideo
in hoc situ etiam linea que transit per centra duorum foraminum ex
cuius verticatione comprehendit visus extremitatem stili erit perpendicularis super superficiem vitri equalem et super superficiem eius
convexam.
185   [4.20] Deinde oportet experimentatorem evellere vitrum et ex-
trahere a centro lamine lineam rectam in superficie lamine que contineat cum diametro lamine super cuius extremitates sunt due linee
perpendiculares in hora instrumenti angulum obtusum, et extrahat
illam donec perveniat ad horam instrumenti. Deinde extrahat a cen-
190   tro lamine lineam in superficie lamine que continet cum prima linea
angulum rectum, et protrahat illam in utramque partem. Tunc hec
linea continet cum diametro lamine angulum acutum, et diameter
lamine erit obliquus super hanc lineam. Deinde superponat vitrum
lamine, et ponat differentiam eius communem super lineam quam ul-
195   timo signavit in superficie lamine, et ponat superficiem vitri equalem

---

165 erit *om. R* / erit linea *transp. E* / *post* linea *rep. et del.* linea *E*   167 eius *inter. E* / *post* ponat
add. experimentator *R* / aqua: aquas *L3*; aquam *R*; vas *C1* / mittat: mutat *S*   168 stili *inter. O*
171 extraxerit *corr. ex* extraherit *O* / a *inter. a. m. C1*   172 medii circuli *transp. SL3O*   173 *post*
stili *add.* non *FP1* / comprehendit recte *transp. C1*   174 circuli medii *inter. O*   175 hoc
mg. F / statu: situ *R*   176 stili non fuerit: non fuerit stili *C1* / non *inter. C1L3* (*a. m. L3*); *om.*
SO / *post* fuerit *rep. et del.* non fuerit *E* / *post* extremitatem *add.* stili *SL3C1OE* (*deinde del. L3C1*)
177 econtra: econtrario *R*; *corr. ex* contra *O* / quod: ut *R*   179 *post* fuerit *scr. et del.* super *C1* /
post stili *add.* etiam *ER* (*deinde del. E*)   180 fuerit in *mg. a. m. C1* / diametri *inter. O* / *post* medii
scr. et del. diametri *O* / ideo *inter. E*   181 etiam *om. L3O*   183 vitri *om. S* / super[2] *om. L3ER*
186 lamine[1] *corr. ex* foramine *P1* / superficie *corr. ex* superficiem *C1* / contineat: continet *L3O*
187 sunt *corr. ex* sint *E*   190 lamine[1] *inter. O* / superficie: superficiem *FP1* / continet: contineat *R*   191 rectum *mg. F; inter. a. m. E*   192 continet: continebit *R* / diameter: diametrum
SO; corr. ex diametrum *C1*   193 obliquus: obliqua *R*   194 lineam quam *transp. deinde*
corr. O   195 signavit *corr. ex* signavi *O* / *post* ponat *scr. et del.* super *E*

ex parte duorum foraminum, et ponat medium differentie communis super centrum lamine.

[4.21] Sic ergo erit centrum vitri super centrum circuli medii, ut prius declaratum est, et linea que transit per centra duorum foraminum transibit per centrum vitri. Et hec linea erit obliqua super superficiem vitri equalem, nam diameter lamine illi equidistans est obliquus super differentiam communem que est in vitro. Et hec linea etiam erit perpendicularis super superficem vitri convexam, quia transit per centrum eius.

[4.22] Deinde extrahat experimentator ab extremitate linee quam primo signavit in lamina lineam perpendicularem in hora instrumenti, et ducat illam ad circumferentiam circuli medii, et hee linee sint nigre. Erit ergo punctus ad quem pervenit, cum ab illo extracta fuerit linea ad centrum circuli medii, quod est centrum vitri, perpendicularis super superficiem vitri equalem et superficiem vitri spericam. Super superficiem autem vitri equalem est perpendicularis, quia est equidistans prime linee signate in lamina super differentiam communem que est in vitro. Super spericam vero, quia transit per centrum eius. Punctus ergo ad quem pervenit linea extracta in hora instrumenti qui est super circumferentiam circuli medii est casus in quo cadit perpendicularis exiens a centro vitri super superficiem vitri planam.

[4.23] Deinde oportet experimentatorem ponere instrumentum in vas et ponere extremitatem stili in puncto quod est extremitas diametri circuli medii, et ponat suum visum super superius foramen, et intueatur horam instrumenti. Tunc non videbit extremitatem stili. Deinde moveat stilum ad partem contrariam illi in qua est casus perpendicularis, et tunc etiam non videbit extremitatem stili. Deinde

196 duorum *inter. a. m.* E / ponat *corr. ex* ponit *a. m.* C1   198 circuli medii *transp.* C1 / medii *om.* O; *inter. a. m.* L3; *mg. a. m.* C1   200 erit *om.* E / *post* erit *scr. et del.* etiam C1 / super *om.* L3   201 diameter: diametrum P1; *alter. ex* diametum *in* diametrum F / illi equidistans *transp. deinde corr.* C1   202 obliquus: obliqua R   203 etiam *om.* L3ER / convexam: communem S / quia: que FE   205 experimentator *corr. ex* experimentatorem S   207 hee . . . sint: sint hee linee R / sint: sunt P1L3O; *corr. ex* sunt SC1   208 nigre *corr. ex* nire F / *post* ergo *add.* linea R; *scr. et del.* punc S / punctus . . . pervenit *om.* R / ab *inter.* O / *post* illo *add.* puncto R   209 linea *om.* R / circuli . . . centrum *mg.* F   210 *post* et *add.* super FP1ER / *post* spericam *add.* et L3   211 *ante* super *inter.* ?? L3 / super superficiem *corr. ex* superficiem *a. m.* L3 / *post* superficiem *scr. et del.* vitri S / autem *inter. a. m.* L3   213 quia: qui S   214 punctus: punctum R / quem: quod R / pervenit *mg.* F   215 qui: quod R / *post* circumferentiam *scr. et del.* instrumenti F / quo: quem R   218 ponere . . . et (219) *rep.* P1   219 et ponere *corr. ex* imponere L3 / in *inter.* O   220 medii *mg. a. m.* E / *post* ponat *add.* experimentator R / suum: suuum S / suum visum *transp.* P1 / superius *corr. ex* superfic F   222 moveat: movet FP1 / stilum *corr. ex* tilum L3 / ad partem *rep.* L3   223 *ante* et *scr. et del.* super superficiem P1 / et . . . perpendicularis (224) *mg.* O / etiam *inter.* P1 / *post* videbit *scr. et del.* in C1

moveat stilum ad illam in qua est casus perpendicularis et per cir-
225 cumferentiam circuli medii, tunc enim, si motus fuerit suavis, videbit
extremitatem stili. Et tunc figat extremitatem stili in suo loco in quo
apparuit. Deinde precipiat alicui cooperire centrum vitri tenui subtili
ligno, et tunc non videbit extremitatem stili, et si abstulerit coopertorium, videbit ipsum.

230 [4.24] Ex hac ergo experimentatione patet quod, cum visus comprehendit extremitatem stili recte, est secundum reflexionem, et quod
reflexio est a centro vitri, et quod forma reflexa est in superficie circuli medii, qui est perpendicularis super superficiem vitri equalem
apud quam fit reflexio perpendicularis, ut prius declaratum est. Et
235 si experimentator aspexerit locum extremitatis stili, inveniet ipsum
inter casum perpendicularis et extremitatem diametri circuli medii,
qui transit per centra duorum foraminum. Linea ergo que exit ab extremitate stili ad centrum vitri, cum extensa fuerit in illa recte in aere,
et extensa fuerit cum illa in aere perpendicularis exiens a centro vitri
240 super superficiem vitri equalem, erit media inter perpendicularem et
lineam que transit per centra duorum foraminum. Et forma extremitatis stili, que extensa est ab extremitate stili ad centrum vitri, extensa
est super hanc lineam et extensa est in rectitudine eius ad centrum
vitri, hec enim linea est perpendicularis super superficiem vitri speri-
245 cam, que est ex parte extremitatis.

[4.25] Deinde cum hec forma fuerit reflexa super lineam que transit per centra duorum foraminum, quia linee radiales que exeunt a
visu in hoc situ non perveniunt ad vitrum preter lineam que transit per centra duorum foraminum, calamus enim qui extenditur in-
250 ter duo foramina secat omnem lineam exeuntem a visu ad vitrum
preter quam lineam que transit per centra duorum foraminum. Vi-

224 *post* ad *add.* partem contrariam E; *add.* partem R / illam: illi E / casus *om.* P1   225 circuli medii *transp.* S / suavis *corr. ex* similis S   226 et . . . stili² *mg.* O; *scr. et del.* E; *om.* R
227 cooperire: cooperue S / *post* tenui *add.* ac P1   228 tunc *om.* FP1 / coopertorium *corr. ex*
coopertum O   230 ex hac *inter.* O   231 *post* stili *add.* et S / recte *om.* R / reflexionem:
refractionem R   232 reflexio: ratio S; refractio R / forma *mg.* F / reflexa: refracta R   233 qui:
que R   234 fit: sit O / fit reflexio *mg. a. m.* L3 / reflexio: refractio R / perpendicularis: ad
perpendicularem R; perpendiculariter O; *corr. ex* perpendiculariter *a. m.* L3 / prius *om.* L3 / declaratum: dictum L3   235 si *inter.* L3O (*a. m.* L3) / aspexerit *corr. ex* aspicit O / ipsum *inter.*
a. m. E   236 diametri *om.* P1 / circuli medii *transp.* C1   237 qui: que ER   238 in illa
om. R / recte *inter. a. m.* E / in² *om.* L3 / in aere *om.* E   239 et . . . illa *mg. a. m.* E / et . . . aere *om.*
R / cum illa *mg.* O / vitri *inter.* O; vitro S   240 superficiem: super F / inter perpendicularem
transp. deinde corr. S   242 extensa est¹ *transp.* C1 / vitri *om.* P1   243 hanc lineam *transp.*
deinde corr. P1 / extensa *corr. ex* extensam C1   244 enim linea *transp. deinde corr.* C1 / vitri
spericam *transp.* L3   245 ex parte *mg.* P1   246 reflexa: refracta R   247 *post* foraminum *scr. et del.* vitri O / quia *om.* R / a . . . situ (248): in hoc situ a visu R   249 *post* foraminum *scr. et del.* cli F / inter: in FP1   250 secat . . . foraminum (251) *mg.* O / omnem lineam:
lineam communem L3 / lineam *mg.* F; in eam R / *post* lineam *scr. et del.* que transit F / exeuntem
a visu: a visu exeuntem ER / exeuntem *mg. a. m.* E   251 quam *inter.* E

sus autem non comprehendit formas nisi ex verticationibus harum
linearum tantum; ergo forme non extenduntur nisi recte; ergo visus
non comprehendit hanc formam nisi ex verticatione huius linee per-
255 pendicularis. Ergo que extenditur recte in aere est perpendicularis
super superficiem aeris contingentis superficiem vitri equalem. Ergo
hec reflexio erit ad partem contrariam parti perpendicularis exeuntis
a loco reflexionis super superficiem aeris, nam linea que transit per
centra duorum foraminum magis distat a perpendiculari que exten-
260 ditur in aere quam linea que exit ab extremitate stili ad centrum vitri
que extenditur in aere. Et hec forma exit a vitro et reflectitur in aere,
et aer est subtilior vitro, et hoc modo fuit reflexio forme de aqua ad
aerem, visus enim comprehendit extremitatem stili in aqua ab isto
loco, scilicet quia comprehendit extremitatem stili quando fuerit in-
265 ter casum perpendicularis et extremitatem diametri circuli medii, qui
transit per centra duorum foraminum. Et illa forma etiam exivit ab
aqua et reflexa est in aere, et aer est subtilior aqua.

[4.26] Deinde oportet experimentatorem evellere vitrum et ponere ipsum super laminam econtra huius situs, scilicet quod ponat
270 convexum eius ex parte duorum foraminum, et ponat differentiam
eius communem super lineam equalem in superficie lamine in qua
posuerat illam in predicto situ, et ponat medium communis differentie super centrum lamine. Et sic linea que transit per centra duorum
foraminum erit obliqua super superficiem vitri equalem et perpen-
275 dicularis super superficiem eius convexam. Et applicet vitrum hoc
situ, et ponat instrumentum in vas, et ponat extremitatem stili super extremitatem diametri circuli medii, ut prius fecerat, et ponat visum suum super superius foramen, et intueatur horam instrumenti,
non enim videbit tunc extremitatem stili. Deinde moveat stilum ad
280 partem casus perpendicularis, et tunc non videbit extremitatem stili.
Deinde moveat illud ad partem contrariam illi in quo est casus perpendicularis per circumferentiam circuli medii, et suaviter, tunc enim

252 autem *mg. a. m. L3; om.* O   253 non *inter.* O / extenduntur: extendentur FP1   255 ergo
. . . perpendicularis *om.* S / perpendicularis . . . superficiem[1] (256) *corr. ex* super superficiem perpendicularis C1   256 aeris *om.* F / aeris . . . superficiem[2] *om.* P1 / *post* contingentis *add.* super F
257 reflexio: refractio R   258 reflexionis: refractionis R   259 a *om.* S   261 *post* forma *scr.
et del.* que C1 / reflectitur: refringitur R   262 *est rep.* P1 / fuit: fuerit S; fiet ER / reflexio: refractio
R / aqua *corr. ex* qua O / ad *corr. ex* in *a. m.* E   263 *post* aqua *add.* et C1 / ab: sub FP1 / isto: illo C1
264 stili *om.* S / fuerit: fuit FP1L3OER   265 casum *corr. ex* visum O / qui: que ER   266 ab: de L3
267 reflexa: reflexio O; refracta R / est[1] *om.* E / est subtilior *transp.* S   268 experimentatorem *corr.
ex* experrimentorem *a. m.* L3   269 econtra: extra ER / huius: huiusmodi R / situs: situm R / quod:
ut R   271 superficie: superficiem O   272 communis differentie *transp.* R   273 sic: si O;
corr. ex si L3   275 *post* vitrum *add.* in R   276 vas: vase L3   277 circuli *om.* FP1 / et *om.* S
278 super *inter. a. m.* L3; *om.* S   279 tunc *inter.* O   280 stili *inter. a. m.* E   281 illud: eundem
R / illi *mg. a. m.* L3 / est *om.* E   282 circuli medii *transp.* R / et *om.* P1

videbit extremitatem stili. Sic ergo linea recta que exit ab extremitate stili ad centrum vitri, cum fuerit extensa recte in corpore vitri, et ex-
285 tensa fuerit cum ipsa perpendicularis exiens a centro vitri super superficiem vitri, erit linea que transit per centra duorum foraminum media inter duas lineas. Et forma extremitatis stili que extenditur super hanc lineam, cum fuerit extensa ad centrum vitri, reflectebatur super lineam que transit per centra duorum foraminum. Erit ergo
290 reflexio ista ad partem perpendicularis exeuntis a loco reflexionis super superficiem vitri, et hec forma exit ab aere, et reflectitur in vitro, et vitrum est grossius aere.

[4.27] Ex omnibus igitur istis experimentationibus patet quod visus comprehendit visibilia que sunt in aqua et ultra corpora diaffona
295 que differunt a diaffonitate aeris secundum reflexionem preter quam illa que sunt super lineas perpendiculares super superficiem corporis diaffoni in quo existit, et quod reflexio formarum ipsorum est in superficiebus perpendicularibus super superficies corporum diaffonorum, omne enim quod experimentatum est per predictum instru-
300 mentum invenitur reflecti in superficie medii circuli, ex quo patuit quod est perpendicularis super superficies corporum diaffonorum et super superficies corporum contingentium superficies eorum. Ex hac ergo experimentatione declarabitur etiam quod forme que comprehenduntur a visu secundum reflexionem que exeunt a grossiori
5 corpore ad subtilius reflectuntur ad partem contrariam illi in qua est perpendicularis exiens a loco reflexionis super superficiem corporis diaffoni, et que exeunt a subtiliori ad grossius reflectuntur ad partem in qua est perpendicularis predicta.

[4.28] Stelle autem comprehenduntur etiam secundum reflexio-
10 nem, nam corpus celi est subtilius corpore aeris, scilicet maioris diaffonitatis. Hoc autem potest experiri experimentatione que ostendet quod stelle comprehenduntur secundum reflexionem et ex qua patebit etiam quod corpus celi est magis diaffonum corpore aeris. Et cum

284 recte *corr. ex* recta S / in *inter. a. m.* C1 / vitri^2 *inter. a. m.* E   285 cum: in *FP1* / super *om.* P1 / super . . . vitri (286) *om.* R   286 erit *inter. a. m.* L3   288 fuerit extensa *transp.* C1 / reflectebatur: refringetur R   290 reflexio: refractio R / reflexionis: refractionis R   291 exit: erit S / aere *alter. in axe a. m.* E / et^2 *inter. a. m.* L3; *om.* O / reflectitur: refringitur R   293 istis *om.* L3   295 reflexionem: refractionem R / quam *om.* S / ipsorum: ipsarum E   296 super^2 *om.* F   297 in quo *mg. a. m.* C1 / existit: existis S / reflexio: refractio R   298 superficies: superficiem E   300 reflecti: refringi R / ex: de P1R   1 est *inter. a. m.* E / superficies: superficiem SE   2 super *inter.* OE (*a. m.* E) / ex: et P1   3 hac: nunc *deinde del.* F / que *inter.* O; *om.* P1   4 reflexionem: refractionem R / exeunt: exivit E / grossiori: grossiore R   5 *post* corpore *add.* diaffono ER / reflectuntur: reflectuntuntur S; refringuntur R / *post* partem *add.* per L3   6 reflexionis: refractionis R   7 subtiliori: subtiliore R / reflectuntur: refringuntur R   9 reflexionem: refractionem R   10 *post* subtilius *add.* a O / scilicet: id est R   11 autem: aut L3 / experiri: experimentari R / ostendet *corr. ex* ostendit O; ostenderet S   12 comprehenduntur: comprehendantur R / reflexionem: refractionem R / et *om.* P1R / qua: quo R / patebit etiam (13) *transp.* L3

quis hoc voluerit experiri, accipiat instrumentum de armillis, et ponat
illud in loco eminenti in quo poterit apparere orizon orientalis, et
ponat instrumenti armillarum suo modo proprio, scilicet quod ponat
armillam que est in loco circuli meridiei in superficie circuli meridiei,
et polus eius sit altior terra secundum altitudinem poli mundi super orizonta loci in quo ponitur instrumentum. Et in nocte preservet
aliquam stellarum fixarum magnarum que transit per verticationem
capitis illius loci aut prope, et preservet illam in ortu suo ab oriente.
Stella autem orta, revolvat armillam que revolvitur in circuitu poli
equinoctialis donec fiat equidistans stelle, et certificetur locus stelle
ex armilla, et sic habebit longitudinem stelle a polo mundi. Et deinde
preservet stellam quousque perveniat ad circulum meridiei, et moveat armillam quam prius moverat donec fiat equidistans stelle, et sic
habebit longitudinem stelle a polo mundi, cum stella fuerit in verticatione capitis. Hoc autem facto, inveniet remotionem stelle a polo
mundi apud ascensionem minorem remotione eius a polo mundi in
hora existentie eius in verticatione capitis, ex quo patet quod visus
comprehendit stellas reflexe, non recte.

[4.29] Stella enim fixa semper movetur per eundem circulum de
circulis equidistantibus equatori diei, et numquam exit ab ipsa ita
quod appareat nisi in longissimo tempore. Et si stella comprehenderetur recte, tunc linee radiales extenderentur a visu ad stellas recte,
et extenderentur forme stellarum per lineas radiales recte quousque
pervenirent ad visum. Et si forma extenderetur a stella ad visum
recte, tunc visus comprehenderet eam in suo loco, et sic inveniret distantiam stelle fixe a polo mundi in eadem nocte eandem. Sed distantia stelle mutatur in eadem nocte a polo mundi, ergo visus non
recte comprehendit stellam. In celo autem non est corpus densum

14 hoc voluerit *transp.* FP1L3 / de . . . instrumenti (16) *mg.* O   15 eminenti: eminente R / *post*
eminenti *scr. et del.* et E / poterit: patuit S   16 quod: ut R   17 meridiei[1] *corr. ex* merediei
O; meridionalis C1ER   18 altior: exaltatus a R / super: supra R   19 orizonta *corr. ex*
orizona L3; horizontem R / preservet: observet R   20 fixarum: fiarum S / verticationem:
verticem R   21 loci *corr. ex* lucis L3 / preservet: observet R / illam *inter. a. m.* L3 / in: ab R / ab:
in R   22 orta *inter.* O   23 *post* stelle *scr. et del.* basi P1   24 ex *corr. ex* extra O; *om.*
L3 / et[2] *om.* FP1R   25 preservet: observet R / perveniat: perveniet C1; pervenerit R / moveat:
revolvat R; *alter. ex* inducat *in* moverat *a. m.* E;   27 verticatione: vertice R   29 apud
ascensionem: in ascensione R / apud . . . mundi[2] *mg.* O   30 *post* hora *scr. et del.* instrumenti
P1 / post eius *scr. et del.* a polo mundi E / in *inter. a. m.* E / verticatione: vertice R / *post* verticatione *scr. et del.* eius L3   31 reflexe: refracte R   32 enim: autem L3   33 diei *corr. ex*
meridiei L3; *om.* R / numquam: nunc quam S; *corr. ex* nunc quam C1   34 quod: ut R   35 extenderentur . . . et (36) *rep.* S / visu *corr. ex* recte E / ad *corr. ex* et *a. m.* C1 / ad . . . recte[2]: recte
ad stellas R / recte[2] *om.* E   36 et *inter. a. m.* L3 / forme *corr. ex* formas S   37 pervenirent
corr. ex pervenient L3 / post visum[1] *rep.* et (36) . . . radiales (36) FP1 / ad[2] . . . recte (38): recte ad
visum R   38 comprehenderet: comprehenderetur FP1; comprehendet L3 / sic *corr. ex* si
O / inveniret: inveniat P1; inveniet L3   39 in . . . mundi (40) *om.* S   40 in *om.* ER
41 recte *om.* P1 / autem: aut S

tersum, nec in aere, a quo possunt forme converti, et cum visus comprehendit stellam et non recte nec secundum conversionem, ergo est
secundum reflexionem, cum hiis solis tribus modis comprehenduntur res a visu. Ex diversitate ergo distantie eiusdem stelle in eadem
nocte a polo mundi patet procul dubio quod visus comprehendit stellas reflexe. Ergo corpus in quo sunt stelle fixe differt in diaffonitate
ab aere.

[4.30] Preterea potest experiri diaffonitas corporis celi per experimentationem lune, nam, cum equaveris locum lune in aliqua hora
prope ortum eius, et post in nocte nota et in loco noto et verificaveris
locum eius a polo mundi. Deinde posueris instrumentum horarum
in illa nocte ante ortum lune, et scias altitudinem lune. Et preservaveris lunam usque ad ortum eius, et perveniat tempus ad minutum idem eiusdem hore quod habet luna, et preservaveris altitudinem lune quam habet in illa hora a verticatione capitis, et preservaveris quod instrumentum elevationis sit divisum per minuta et per
minora minutis si possibile est. Tunc invenies distantiam lune a verticatione capitis in illa hora per instrumentum minorem spatio remotionis a verticatione capitis in illa hora per computationem. Ergo lux
lune non extenditur per duo foramina instrumenti per quod sumpta
est elevatio recte, tunc enim distantia eius a verticatione capitis esset
eadem cum illa que inventa est per computationem, sed distantia
inventa per instrumentum differt a distantia per computationem.
Ergo lux lune non extenditur a celo ad aerem per lineas rectas; ergo
secundum reflexionem. Ex hiis ergo experimentationibus patet quod
comprehendit visus omnes stellas que sunt in celo reflexe. Ergo universum celum differt a diaffonitate aeris. Restat ergo declarare quod
corpus celi differt in subtilitate ab aere.

42 nec *inter. E* / possunt: possint *R* / forme converti *transp. C1* / converti: reflecti *R* / comprehendit: comprehendat *R*   43 *ante* stellam *scr. et del.* illam *C1* / et *om. E* / et non *om. R* / conversionem: reflexionem *R* / est *om. R*   44 reflexionem: refractionem *R* / hiis solis *transp. L3O* /
comprehenduntur: comprehendantur *R*   45 eiusdem *corr. ex* eius *O*   46 comprehendit:
comprehendat *R*   47 reflexe: refracte *R* / differt: differunt *L3O* / in²: a *S*   49 experiri:
experimentari *R*   50 aliqua *corr. ex* aqua *C1*   51 post *corr. ex* potest *O* / post nocte *scr.
et del.* noc *F* / loco: suo *P1* / et³ *om. R* / verificaveris: verificationis *O*; *corr. ex* verificationis *L3*
52 posueris *corr. ex* posuerit *O*   53 *post* ante *scr. et del.* horam *C1* / et . . . lune² *inter. O* / scias:
sciveris *R* / preservaveris: observaveris *R*   54 lunam *corr. ex* lum *F* / post tempus *add.* ad
instrumentum *E*; *add.* in instrumento *R* / ad² . . . hore (55): eiusdem hore ad minutum idem *E*
55 *post* eiusdem *scr. et del.* deinde *O* / habet: habes *C1*; *om. S* / preservaveris: observaveris *R*
56 verticatione: vertice *R* / preservaveris: observaveris *R*   57 quod: ut *R* / sit: si *P1*   58 est
inter. a. m. C1 / verticatione: vertice *R*   59 per *om. S* / per . . . hora (60) *mg. a. m. E* / minorem: minor est *P1*; *corr. ex* minor est *F*   60 verticatione: vertice *R*   61 quod: que *R*
62 verticatione: vertice *R*   63 eadem: eam *S* / illa: linea *O*; *corr. ex* linea *a. m. L3* / inventa
est *transp. ER*   64 instrumentum: computationem *ER* / computationem: instrumentum *ER*
65 *post* lux *rep. et del.* lux *F*   66 reflexionem: refractionem *R*   67 comprehendit visus
transp. R / reflexe: refracte *R*

70   [4.31] **[PROPOSITIO 1]** Et hoc declarabitur per experimentationem predictam. Sit ergo circulus meridiei in loco experimentationis
circulus ABG [FIGURE 7.4.1, p. 440], et cenit capitis B, et polus mundi
D, et centrum mundi E. Et continuemus B cum E, et sit locus visus
Z, et circulus equidistans equinoctiali cuius distantia a polo mundi
75   est illa in qua invenitur stella in hora certificationis distantie prime
circulus HT. Et sit locus stelle in illa hora H, et sit circulus equidistans equinoctiali cuius distantia a polo est illa in qua invenitur stella
in secunda hora circulus KB. Iste ergo circulus erit ille in quo requiescet stella secundum verticationem, nam, cum stella fuerit in vertica-
80   tione capitis aut valde prope, tunc visus comprehendet illam recte,
quia linea recta que transit per visum et per verticationem capitis est
perpendicularis super concavum spere celi et perpendicularis super
convexum aeris. Et cum sit perpendicularis super utrumque corpus,
ergo visus comprehendet stellam que est super hanc lineam recte,
85   sive hec duo corpora celi et aeris fuerint diverse diaffonitatis aut consimilis.
        [4.32] Cum ergo stella fuerit in verticatione capitis aut prope, vi-
sus comprehendit illam in suo vero circulo equidistanti equinoctiali
super quem movebatur ab initio noctis quousque pervenit ad circu-
90   lum meridiei. Circulus ergo KB est ille in quo erat stella in experimentatione. Et sit circulus verticationis qui transit per stellam in hora
experimentationis prima circulus BHK, et secet iste circulus circulum
KB in puncto K. Et quia distantia stelle a polo mundi fuit in prima experimentatione minor quam in secunda, erit circulus HT propinquior
95   polo circulo KB; ergo punctus H est propinquior cenit capitis quam
punctus K.   .
        [4.33] Et continuemus duas lineas HZ, KZ. Quia ergo stella com-
prehenditur a visu in hora experimentationis prima in puncto H, et

70 per *rep. C1*   72 *post* et[1] *add.* zotit id est *E* / cenit: zenith *R*; xot *S*; zot *FP1*; cenith *L3*; *alter. ex*
zot in cenith *O* / *post* B *scr. et del.* cum E et sit *S*   74 Z *inter. C1*   75 *post* est *add.* in *P1* / *post*
distantie *scr. et del.* prius *C1*   76 HT: habet *S* / sit[1,2] *corr. ex* si *O* / illa hora *transp. C1* / H *om. S*
77 *post* polo *add.* mundi *FP1* (*mg. F*)   78 requiescet *corr. ex* requiescit *E*   79 verticationem:
verificationem *SFP1C1O* / verticatione *corr. ex* vertificatione *S*; vertice *R*   80 *post* visus *add.*
non *L3O* (*deinde del. L3*)   81 verticationem: verticem *R*   82 concavum ... super[2] *om. S*
83 cum *om. C1*   84 comprehendet: comprehendit *L3ER* / hanc lineam *transp. ER*   85 diverse
diaffonitatis *transp. C1* / aut: sive *R*   87 verticatione: vertice *R*   88 equidistanti: equidistante *R*   89 quem: quam *SC1O*; *alter. ex* quod *in* quam *a. m. E* / pervenit: perveniat *C1* / ad *om. S*
90 KB: KBG *R* / *post* in *add.* prima *SL3C1O* (*deinde del. L3*)   91 verticationis *corr. ex* verificationis
O / qui: que *S*   92 prima: prime *R* / iste: ille *ER*   93 KB: KBG *R* / *post* K *add.* et circulum HT
in puncto H *R* / fuit *om. S*   94 HT *inter. O*   95 KB: KBG *R* / punctus: punctum *R* / propinquior: propinquius *R* / cenit: zenith *R*; zot *SFP1*; cenith *L3*; *alter. ex* zot *in* cenith *O*; *alter. ex* et
punctus *in* cenith *E*   96 punctus: punctum *R* / K: capitis *L3O* / *post* K *scr. et del.* ergo punctus
H *S*   97 duas lineas *transp. deinde corr. C1* / KZ *mg. a. m. L3*   98 prima: prime *R*

tunc erat in superficie circuli BHK verticalis, et stella erat in illa hora
in circumferentia circuli KB, ergo stella erat in illa hora in puncto K, et
comprehendebatur a visu in puncto H et per rectitudinem linee ZH,
visus enim nichil comprehendit nisi per verticationes radialium linearum per quas forme perveniunt ad visum. Visus ergo comprehendit stellam in puncto H, quia forma pervenit ad illum in rectitudine
linee HZ. Et cum visus comprehendit illam in rectitudine linee HZ, et
linea recta que est inter stellam et visum est linea KZ, manifestum est
ergo quod visus non comprehendit stellam que est in puncto K recte;
ergo reflexe.

[4.34] Sit ergo locus reflexionis M, et continuemus KM, et protrahamus illam recte usque Z. Forma ergo stelle que pervenit ad Z, ex
qua visus comprehendit stellam, extenditur a stella per lineam KM,
et reflectitur per lineam MZ. Et non reflectuntur forme nisi cum occurrerit corpus diverse diaffonitatis a diaffonitate corporis in quo
existerit. Ergo corpus in quo est stella, scilicet celum, est diaffonum
differens a diaffonitate ab aere, et quod locus reflexionis est apud
superficiem que transit inter duo corpora que differunt in diaffonitate. Punctus ergo M est punctus in concavitate celi. Et continuemus
lineam inter E, M, et sit diameter spere celi. Erit ergo linea EM perpendicularis super superficiem celi concavam contingentem aerem
et super superficiem aeris convexam. Et cum forma stelle que est in
puncto K extenditur per lineam MK et reflectetur in aere per lineam
MZ, patet quod hec reflexio est ad partem in qua est perpendicularis
EM, que transit per punctum reflexionis, que est perpendicularis
super superficiem aeris. Et cum reflexio in aere est ad partem perpendicularis exeuntis per locum reflexionis, ergo corpus aeris est grossius
corpore celi.

99 erat[1]: erit *E*/BHK: HBK *L3*/verticalis: verticalius *O*/illa hora *transp. L3*   100 KB: HB *O*;
KBG *R*/erat: erit *FP1*   101 comprehendebatur: comprehenditur *ER*/a . . . puncto *corr. ex*
in puncto a visu *O*/ZH: HZ *S*   102 enim *inter. L3*; *om. O*/radialium linearum *transp. L3R*
104 in[2] *inter. O*   105 et[1] . . . HZ[2] *inter. O*/comprehendit: comprehendat *R*/illam in rectitudine *corr. ex* illam rectitudinem *C1*/linee[2] *om. R*   106 linea[1]: line *S*/est[2]: sit *R*   108 reflexe:
refracte *R*   109 reflexionis: refractionis *R*   110 illam recte: ab M rectam *R*/pervenit
corr. ex pervenitur *S*   111 *post* extenditur *scr. et del.* a linea per KM *C1*   112 reflectitur:
refringitur *R*/non *om. S*/reflectuntur: refringuntur *R*/occurrerit: occurrit *FP1R*; occurent *L3O*
114 existerit: existit *ER*/corpus *om. S*/*post* stella *scr. et del.* est *P1*/est[2]: scilicet *S*   115 differens: different *S*/a: in *R*/quod: quia *R*/reflexionis: refractionis *R*   116 inter: in *R*/corpora
corr. ex puncta *a. m. E*   117 punctus[1,2]: punctum *R*   119 concavam contingentem *transp.*
deinde corr. E/*post* concavam *add.* et *F*   121 extenditur: extendatur *R*/reflectetur: reflectitur
SE; refringatur *R*   122 hec reflexio *transp. L3*/ reflexio: refractio *R*/partem: lineam *R*; *corr.*
ex lineam *a. m. E*/*post* qua *scr. et del.* extenditur *E*   123 *post* transit *add.* pertransit *P1*/re-
flexionis: refractionis *R*   124 *post* superficiem *scr. et del.* aque *P1*/reflexio: refractio *R*/est:
sit *R*   125 *ante* exeuntis *scr. et del.* super superficeim aeris *S*/*post* exeuntis *add.* transeuntis
SFP1C1OE/per locum reflexionis: a loco refractionis *R*

[4.35] Patet ergo quod hoc quod invenimus per experimentationem de apparitione stellarum signat demonstrative quod visus non
comprehendit stellas nisi reflexe, et quod corpus aeris est grossius
130 corpore celi, et quod corpus celi est subtilius corpore aeris. Ex hiis
ergo omnibus patet quod omnia que comprehenduntur a visu ultra
corpora diaffona quorum diaffonitas differt a diaffonitate aeris, si visus fuerit obliquus a perpendicularibus egredientibus ex ipsis super
superficiem diaffonorum corporum in quibus existunt, comprehen-
135 duntur reflexe.

QUINTUM CAPITULUM
De ymagine

[5.1] Ymago est forma rei visibilis quam visus comprehendit ultra
diaffonum corpus quod differt in sui diaffonitate a diaffonitate aeris,
140 cum visus fuerit obliquus a perpendicularibus exeuntibus ab illo visibili ad superficiem illius corporis diaffoni, nam forma quam comprehendit visus in corpore diaffono de re visa que est ultra ipsum corpus non est ipsa res visa, quoniam visus tunc non comprehendit rem
visam in suo loco nec in sua forma sed in alio loco et in alio modo,
145 scilicet reflexione, et cum hoc comprehendit illam rem in sui oppositione. Hec autem forma dicitur ymago. Hec autem comprehenditur
ratione et experientia.
[5.2] Ratione vero quoniam ex predicto capitulo patet quod visum
quod est in diaffono corpore diverse diaffonitatis ab aere comprehen-
150 ditur a visu reflexe, cum visus fuerit declinis a perpendicularibus
exeuntibus a re visa super superficiem corporis diaffoni. Et cum visus comprehenderit huiusmodi visum reflexe, nec est in oppositione
eius, nec comprehendit ipsum recte, nec sentit ipsum comprehendere

127 quod *inter. a. m.* E/invenimus: inveniemus E; invenitur C1   128 de apparitione *om.* R/
apparitione *corr. ex* aspectione E/signat: significat C1R   129 reflexe: refracte R   132 diaffonitas: diaffonitatis SC1E   134 superficiem: superficies L3/diaffonorum *corr. ex* diaffona-
rum F/existunt: consistunt ER   135 reflexe: refracte R   136 quintum ... ymagine (137):
de ymaginibus capitulum quintum R; *om.* S   137 ymagine: ymaginibus L3; ymaginebus O
138 est *inter. a. m.* L3; *om.* O/*post* comprehendit *inter.* comprehendit O   139 diaffonum corpus
transp. L3/*post* corpus *scr. et del.* visu O/quod ... diaffonitate[1] *inter.* O/sui: sua R/*post* aeris *scr. et
del.* quod differt O   140 visus fuerit *transp.* L3/exeuntibus *corr. ex* exeuntis P1   141 illius
mg. C1/*post* diaffoni *add.* illius C1/quam: aquam L3   142 in *om.* L3/*post* est *rep. et del.* que
est O   144 nec: neque R   145 reflexione: reflexive E; refracte R/hoc *inter. a. m.* C1/com-
prehendit: comprehenderit L3/*post* comprehendit *scr. et del.* hoc C1/in sui: visam L3/sui: sua R
146 *ante* hec[1] *add.* visui L3/hec[1] ... ymago *rep.* S/dicitur *inter.* O/hec[2]: hoc R   148 vero *om.* R
149 est *corr. ex* extra L3   150 reflexe: refracte R/declinis: declivis R   152 comprehenderit:
comprehendit R/reflexe: refracte R   153 nec[1]: non R/*post* nec[1] *scr. et del.* est P1/recte ...
ipsum[2] *rep.* S/ipsum[2]: se R

ipsum reflexe, patet quod ipse comprehendit ipsum extra suum lo-
155 cum.

[5.3] Per experientiam vero sic potest cognosci. Nam si aliquis acceperit vas habens horas erectas perpendiculares, in cuius medio posuerit aliquod visum manifestum, ut obulum aut denarium, et steterit
a longe quousque videat rem visam in profundo vasis, deinde elon-
160 gaverit se a re visa quousque non videat rem paulatim paulatim, tunc
initio occultationis stet in suo loco, et precipiat alii infundere aquam
in vas, ipso existente in suo loco, nec moveat visum nec mutet situm,
tunc enim, cum aspexerit aquam que est in vase, videbit rem visam
postquam non viderat eam. Et videbit eam in eius oppositione, ex
165 quo patet quod forma quam videt in aqua non est in loco visi, nam
forma quam vidit in aqua que est in vase non est in loco visi, nam si
forma esset in loco visi, tunc visus comprehenderet rem visam, non
existente aqua in vase. Visus enim in secundo statu comprehendit
rem visam in sui oppositione ipsa non existente. Hoc igitur modo de-
170 clarabitur utroque modo, ratione videlicet et experientia, quod ymago
rei vise quam visus comprehendit reflexe non est in loco rei vise.

[5.4] Deinde dico quod ymago cuiuslibet puncti quod visus comprehendit reflexe est in puncto quod est differentia communis linee
per quam forma pervenit ad visum et perpendiculari exeunti ab illo
175 puncto viso super superficiem diaffoni corporis. Hoc autem declarabitur per experientiam hoc modo. Accipiat aliquis circulum ligneum,
cuius diameter non sit minor uno cubito, et adequet superficiem eius
quantumcumque poterit. Et inveniat centrum eius, et extrahat in ipso
diametros sese intersecantes quotcumque voluerit, et signentur ferro
180 ut appareant, et impleat lineas illas corpore albo, ut cerusa mixta lacte
nivio, et punctum centri sit nigrum. Hoc autem perfecto, accipiat

154 reflexe: refracte *R* / ipse *om. R* / comprehendit: comprehendet *S*   156 aliquis: quis *FP1* / acceperit: accipit *O*; accipiat *L3*   157 *post* erectas *scr. et del.* ei *P1*   158 obulum: obolum *C1*;
corr. ex obtulum *S*   159 videat: viderit *C1ER* / elongaverit: elongavit *S*   160 paulatim[1]:
visam *P1* / paulatim[2] *om. FER* / *post* tunc *add.* in *FP1ER*   161 alii: alteri *ER*   162 moveat:
movet *S*   164 viderat *corr. ex* videat *OE* (*a. m.* E); viderit *L3*   165 videt in aqua: in aqua
videt *C1* / nam . . . visi (166) *om. L3R*   167 *post* visi *add.* esset *L3*   168 *ante* aqua *scr. et
del.* in *C1*   169 sui: sua *R* / ipsa non existente *om. FP1* / *post* existente *add.* in sui oppositione
C1ER (in *inter. a. m.* E; sui: sua *R*)   170 ratione videlicet *transp. C1* / et *inter. O*   171 quam:
cum *E* / visus *om. FP1* / reflexe: refracte *R* / est *inter. a. m.* L3   172 dico *inter. O*   173 reflexe: refracte *R*   174 *post* illo *rep. et del.* illo *O*   175 puncto viso *transp. FP1* / declarabitur:
declaratur *FP1*   176 accipiat *corr. ex* accipiatat *P1*   177 diameter: diametrum *L3* / *post*
cubito *add.* altitudo duorum vel trium digitorum *R* / *post* adequet *add.* super *FL3C1O* (*deinde
del. L3*) / superficiem: superficies *R*   178 quantumcumque: quodcumque *S*; quamcumque *E*;
corr. ex quodcumque *O* / inveniat: inveniet *FP1*   179 intersecantes *corr. ex* inter se secantes
L3 / quotcumque: quomodocumque *R*; quocumque *SFE*; *corr. ex* quocumque *P1L3* (*a. m. L3*)
180 impleat: impleant *FP1* / cerusa: cerula *FP1*; cerusia *L3* / *post* mixta *inter.* cum *C1*   181 nivio:
nimio *SO*; niveo *L3E*; *om. R* / perfecto *corr. ex* perfectio *L3*

vas amplum, ut pelvim, habens horas elevatas, et ponat vas in loco luminoso. Et infundat in vas aquam claram, et sit altitudo aque minor diametro circuli et maior semidiametro eius. Et mensuretur hoc
185 ipso circulo quousque aqua transeat centrum circuli aliquibus digitis, duobus scilicet diametris aut pluribus signatis in ipso vase, scilicet quod sit aqua cooperiens aliquam partem utriusque diametri et quod remaneat altera pars extra aquam.

[5.5] Et expectet donec aqua quiescat in vase, et tunc mittat circu-
190 lum ligneum in vas, et erigat circulum super horam ipsius, et ponat superficiem ipsius in qua sunt linee signate ex parte visus; deinde moveat circulum donec aliquis diametrorum suorum sit perpendicularis super superficiem aque. Deinde demittat visum suum, et erigat vas quousque visus suus appropinquet equidistantie superficiei aque
195 et extra horam vasis et supra superficiem aque in tantum ut possit videri centrum circuli, experientia enim secundum hunc modum erit manifestior.

[5.6] Hoc igitur facto, intueatur centrum circuli et diametrum circuli perpendicularem super superficiem aque, tunc enim inveniet
200 centrum circuli in rectitudine diametri perpendicularis. Deinde intueatur diametrum circuli declinem, cuius pars est preminens aque, tunc enim inveniet ipsum incurvatum, cuius incurvatio erit apud superficiem aque. Et illa pars que est intra aquam continet cum illa que est extra aquam angulum obtusum. Et inveniet angulum ex parte
205 diametri perpendicularis, et inveniet illud quod est intra aquam rectum continuum, ex quo patet quod forma puncti que est centrum circuli, scilicet forma quam visus comprehendit, non est apud centrum circuli, nam si esset apud centrum circuli, tunc esset in rectitudine diametri declinis, nam in rei veritate talem habet situm.
210 [5.7] Cum ergo visus comprehendit hoc punctum extra rectitudinem diametri declinis, et angulus quem continent partes diametri

182 *post* vas[1] *add.* magnum vel *C1*/habens: habentem *L3* 184 *post* hoc *scr. et del.* in *L3* 185 aliquibus: aliquot *R; corr. ex* aliqui *O* 186 duobus: duabus *R*/*post* diametris *add.* scilicet *S* 187 quod[1]: ut *R*/sit: si *E*/quod[2] *om. R* 188 remaneat: remanet *E* 189 expectet: exspectet *S*/*post* donec *scr. et del.* pervenit usque diametri *O* 190 *post* erigat *scr. et del.* vas *O* 191 qua *corr. ex* aqua *L3* 192 aliquis . . . suorum: aliqua suarum diametrorum *R* 193 *post* deinde *scr. et del.* mo *P1*/demittat *corr. ex* emittat *O* 194 suus: simul *R; om. C1*/appropinquet: appropinquat *C1*; propinquat *E* 195 *post* aque *scr. et del.* in qua *P1*/possit: posset *O* 196 videri: videre *C1ER*/experientia . . . circuli[1] (198) *rep. SO* (*mg. O*) 198 *post* circuli[1] *scr. et del.* experiei *E*/et . . . circuli[2] *mg. O* 200 in . . . circuli (201) *mg. a. m. L3*/intueatur *corr. ex* intuetur *O* 201 declinem: declivem *R*/est: eius *FP1*/est . . . aque: eminet supra aquem *R* 202 ipsum incurvatum: ipsam incurvatam *R*/incurvatum . . . apud *mg. F* 205 illud: illum *FP1* 206 *ante* continuum *add.* et *P1R*/que: quod *ER* 207 scilicet . . . circuli[1] (208) *mg. a. m. L3*/apud *inter. O* 208 nam . . . circuli[2] *mg. a. m. E* 209 declinis: declivis *R*/*post* declinis *scr. et del.* eius qui est forma declinis *O* 211 *post* diametri[1] *scr. et del.* illius *C1*/declinis: declivis *R*/declinis . . . perpendicularem (212) *mg. O*/angulus: anguli *R*

declinis sequitur diametrum perpendicularem, tunc punctus qui est
forma centri est elevatus a centro. Et quia visus comprehendit hoc
punctum in rectitudine diametri perpendicularis super superficiem
215   aque, erit hoc punctum, quod est forma puncti quod est in centro,
extra centrum et elevatum a centro, et cum hoc est in rectitudine perpendicularis exeuntis a centro super superficiem aque. Et declarabitur ex incurvatione diametri declinis apud superficiem aque, et
rectitudine eius quod est intra aquam ex diametro et continuatione
220   eius quod omne punctum partis que est intra aquam ex diametro
declini est elevatum a suo loco.

[5.8] Deinde oportet quod experimentator revolvat circulum ligneum quousque diameter declinis fiat perpendicularis super superficiem aque et diameter qui erat perpendicularis fiat declinis. Deinde
225   demittat visum suum, et intueatur centrum circuli, et tunc inveniet
formam centri in rectitudine diametri qui nunc est perpendicularis
super superficiem aque extra cuius rectitudinem erat forma centri
quando erat declinis, et inveniet formam extra rectitudinem diametri
qui est nunc declinis qui prius erat perpendicularis super superfi-
230   ciem aque. Et inveniet diametrum declinem incurvatum apud superficiem aque, et angulus incurvationis erit ex parte diametri declinis.
Et si in circulo fuerint plures diametri, et revolverit experimentator
circulum quousque unusquisque eorum fuerit perpendicularis successive super superficiem aque, et fuerit diameter qui sequitur illum
235   diametrum declinis, et aliqua pars eius fuerit extra aquam, tunc inveniet formam puncti quod est centrum circuli semper in rectitudine
diametri perpendicularis elevatam a rectitudine diametri declinis. Et
semper inveniet illud quod est intra aquam rectum.

212 declinis: declivis *R*/sequitur: sequuntur *R*/punctus qui: punctum quod *R*   213 elevatus: elevatum *R*   215 *post* aque *scr. et del.* et rectitudine eius quod est intra aquam *C1*/*post*
est[1] *scr. et del.* in *L3*   216 extra . . . et[1] *om. R*/et[1] *om. E*/a centro *mg. a. m. E*/cum *inter. a. m.*
L3/hoc *om. E*   217 super *inter. O*   218 incurvatione *corr. ex* decurvatione *C1*/declinis:
declivis *R*/et: in *O*; *corr. ex* in *a. m. L3*   220 ex diametro *om. S*   221 declini: declinis
C1; declivi *R*   222 quod *om. ER*/experimentator revolvat: experimentatorem revolvere *ER*
223 declinis: declivis *R*   224 qui: que *R*/declinis: declivis *R*   225 demittat: dimittat
FP1OR; damittat *L3*/intueatur: intuetur *FP1*/circuli *om. R*/et[2] *om. L3*/tunc *om. FP1*/*post* inveniet
scr. et del. centrum circuli *O*   226 centri *inter. O*/qui: que *R*   227 erat: erit *FP1*   228 declinis: declivis *R*   229 qui[1,2]: que *R*/declinis: declivis *R*   230 inveniet *corr. ex* invenit
F/declinem: declivem *R*   231 declinis: declivis *R*   232 in . . . fuerit: fuerint in circulo
R/fuerint *mg. O*/revolverit *corr. ex* volverit *a. m. E*   233 circulum *inter. O*/unusquisque
eorum: unaqueque earum *R*/successive . . . aque (234): super superficiem aque successive *R*
234 qui: que *R*/illum: illam *R*   235 declinis: declivis *R*   237 *post* perpendicularis *add.*
et *SC1ER*/elevatam: elevatum *S*/diametri[2] *inter. a. m. E*/diametri declinis *transp. C1*/*post* diametri[2] *scr. et del.* perpendicularis *P1*/declinis: declivis *R*   238 illud: illum *FP1*/*post* aquam
scr. et del. pi *S*

[5.9] Ex omnibus igitur istis patet quod forma cuiuslibet puncti
240 comprehensi a visu in corpore diaffono grossiore corpore aeris comprehenditur extra suum locum et elevatum a suo loco et in rectitudine perpendicularis exeuntis ab illo puncto super superficiem corporis
diaffoni, cum linea que continuat centrum visus cum illo puncto non
fuerit perpendicularis super superficiem corporis diaffoni. Omne
245 autem punctum comprehenditur a visu in eius oppositione et in rectitudine linee recte per quam extenditur forma ad visum; puncta ergo
que visus comprehendit reflexive comprehenduntur in eius oppositione et in rectitudine linee recte per quam forma pervenit ad visum.
[5.10] Hoc autem declarabitur per experimentationem comprehen-
250 sionis rerum visibilium secundum reflexionem per illud instrumentum predictum, nam si experimentator clauserit secundum foramen
quod est in instrumento, tunc non comprehendet rem visam quam
comprehendebat secundum reflexionem, et cum clauserit secundum
foramen, nichil aliud fecit nisi secare lineam rectam ymaginabilem
255 que exit a centro visus ad locum reflexionis, ex quo patet quod forma
que extenditur a visu in corpore diaffono in quo est res visa et reflectitur in corpore diaffono in quo est visus extenditur per lineam rectam
que exit a centro visus ad locum reflexionis. Et omne punctum quod
comprehenditur a visu in corpore diaffono magis grosso quam cor-
260 pus sit aeris, si centrum visus fuerit extra perpendicularem exeuntem
ab illo puncto super corpus diaffonum, comprehenditur in puncto
quod est differentia communis linee super quam pervenit forma ad
visum et perpendiculari exeunti a puncto viso super superficiem corporis diaffoni quod est ex parte visus.
265 [5.11] Si autem experimentator voluerit experiri ymaginem rei
vise cuius forma reflectitur a corpore subtiliore ad corpus grossius,
accipiat frustrum vitri cuius superficies sint equate equidistantes
habens in longitudine octo digitos, et in altitudine quatuor, et in spis-

239 *ante* ex *add.* et *L3O/* ex *inter.* O   240 diaffono . . . corpore^2 *om.* P1   241 locum *om.* E/ele-
vatum *corr. ex* levatum E/*post* elevatum *add.* est FP1   243 centrum *rep.* S/illo puncto *transp.* FP1
244 super superficiem *corr. ex* superficiem O   246 recte *om.* C1/extenditur . . . quam (248) *mg. a. m.*
C1/forma *om.* C1   247 visus comprehendit *transp.* R/reflexive: reflexe L3; refracte R   248 recte
om. C1   250 secundum *inter.* F/reflexionem: refractionem R/illud *inter. a. m.* E   251 foramen
corr. ex formam C1   252 in *om.* FP1O/comprehendet: comprehendit FP1   253 reflexionem:
refractionem R   254 fecit: facit FP1R   255 ad locum *corr. ex* a loco L3/reflexionis: refractionis
R/ex . . . reflexionis (258) *rep.* S (visus extenditur^1 *corr. ex* res visa et reflectitur extenditur)   256 est
res visa: res visa est C1R/res . . . est (257) *mg.* F (et *inter.*; diaffono [257]: diaffonos)/reflectitur: refringitur R   258 reflexionis: refractionis R/*post* et *add.* quod R; *scr. et del.* omni P1   259 corpus *corr.*
ex corporis O/corpus sit (260) *transp.* R   263 perpendiculari: propendiculari S   264 parte *corr.*
ex punctus S   266 reflectitur: refringitur R/*post* corpore *add.* grossiore ad corpus subtilius et a *mg.*
a. m. E/subtiliore: subtiliori P1L3; grossiore C1/ad *inter.* O/grossius: subtilius C1   267 frustrum:
frustum R/sint: sunt FP1/*post* equate *add.* et R   268 *post* habens *scr. et del.* gross P1/in^2 *om.* C1/al-
titudine: latitudine S; *alter. in* latitudine *a. m.* L3/et^2 . . . quatuor (269) *om.* FP1/in^3 *om.* L3

situdine quatuor. Et accipiat circulum ligneum predictum, et signet in
270   dorso eius cordam in longitudine decem digitorum, et dividat illam
in duo equalia, et continuet locum divisionis cum centro circuli linea
recta que transit in utramque partem. Hec ergo linea erit perpendicularis super primam lineam. Deinde continuet alteram extremitatem
corde cum centro circuli linea recta que etiam transeat in utramque
275   partem, et hii duo diametri sint signati ferro, quorum perpendicularem impleat corpore albo et aliud alterius modi corpore.

[5.12] Deinde ponat vitrum longum super dorsum instrumenti
circuli lignei, et superponat alteram extremitatem longitudinis eius
medietati corde, et distinguat de vitro tres digitos, ex quibus duo
280   erunt ex parte diametri declinis extra circulum, et remanebit de longitudine vitri unus digitus qui erit ultra diametrum perpendicularem
supra cordam. Et sit corpus vitri ex parte centri, et applicet vitrum
secundum hunc situm circulo ligneo applicatione fixa. Sic ergo diameter perpendicularis super cordam erit perpendicularis super ex-
285   tremitates vitri equidistantes, et alter diameter erit declinis super has
duas superficies.

[5.13] Deinde oportet quod experimentator ponat horam circuli in
qua est extremitas vitri eminens ex parte sui visus, et ponat alterum
visum in differentia communi circumferentie et extremitati vitri, que
290   est extremitas diametri declinis, et appropinquet suum visum vitro
quantum potuerit ita quod non possit per illum videre ex superficie
aliquid preter extremitatem diametri declinis. Reliquus autem visus
sit in parte in qua est vitrum et circulus. Deinde cooperiat illud quod
opponitur alteri visui ex superficie vitri cum bombace, quem appli-
295   cet super aliquam partem vitri ita quod comprehendat diametrum
declinem, qui est ultima linea per unum visum qui contingit vitrum,

269 *post* in *scr. et del.* cor *P1* · 270 decem digitorum *transp. FP1* (digitorum: digditorum *mg.*
F; corr. ex digito *P1*)/dividat *corr. ex* dividet *C1*   271 in *corr. ex* et *FP1*/et . . . locum *inter. O*/
continuet: continet *L3; corr. ex* continet *E*   272 transit: transeat *R*/erit *rep. P1*   273 pri-
mam lineam *transp. FP1ER*/alteram extremitatem *transp. P1*   274 *post* que *scr. et del.* est
L3/etiam *om. C1*/transeat *corr. ex* transi *E*/in *om. SO; inter. a. m. L3*   275 hii duo: he due
R/sint: sunt *SFP1L3O*/quorum perpendicularem: quarum alteram *R*   276 aliud: aliam
R/*post* modi *scr. et del.* sunt *P1*/corpore: colore *FP1; corr. ex* colore *a. m. E*   278 superponat:
supponat *S*/alteram *corr. ex* alteam *S*   280 erunt: erant *L3C1*/declinis: declivis *R*   281 vitri
om. P1   282 supra: super *R*   284 *post* perpendicularis[1] *scr. et del.* erit *O*/super[1]:
supra *C1*/extremitates *corr. ex* extremitatem *C1*   285 alter: altera *R*/declinis: declivis *R*
287 oportet quod *mg. a. m. C1*/quod: ut *R*   288 qua: quo *C1*   289 *post* extremitati
scr. et del. visus *P1*   290 declinis: declivis *R*/suum visum *transp. ER*/suum . . . vitro: vitro
suum visum *C1*   291 potuerit: poterit *R*/quod: ut *R*/*post* superficie *scr. et del.* videre *O*
292 declinis: declivis *R*   293 sit *om. P1*/in[1]: ex *L3O*   294 ex *corr. ex* cum *L3*/quem:
quam *ER*/applicet: amplicet *S*; appticæs *L3*   295 *ante* super *scr. et del.* aliquod vitrum *P1*/
quod: ut *R*   296 declinem: declivem *R*/qui[1]: que *ER*/*post* linea *scr. et del.* perpendicularis
P1/unum visum: universum *FP1*/contingit *corr. ex* continget *E*

et non videat ultra hanc lineam, et videat lineam albam perpendicularem utroque visu.

[5.14] Ipso autem existente in hoc situ, intueatur centrum circuli,
300 et inveniet illud in rectitudine linee albe que est perpendicularis super superficiem vitri. Et intueatur diametrum declinem apud cuius extremitatem tenet visum suum, et tunc videbit eum incurvatum apud superficiem vitri que est ex parte centri, et inveniet angulum incurvationis ex parte circumferentie. Visus autem comprehendit partem
5 huius diametri declinis que est sub vitro in rectitudine, nam visus tangit superficiem vitri, et diametri perpendicularis una pars est sub vitro, alia extra vitrum ex parte centri, alia extra vitrum ex parte extremitatis diametri.

[5.15] Pars igitur que est sub vitro comprehenditur a visu extra
10 vitrum secundum reflexionem, et pars que est ex parte extremitatis diametri comprehenditur a visu extra vitrum, qui est visus extra vitrum recte et sine reflexione. Pars autem que est ex parte centri comprehenditur ab utroque visu secundum reflexionem, nam linee que exeunt a centro visus contingentis vitrum et extenduntur in corpore
15 vitri, quando perveniunt ad superficiem vitri que est ex parte extremitatis centri, omnes erunt declines super superficiem vitri. Pars ergo que est ex parte centri ex diametro perpendiculari comprehenditur a visu contingenti vitrum secundum reflexionem.

[5.16] Linee vero que exeunt a reliquo visu ad superiorem super-
20 ficiem vitri erunt declines super superficiem vitri superiorem, et cum extenduntur ad aliam superficiem vitri que est ex parte centri erunt etiam declines. Reliquus ergo visus etiam comprehendit partem diametri perpendicularis que est ex parte centri duabus reflexionibus; partem autem que est sub vitro una sola reflexione, partem vero su-

297 lineam^2 corr. ex linear F   299 autem om. FP1 / hoc: tali C1 / intueatur: intuetur FP1; corr. ex intueaur S   300 inveniet corr. ex invenit O / illud: illum FP1C1 / albe inter. a. m. E   1 post et scr. et del. inveniet illud O / declinem: declivem R / cuius: eius E; corr. ex eius O / extremitatem corr. ex extremi-tas F   2 tunc inter. P1 / eum incurvatum: eam incurvatam L3R   3 ex: in S   4 post autem add. huius diametri FP1 / comprehendit: comprehendet ER   5 huius om. SFP1 / diametri om. FP1 / declinis: declivis R / nam: et quia R / visus alter. in res visa O   6 ante tangit scr. et del. visa res O / et mg. a. m. L3   7 centri: vitri S / centri . . . parte^2 om. FP1 / alia: altera ER / ex parte^2 inter. a. m. E; om. L3   9 est sub vitro: sub vitro est R   10 vitrum: visum P1; alter. ex visum in vi um L3 / reflexionem: refractionem R   11 vitrum^1: visum S; corr. ex visum P1 / qui . . . vitrum^2 mg. a. m. E / est visus transp. SFC1R / visus om. P1   12 ante recte add. et L3 / reflexione: refractione R   13 reflexionem: refractionem R   14 vitrum et transp. L3 / et om. O   15 vitri^1: cum FP1 / que est om. P1   16 centri om. L3 / declines: declives R   17 perpendiculari: perpendicularis ER   18 contingenti: contingente ER / reflexionem: refractionem R   19 a inter. O   20 declines: declives R / super . . . superiorem mg. a. m. E   21 ad: super R; corr. ex super a. m. E / aliam superficiem transp. ER / post parte scr. et del. vitri S   22 declines: declives R / post declines add. et C1 / visus corr. ex declinis F / etiam: et E; om. P1R   23 reflexionibus: refractionibus R; corr. ex super O   24 reflexione: refractione R / partem^2 . . . reflexione (25) om. R / vero: autem S

25 periorem absque reflexione, et cum hoc toto uterque visus comprehendit hunc diametrum rectum. Et si experimentator cooperuerit
alterum visum et aspexerit per visum qui est ex parte vitri, comprehendet perpendicularem rectum. Et si elevaverit visum suum a vitro
et intuens fuerit diametrum perpendicularem ultra vitrum, compre-
30 hendet ipsum rectum cum hoc quod comprehendit ipsum secundum
reflexionem.

[5.17] Causa autem huius est quoniam omne punctum diametri
perpendicularis, quando comprehenditur a visu secundum reflexionem, comprehenditur non suo loco, sed tamen comprehendit ipsum
35 in loco qui est in rectitudine perpendicularis que exit ab illo super
superficiem vitri, et iste diameter est perpendicularis que exit a quolibet puncto eius ad superficiem vitri, et nullum punctum comprehenditur reflexive nisi super ipsum. Cum igitur visus comprehendit
hunc diametrum rectum et comprehendit formam centri in rectitudi-
40 nem huius diametri, forma centri quam visus comprehendit ultra vitrum, quando visus tangit vitrum, est in rectitudine perpendicularis
exeuntis a centro super superficiem vitri.

[5.18] Et cum comprehenderit diametrum declinem incurvatum,
comprehendit partem eius que exit a vitro, que est ex parte centri,
45 non in suo loco. Punctus centri non comprehenditur a visu nisi preter
suum locum, et cum angulus incurvationis fuerit ex parte circumferentie, tunc punctum quod est forma centri est sub centro, ex quo
patet quod ymago cuiuslibet puncti comprehensi a visu ultra corpus
diaffonum subtilius corpore diaffono quod est in parte visus est in
50 rectitudine linee que exit ab illo puncto perpendiculariter super superficiem corporis diaffoni quod est in parte visus, et est remotior
a superficie corporis diaffoni quod est in parte visus quam ipsum
punctum. Et omne punctum comprehensum a visu est in rectitudine linee per quam pervenit forma ad visum, et ymago cuiuslibet

25 uterque *corr. ex* in *O; om.* R   26 hunc: hanc R; *inter. a. m.* L3 / *post* hunc *scr. et del.* visum O /
rectum: rectam R   27 per *inter.* O / qui: que C1 / est *om.* R / comprehendet *corr. ex* comprehendit O   28 rectum: rectam R   30 ipsum rectum: ipsam rectam R / quod *mg. a. m.* C1 / ipsam²:
ipsam R   31 reflexionem: refractionem R   32 quoniam: quod R   33 perpendicularis:
perpendiculares C1 / quando comprehenditur *rep.* P1 / secundum *mg.* C1 / reflexionem: refractionem R   34 *post* non *add.* in ER / comprehendit: comprehenditur R / ipsum *om.* R   35 super
om. FP1   36 iste: ista R   38 reflexive: refracte R / ipsum: ipsam R   39 hunc: hanc R / rectum: rectam R / comprehendit: comprehen L3   40 quam visus *om.* L3 / comprehendit ... vitrum
inter. a. m. L3   41 quando *corr. ex* quam L3 / tangit: transit C1 / *post* tangit *scr. et del.* illum F / est:
et C1 / *post* in *scr. et del.* superficie P1   42 *post* vitri *scr. et del.* convexam P1   43 declinem:
declivem R / incurvatum: incurvatam R   44 comprehendit: comprehendet ER / vitro: centro
R / ex: in S   45 *post* loco *add.* et R / punctus: punctum R / *post* punctus *add.* ergo C1   46 suum:
suuum R   48 ymago: forma FP1 / *post* a *scr. et del.* vi P1   50 ab *inter.* O / perpendiculariter:
perpendicularis R   51 *post* in *scr. et del.* ea C1 / et ... visus (52) *inter. a. m.* L3   52 a *om.* L3
53 et *om.* FP1 / et ... punctum *inter.* O   54 pervenit forma *transp.* R

55 puncti comprehensi a visu ultra corpus diaffonum subtilius corpore
diaffono quod est in parte visus est in differentia communi linee per
quam forma pervenit ad visum et perpendiculari que exit a puncto
viso super superficiem corporis diaffoni quod est in parte visus.

[5.19] Ex omnibus ergo istis declaratis in hoc capitulo patet quod
60 ymago cuiuslibet puncti cuiuslibet visi comprehensi a visu ultra corpus diaffonum diverse diaffonitatis a diaffonitate corporis quod est
in parte visus, cum visus fuerit declinis a perpendicularibus exeuntibus ab illa re visa super superficiem corporis diaffoni quod est in
parte visus, est in differentia communi linee per quam forma illius
65 puncti pervenit ad visum et perpendicularis que exit ab illo puncto
super superficiem corporis diaffoni quod est in parte visus, sive corpus diaffonum quod est in parte visus sit subtilius corpore diaffono
quod est in parte rei vise aut grossius.

[5.20] Quare autem visus comprehendit rem visam in loco yma-
70 ginis et quare ymago est in loco sectionis inter lineam per quam forma
pervenit ad visum et inter perpendicularem que exit a puncto viso ad
superficiem corporis diaffoni postea dicetur. Quod autem visus comprehendit formam puncti visi quam comprehendit reflexive etiam in
rectitudine linee per quam forma pervenit ad visum manifestum est,
75 et causa eius declarata est in predictis tractatibus, et est quoniam visus nichil comprehendit nisi in rectitudine linearum radialium, non
enim patitur nisi in verticationibus istarum linearum.

[5.21] Quare autem comprehendit formam per perpendiculares
exeuntes a re visa super superficiem corporis diaffoni est quia, ut in
80 secundo declaravimus, quando lux extenditur in corpore diaffono,
extenditur per motum velocissimum. Et in quarto capitulo huius
tractatus declaravimus quod motus lucis in corpore diaffono super
lineam declinem super superficiem illius corporis est compositus ex

55 comprehensi: apprehensi FP1   56 in¹: ex ER/post visus scr. et del. quam ipsum punctum
et omne punctum C1/est² inter. O/linee corr. ex linea E   57 perpendiculari corr. ex perpendicularis C1   58 in: ex ER/post visus add. et L3O; rep. et del. est² (56) . . . visus (58) E (linee
[56]: linea; perpendiculari [57]: perpendicularis; in [58]: ex)   59 ex corr. ex in O/ergo om.
L3/post ergo add. ex C1   60 cuiuslibet om. P1ER/visi om. P1/ultra: intra FP1   62 in om.
L3/cum: et C1/cum visus mg. a. m. L3; om. SO/declinis: declivis R   63 visa om. R/super
corr. ex superficie F   64 communi corr. ex communis L3   65 et: erit O; corr. ex erit L3/
perpendicularis: perpendiculari R   68 aut: sive R   69 autem visus transp. C1/compre-
hendit: comprehendat R/ymaginis: ymaginationis SL3; alter. in ymaginationis a. m. E; corr. ex
ymaginationis C1   70 est: sit R/forma inter. a. m. E   71 et inter. a. m. E   72 comprehendit: comprehendat R   73 reflexive: refracte R/etiam: et L3   75 eius: huius L3/post
in scr. et del. ipsis C1/predictis tractatibus transp. C1   76 nisi inter. L3; om. O   77 in om.
E/istarum linearum transp. L3O/post linearum scr. et del. rectarum O   78 comprehendit:
comprehendat R/per inter. a. m. L3E; om. SO   79 exeuntes corr. ex exeuntis E/post visa scr. et
del. ea L3/post est scr. et del. enim ut F/in om. SL3O   80 post secundo add. tractatu FP1; add.
libro R/extenditur . . . diaffono mg. a. m. L3   81 per inter. a. m. L3; om. O/capitulo alter.
in libro a. m. C1   83 declinem: declivem R

motu super perpendicularem exeuntem a puncto in quo extenditur
85   lux super superficiem illius corporis diaffoni et ex motu super lineam
que est perpendicularis super hanc perpendicularem. Forma autem
que extenditur a puncto viso reflexive ad locum reflexionis, que est
forma lucis existens in puncto viso mixta cum forma coloris, semper
extenditur super lineam declinem super superficiem corporis diaf-
90   foni. Hec igitur forma extenditur ad locum reflexionis motu composito
ex motu super perpendicularem que exit a puncto viso super superficiem corporis diaffoni et ex motu super lineam que est perpendicularis super hanc perpendicularem.

[5.22] Est ergo motus forme que movetur super perpendicularem
95   que est super superficiem corporis diaffoni, et deinde translata est ab
hac perpendiculari alio motu, aut super perpendicularem que existit
super primam perpendicularem, et translata est post motum ipsius
super primam perpendicularem motu composito ex predictis duobus
motibus. Hoc autem punctum comprehenditur a visu in rectitudine
100   linee per quam forma pervenit ad visum. Forma ergo existens in loco
reflexionis pervenit ad ipsum per motum forme que movetur super
lineam perpendicularem super superficiem corporis diaffoni, deinde
translata est ab hac perpendiculari per motum in rectitudine linee per
quam forma pervenit ad visum.

105   [5.23] Forma autem que est super perpendicularem existentem
super superficiem corporis diaffoni deinde movetur in rectitudine
linee per quam forma extenditur ad visum est forma que extenditur
a puncto viso in rectitudine perpendicularis exeuntis ex ipso super
superficiem corporis diaffoni donec perveniat ad punctum sectionis
110   inter hanc perpendicularem et lineam per quam forma extenditur ad
visum. Forma igitur puncti quam visus comprehendit reflexive ultra
corpus diaffonum est per motum forme que pervenit ad visum a loco
ymaginis. Visus autem comprehendit hanc formam ex loco ymaginis, quia est per motum forme quam visus comprehendit recte et sine

84 super *om. S*   85 illius corporis *transp. FP1*   87 reflexive: refracte *R* / reflexionis: refractionis *R*   89 declinem: declivem *R*   90 reflexionis: refractionis *R* / *post* reflexionis
scr. et del. ex C1 / *post* motu *scr. et del.* super perpendicularem *F*   91 ex . . . perpendicularem *corr. ex* super perpendicularem ex motu *C1*   94 *post* que *scr. et del.* non *F* / *post*
movetur *add.* aut *L3R* (*inter. a. m. L3*); *add.* est *OE* (*mg. a. m. E*; *deinde del. O*)   95 et *om.*
L3 / ab *inter. C1*   96 hac perpendiculari *transp. C1* / que . . . perpendicularem (97) *inter.*
L3 / existit: est *L3*   97 *post* est *scr. et del.* illa forma super *E* / *post* . . . primam (98) *mg.*
a. m. E   98 motu composito: composito motum *C1*   101 reflexionis: refractionis
R / *post* motum *scr. et del.* notum *O*; *rep. et del.* per motum *P1*   102 lineam: linem *S* / *post*
lineam *scr. et del.* per *F*   105 existentem: exeuntem *F*   106 *post* diaffoni *add.* et *R*
108 in . . . ipso *om. R* / *post* perpendicularis *add. ex S* / ex: ab *L3O*   110 *post* extenditur
add. forma *C1*   111 re- flexive: refracte *R* / reflexive . . . diaffonum (112) *corr. ex* ultra
corpus diaffonum reflexive *O*   112 a . . . motum (114) *mg. a. m. L3*   113 autem *om.*
O / *post* autem *scr. et del.* que pervenit *C1*   114 est *om. L3*

115 reflexione, et est locus qui distat a visu quantum punctus ymaginis, cuius situs in respectu visus est situs forme que est in loco ymaginis, unde visus comprehendit illud punctum secundum reflexionem in loco ymaginis.

[5.24] Hec ergo est causa propter quam visus comprehendit rem
120 visam ultra corpus diaffonum in loco ymaginis et propter quam yma- go cuiuslibet puncti rei vise comprehense secundum reflexionem est in loco in quo linea per quam forma pervenit ad visum secat perpen- dicularem exeuntem ab illo puncto super superficiem corporis diaf- foni.

125 [5.25] Hoc autem declarato, dicamus quod nullum visum compre- hensum a visu ultra aliquod corpus diaffonum quod differt in diaf- fonitate a corpore quod est in parte visus, si corpus fuerit ex corpori- bus communibus, nichil habet nisi unam solam ymaginem. Corpora autem diaffona assueta sunt celum, et aer, et aqua, et vitrum, et
130 lapides diaffoni, et superficies celi que est ex parte visus est sperica concava, unde omnis superficies equalis plana que secat eam facit in ea lineam circularem cuius concavitas est ex parte visus. Superficies autem aeris que tangit illam est sperica convexa, unde, si secetur a superficie equali, fiet in ipsa linea circularis cuius convexum est ex
135 parte celi. Superficies vero aque que est ex parte visus est sperica convexa, et si secetur a superficie equali, fiet in ipsa linea circularis cuius convexum est ex parte visus.

[5.26] Vitrorum autem et lapidum diaffonorum figure assuete sunt rotunde aut plane, unde, si secentur a planis superficiebus, fient
140 in illis aut circuli aut linee recte. Et universaliter dicimus quod omne punctum comprehensum a visu ultra quodcumque corpus diaffonum cuius superficies que opponitur visui est una superficies et quod, si secetur a superficie equali, fiet in superficie eius linea recta aut circu- laris non habet hoc punctum nisi unam ymaginem, nec comprehen-
145 ditur a visu nisi unum punctum tantum.

115 reflexione: refractione R/post distat add. tantum R/punctus: punctum R; corr. ex punc- tum S   117 illud: illum FP1/reflexionem: refractionem R   118 post ymaginis rep. et del. unde (117) . . . punctum (117) S   119 ergo: autem R; corr. ex autem E/ergo est transp. SC1/propter: per L3   121 reflexionem: refractionem R   122 post linea scr. et del. per- veniet C1   123 ab: a ER/illo puncto transp. ER   125 nullum: omne R/comprehensum: comprehenditur P1   126 aliquod om. C1/post diaffonum scr. et del. aliquod C1/in: a C1 127 post in scr. et del. cor C1   128 nichil: non R/solam om. R   130 ex corr. ex in C1/post sperica add. et C1ER   131 equalis inter. a. m. L3; om. R/plana om. FP1L3/post plana rep. unde omnis superficies SO   132 post ea add. secundum C1/post circularem scr. et del. con E 133 post tangit add. lineam L3   135 est² inter. O/sperica mg. F   136 a: ex C1/circularis corr. ex circulis a. m. C1   138 autem inter. O   139 secentur: secetur S   140 aut: autem O; corr. ex autem L3   142 una: unica R/post superficies² scr. et del. equali O/quod inter. P1; om. R   143 fiet: fiat R/in corr. ex a E   144 hoc: hunc L3/hoc punctum om. R   145 tantum inter. O

[5.27] **[PROPOSITIO 2]** Sit ergo visus A [FIGURE 7.5.2, p. 441] et punctum visibile B. Et corpus diaffonum ultra quod est B sit illud in cuius superficie est G, et sit diaffonitas huius corporis grossior diaffonitate corporis quod est ex parte visus. Et sit superficies eius que
150   est ex parte visus equalis, et extrahamus a puncto A perpendicularem AGC. Punctum ergo B aut erit super lineam AGC, aut erit extra ipsam.

[5.28] Si ergo punctum B fuerit in linea GC, tunc visus A comprehendet B recte et sine reflexione, nam forma B, quando extenditur
155   per BG, exit ad corpus quod est in parte A in rectitudine BG, nam BG est perpendicularis super superficiem corporis diaffoni quod est ex parte visus. Visus ergo A comprehendit B in suo loco et in rectitudine AGB.

[5.29] Dicimus ergo quod punctum B extra hanc lineam numquam
160   reflectitur ad A, quod si sit possibile, reflectitur forma B ad A ex T. Et extrahamus superficiem in qua est perpendicularis AGB et punctum T. Faciet ergo in superficie corporis diaffoni lineam rectam. Sit ergo GTD, et extrahamus a puncto T perpendicularem super lineam GD, et sit KTL, Erit ergo KTL perpendicularis super superficiem corporis
165   diaffoni. Et continuemus BT, et extrahamus illam ad H.

[5.30] Erit ergo angulus KTH ille quem continet linea per quam extenditur forma et perpendicularis exiens a loco reflexionis super superficiem corporis diaffoni. Quia ergo corpus quod est ex parte A est subtilius illo quod est ex parte B, cum B pervenit ad T, reflectetur
170   ad partem contrariam illi in qua est perpendicularis TK. Non ergo pervenit forma reflexa ad lineam AB, sed est ex parte reflexa ad punctum A, quod est impossibile. Non ergo reflectetur forma B ad A ex

147 in . . . sit (148) *corr. ex* inter superficiem H et *a. m. L3*   148 est: sit *L3*/G *om. O*/sit *corr. ex* si *O*/corporis *corr. ex* operis *F*/grossior: grossiore *FP1*/*post* grossior *add.* a E   149 et . . . visus (150) *om. L3*/que: quod *FP1*   150 est: sit *P1*/*post* extrahamus *add.* super ipsam *R*/a: ex *L3*   151 super *rep. O*/AGC: AT *C1*; HTA *O*; *corr. ex* ABC *a. m. E*/erit² *om. ER*/erit extra *transp. deinde corr. F*   153 GC: HT *O*   154 reflexione: refractione *R*   155 per *om. S*/BG¹,²,³: BH *O*/exit: erit *S*/*post* ad *scr. et del.* a corpore *C1*/nam BG *inter. O*   156 est perpendicularis *transp. F*   158 AGB: ABH *O*; ABG *L3*; *corr. ex* ABG *SE*   159 extra . . . lineam *inter. a. m. L3*; *om. O*/numquam: numquid *S*   160 reflectitur¹: refringetur *R*/reflectitur²: refringetur *R*; *alter. in* reflectatur *a. m. C1*/forma *om. L3*/*post* ex *add.* puncto *R*/T: P *R*   161 *post* superficiem *scr. et del.* vitri *C1*/AGB: AHB *O*; ABG *C1*; *corr. ex* ABG *a. m. E*   162 T: P *R*; *inter. a.. m. L3*   163 GTD: GBTD *S*; HTD *L3*; GPD *R*/T: P *R*/super lineam *inter. a. m. E*/GD: HGD *S*; GTD *C1*; DTG *E*; DPG *R*; *alter. ex* HGD *in* DTG *L3*   164 KTL¹: KPL *R*/erit ergo *transp. C1*/erit ergo KTL *inter. L3*/KTL²: KPL *R*; *corr. ex* HTL *O*   165 BT: HT *SL3O*; BP *R*/illam *om. R*/H *corr. ex* T *O*   166 KTH: KTL *P1*; KPH *R*   167 reflexionis: refractionis *R*/*post* reflexionis *inter.* BC *O*; *scr. et del.* BC *L3*/super *om. FO*   169 pervenit: pervenerit *R*/T: P *R*; *corr. ex* TE *E*/reflectetur: refringetur *R*   170 TK: PK *R*   171 pervenit: perveniet *SC1ER*/reflexa¹,²: refracta *R*/lineam . . . ad² *rep. mg. a. m. E*/ex parte *om. R*   172 reflectetur: refringetur *R*/*post* ex *scr. et del.* parte *P1*

T, nec ex alio puncto. A ergo non comprehendet B nisi ex rectitudine
AGB; non ergo comprehendit ipsum nisi ex puncto uno tantum, et
175 hoc voluimus declarare.

[5.31] **[PROPOSITIO 3]** Si ergo B fuerit extra AGC [FIGURE 7.5.3,
p. 441], extrahamus superficiem in qua est AGC linea et punctum B.
Ergo erit perpendicularis super superficiem corporis diaffoni, et fiat
in superficie huius corporis linea GD. Erit ergo GD recta. Non ergo
180 reflectetur forma B ad A nisi in superficie in qua est GD, non enim
transit per duo puncta A, B superficies perpendicularis super superficiem corporis diaffoni aut superficies transiens per perpendicularem
AC, et non transit per perpendicularem AC et per punctum B superficies equalis nisi una sola tantum. Forma ergo B non reflectitur ad A
185 nisi ex linea GD.

[5.32] Reflectatur ergo forma B ad A a puncto E, et continuemus
duas lineas BE, EA, et extrahamus ex E perpendicularem super lineam GED. Sit ergo HEZ. Erit ergo HEZ perpendicularis super duas
superficies duorum corporum diaffonorum. Et extrahamus BE recte
190 ad T. Erit ergo ET inter duas lineas EH, EA, nam corpus diaffonum
quod est ex parte A est subtilius illo quod est ex parte B. Forma ergo
B, que extenditur per lineam BE, cum pervenerit ad E, reflectetur ad
partem contrariam parti perpendicularis ZEH; ideo erit linea ET inter
duas lineas EH, EA.

195 [5.33] Et extrahamus ex B perpendicularem super lineam GD, scilicet BK. Erit ergo BK perpendicularis super superficiem corporis
diaffoni quod est ex parte B. Et extrahamus AE recte ut secet angulum BEK, et secet lineam BK in M. M ergo erit ymago puncti B, et
angulus TEA erit angulus reflexionis. Dico ergo quod B non habebit
200 aliam ymaginem preter M, nec forma eius reflectetur ad A ex alio
puncto quam ex E.

173 T: P R/nec: neque R/post non scr. et del. qui F/comprehendet: comprehendit R; corr. ex comprehendit a. m. E/ex²: in R/post rectitudine add. linee R   174 comprehendit alter. in comprehendet a. m. E/ex om. R/et . . . declarare (175) om. R   176 ergo: vero R/B om. P1   177 B: G
S; BG L3O   178 ergo om. S   179 in inter. a. m. L3; om. O/superficie: superficies O; corr. ex
superficies L3/post GD¹ add. sectio communis R/erit inter. E; om. R/post GD² add. est R   180 reflectetur: refringetur R/post enim scr. et del. qui P1   182 aut: nisi R; corr. ex ut O/per inter. a.
m. L3E; om. SC1O   183 et¹ inter. F/et¹ . . . AC² mg. O/non . . . et² om. R/per¹ inter. a. m. C1E; om.
SO/per² om. L3/post B add. et per perpendicularem AC non transit R   184 reflectitur: reflectetur SE; refringitur R   186 reflectatur: reflectat FP1; refringatur R/reflectatur . . . A inter. O/a
inter. E   187 E corr. ex EA S   188 erit ergo HEZ mg. E/HEZ²: EHZ SL3O; corr. ex ERZ F; corr.
ex EZ P1/duas corr. ex lineas O   190 T: P R/ET: EP R/nam . . . EH (194) mg. C1   191 est³
om. C1   192 reflectetur: refringetur R   193 ET: EP R   195 post extrahamus scr. et del. q O
196 erit ergo BK om. S/corporis diaffoni (197) transp. ER   198 et¹ . . . BK om. FP1/post secet scr.
et del. angulum C1/lineam corr. ex lineas L3/M² inter. O; om. F   199 TEA: PEA R/reflexionis:
refractionis R/B corr. ex D O   200 nec . . . est¹ (202) om. R/A corr. ex EA L3

[5.34] Huius demonstratio est quoniam demonstratum est quod
B non comprehenditur a visu nisi per perpendicularem BK. Si ergo
B aliam habuerit ymaginem, erit in linea BK et inter duo puncta B, K,
205 corpus enim quod est ex parte B est grossius illo quod est ex parte A.
Sit ergo illa alia ymago, si possibile est, punctum N. Erit ergo N aut
inter duo puncta M, K aut inter duo puncta M, B.

[5.35] Et continuemus AN. Secabit ergo lineam GD in puncto O.
Et continuemus BO, et transeat usque ad L. Erit ergo O punctum
210 reflexionis, quia linea AON est illa per quam extenditur forma, que
est apud N, ad A, et erit angulus LOA angulus reflexionis. Et extrahamus ex O perpendicularem super lineam GD, et sit FOQ. Erit ergo
linea FOQ perpendicularis super superficiem corporis diaffoni, et
erit angulus LOF angulus quem continet perpendicularis et linea per
215 quam extenditur forma ad locum reflexionis.

[5.36] Si autem N fuerit inter duo puncta M, K, tunc O erit inter
duo puncta E, K; angulus ergo EBK erit maior angulo OBK. Angulus
ergo TEH est maior angulo LOF. Et angulus TEA est angulus reflexionis ex angulo TEH, et angulus LOA est angulus reflexionis ex angulo
220 LOF. Angulus ergo TEA est maior angulo LOA, ut declaratum est in
tertio capitulo huius tractatus; angulus ergo AEH est maior angulo
AOF quod est impossibile.

[5.37] Si autem N fuerit inter duo puncta M, B [FIGURE 7.5.3a, p.
441], tunc punctus E erit inter duo puncta O, K, et erit angulus EBK
225 minor angulo OBK. Erit ergo angulus TEH minor angulo LOF; erit
ergo angulus TEA, qui est angulus reflexionis, minor angulo LOA,
qui est angulus reflexionis. Angulus ergo AEH est minor angulo

202 est[1] *inter. a. m.* L3; *om.* SO/*post* quoniam *add.* enim R/est[2] *scr. et del.* S   203 B *om.* S/*post* com-
prehenditur *scr. et del.* a C1/per *mg.* L3; *om.* SO; super C1ER (*inter.* E)   204 B[1] *om.* S/et *om.* FP1/et
... K *mg. a. m.* E   205 est[2]: erit C1E   206 N[1]: A L3/N[2] *om.* R/aut: autem O   207 *post* B
add. sit inter MK R; *inter.* aut inter MK *a. m.* E   209 transeat *corr. ex* transit O/L: B E; *corr. ex*
AL C1/O *inter.* C1   210 reflexionis: refractionis R/AON: AOL S; BOL R; *alter. ex* AO in AOB
P1/*post* est *scr. et del.* perpendicularis P1/extenditur: ostenditur SL3C1O   211 N: B R/ad A *om.*
R/reflexionis: refractionis R   212 O *corr. ex* A O/FOQ *om.* FP1   213 et ... angulus[1] (214) *scr.*
et del. E   214 LOF angulus *om.* S/*post* LOF *add.* sicut R/et *inter.* E   215 *post* extenditur *rep. et*
del. extenditur O/reflexionis: refractionis R   216 autem: igitur R   217 erit: est R   218 ergo
om. F/TEH: PEH R; *corr. ex* DEH O/maior *corr. ex* minor L3/maior ... est[2] *mg. a. m.* C1/LOF: LOT
SL3; *corr. ex* OBK E; LOK *inter.* O/et *inter.* O/TEA: PEA R/reflexionis: refractionis R   219 *ante* ex[1]
add. et L3/TEH: PEH R/LOA *corr. ex* LEA *a. m.* E/reflexionis: refractionis R   220 *post* LOF *add.*
angulus LOF L3O (*deinde del.* L3)/TEA: PEA R/*post* maior *scr. et del.* recto P1   221 capitulo: capite
R/AEH: AEB SC1/maior: minor SL3OE/*post* maior *scr. et del.* recto P1   222 AOF: AEB S; AOB O;
AEF E   224 punctus: punctum R/E: O L3/*post* duo *rep. et del.* duo F/puncta *mg.* F/angulus EBK
transp. FP1   225 TEH: PEH R; *corr. ex* TH P1/minor[2] ... TEA (226) *mg. a. m.* E   226 TEA: PEA
R/reflexionis: refractionis R/minor ... angulus[2] (227) *mg.* O/*post* angulo *scr. et del.* AOF quod est
impossibile C1/LOA ... reflexionis (227) *om.* P1   227 reflexionis: refractionis R/*post* reflexionis
rep. et del. angulus (221) ... reflexionis (227) F/minor: maior SFP1OER; *corr. ex* maior L3

AOF, quod est impossibile. Ergo impossibile est quod punctum N sit ymago puncti B, nec aliud punctum ab M; ergo punctum B respectu visus A nullam habet ymaginem preter quam punctum M, et hoc de-clarare debuimus.

[5.38] **[PROPOSITIO 4]** Et iterum sit corpus grossius ex parte visus et subtilius ex parte rei vise, et sit differentia communis inter hanc superficiem et superficiem corporis diaffoni linea GD [FIGURE 7.5.4, p. 441]. Extrahamus ex B perpendicularem super lineam GD, et sit BK. Erit ergo BK perpendicularis super superficiem corporis diaf-foni. Et reflectatur forma B ad A ex E, et continuemus BE, EA. Et extrahamus perpendicularem HE, et extrahamus BE recte ad T.

[5.39] Erit ergo AE linea media inter duas lineas ET, EH, nam pri-ma linea per quam extenditur forma ad locum reflexionis est linea BET. Reflexio autem est ad partem perpendicularis EH, nam corpus quod est ex parte A est grossius illo quod est ex parte B; linea ergo AE est media inter duas lineas ET, EH. Et extrahamus directe AE ad partem E quousque occurrat linee BK, secabit enim HEZ. Occurret ergo illi in puncto M. M ergo erit ymago puncti B, nam corpus quod est ex parte B est subtilius illo quod est ex parte A. Dico ergo quod B non habet ymaginem nisi M.

[5.40] Habeat ergo N, si possibile est. N ergo erit in perpendiculari BK et infra punctum B, quia corpus quod est ex parte B est subtilius illo quod est ex parte A. Est ergo aut inter duo puncta M, B aut infra M. Et continuemus AN. Secabit ergo lineam GD in O; O ergo est punctum reflexionis. Et continuemus BO; et transeat usque ad L et extrahamus ex O perpendicularem FOQ. Linea ergo BO est illa per quam extenditur forma ad locum reflexionis; ergo linea AO erit inter duas lineas OL, OF, reflexio enim est ad partem perpendicularis.

228 AOF: AOB *alter. ex* AOFB *O*/ergo impossibile *inter.* L3O (*a. m.* L3)/est² *om.* L3/quod: ut *R* 229 nec: neque *R*/aliud: alium *FP1*/ab: est preter *R*/M *inter.* O/B² *inter.* S   231 debuimus: voluimus *R*   233 *post* parte *scr. et del.* centri *P1*   235 *ante* extrahamus *add.* et *FP1ER*/*post* B *add.* lineam *R*/GD: DG *C1; corr. ex* AG *E*   236 super *rep.* S   237 reflectatur: re-fringatur *R*/*post* continuemus *add.* lineas *ER*   238 *post* extrahamus¹ *scr. et del.* ex B *C1*/*post* perpendicularem *scr. et del.* perpendicularis *E*/T: P *R*   239 AE *corr. ex* DE *O*/*post* linea *scr. et del.* DO *O*/lineas *om.* P1/ET: et SO; EP *R*   240 forma *mg.* F/reflexionis: refractionis *R* 241 BET: BEP *R*/reflexio: refractio *R*/autem: enim *ER*   242 A: E *O; corr. ex* E L3   243 AE¹ *inter. a. m.* L3/*post* est *scr. et del.* mdia *P1*/ET: EP *R*/directe AE *transp.* R   244 E *corr. ex* AE *F*/secabit: secat *R*/enim *scr. et del.* L3/HEZ: EHZ *SFO; corr. ex* EHZ L3/occurret: occurrit *C1*; occurrat *R*   245 illi *mg.* F   246 B¹ *inter.* E/B¹ . . . illo: visus P1   248 ergo *rep.* SE; enim *R*/est *inter.* O/ergo erit *transp.* L3   249 ex: in *R*/*post* B² *scr. et del.* est subtilius illo quod est ex parte B P1   251 M: AN S/GD: DG *R; corr. ex* AG E   252 reflexionis: refractionis *R*   253 *post* est *add.* in *FP1*/illa: linea *C1ER*/per . . . linea (254) *mg. a. m.* E 254 reflexionis: refractionis *R*/AO: OA *C1E*; OST S/inter *corr. ex* in *C1*   255 OL *corr. ex* OB *C1*/reflexio: refracti *R*/*post* enim *scr. et del.* N *E*/partem: partes *E*

[5.41] Si ergo N fuerit inter duo puncta M, B, tunc punctum O erit inter duo puncta E, K. Angulus ergo OBK est minor angulo EBK; ergo angulus LOF est minor angulo TEH. Ergo angulus LOA, qui est angulus reflexionis, est minor angulo TEA, qui est angulus reflexio-
260   nis. Et angulus AOF, qui remanet post angulum reflexionis, est minor angulo AEH, qui remanet post angulum reflexionis, sicut declaravimus in tertio capitulo huius tractatus. Sed angulus AOF est equalis angulo ANK, et angulus AEH est equalis angulo AMK; ergo angulus ANK est minor angulo AMK, quod est impossibile.

265   [5.42] Si autem N fuerit infra M [FIGURE 7.5.4a, p. 442], tunc E erit inter duo puncta O, K, et erit angulus OBK maior angulo EBK; angulus ergo LOF erit maior angulo TEH. Ergo angulus LOA est maior angulo TEA. Et angulus AOF est maior angulo AEH; ergo angulus ANK est maior angulo AMK, quod est impossibile. N ergo non
270   est ymago B, nec aliud punctum preter quam M; B ergo non habet ymaginem nisi M, et hoc est quod voluimus.

        [5.43] [**PROPOSITIO 5**] Ad duas autem lineas circulares convexam et concavam premittemus hoc: cum due corde sese secuerint in circulo, angulus sectionis erit equalis angulo qui est apud circumferen-
275   tiam quem cordant duo arcus quos distinguunt ille due corde, et si due linee secuerint circulum et secuerint se extra circulum, angulus sectionis erit equalis angulo qui est apud circumferentiam quem cordat excessus maioris illorum duorum arcuum quos distinguunt ille due linee super reliquum.
280   [5.44] Verbi gratia, in circulo ABG [FIGURE 7.5.5, p. 442] secent se due corde AG, BD in E. Dico ergo quod angulus AEB est equalis

256 N fuerit *transp. ER* / duo puncta *transp. F* / tunc *inter. a. m. L3*   257 angulus ergo *transp. C1*; ergo erit angulus *R* / est *om. R*   258 ergo[1] . . . TEH *mg. O* / TEH: DEH *FP1*; PEH *R* / ergo[2] . . . LOA *inter. a. m. E*   259 reflexionis[1,2]: refractionis *R* / est[1] . . . reflexionis[2] *mg. a. m. E* / TEH: DEH *FP1*; PEH *R*   260 reflexionis: refractionis *R* / post est *scr. et del.* in parte *P1*   261 reflexionis: refractionis *R* / sicut . . . tractatus (262) *om. R*   263 ANK *corr. ex* AMK *L3O* / et . . . AMK (264) *inter. a. m. L3* / AEH: AEF *P1*   264 ANK *corr. ex* AMK *C1* 265 autem *inter. O* / post autem *scr. et del.* EM *S* / E *corr. ex* M *O* / E erit (266) *transp. ER*   266 inter *om. L3* / maior: minor *L3O* (*inter. O*) / angulo EBK *inter. O* / post EBK *add.* ergo angulus ANK est maior angulo AMK *SFP1*   267 erit: est *L3O* / maior: minor *O*; *corr. ex* minor *L3* / TEH: PEH *R*   268 *post* angulo[1] *scr. et del.* AEH ergo angulus ANK *O* / TEA: PEA *R* / TEA . . . AOF *inter. O* / angulo[2] *om. SC1* / AEH: AEF *P1*; *corr. ex* AMK *O* / ergo . . . AMK (269) *om. SFO*   269 *ante* ANK *scr. et del.* AMK *P1*   270 *post* punctum *scr. et del.* MB *O* / M: E *E* / ergo *om. S* / post ergo *scr. et del.* erit non est *P1*   271 M: N *L3O* / voluimus: volumus *C1* / post voluimus *add.* declarare *C1R*   273 *post* premittemus *scr. et del.* ergo *L3* / hoc: hec *L3R*   274 *post* apud *scr. et del.* circulos *P1*   275 quem: qui *FP1*; quam *ER* / duo: duos *O*; ii *L3* / duo . . . cordat (277) *mg. a. m. L3* / quos: qu *F*   276 secuerint[1] . . . et *om. S* / et . . . circulum[2] *mg. C1*   277 quem: quam *ER*   278 *ante* excessus *add.* continet *L3O* (*deinde del. L3*)   279 super: supra *R*   280 ABG: ABCD *R* / secent: secabunt *L3O* 281 AG: AC *R* / BD: HD *S*

angulo qui est in circumferentia quem respiciunt duo arcus AB, GD, et quod angulus BEG est equalis angulo in circumferentia quem respiciunt duo arcus AD, GB.

285   [5.45] Probatio huius: extrahemus ex B lineam HBZ equidistantem linee AG. Arcus ergo GZ est equalis arcui AB, et arcus GD est communis; ergo arcus DZ est equalis duobus arcubus AB, GD. Sed arcus DZ respicit angulum DBZ; ergo DZ respicit arcus equales duobus arcubus AB, GD. Et angulus DBZ est equalis angulo AEB; ergo
290 angulus AEB equalis angulo qui est in circumferentia quem respiciunt duo arcus AB, GD, et hoc est quod voluimus.

    [5.46] Item continuemus DZ. Erit ergo angulus HBE equalis duobus angulis BDZ, BZD, et duo anguli BZD, BDZ respiciuntur a duobus arcubus DB, BZ; angulus ergo HBE est equalis angulo quem respicit
295 arcus DB, BZ. Et arcus AB est equalis arcui ZG, et arcus DABZ est equalis duobus arcubus DA, BG; ergo angulus HBE est equalis angulo quem respiciunt duo arcus DA, BG. Et angulus HBE est equalis angulo BEG; ergo angulus BEG est equalis angulo qui est in circumferentia quem respiciunt duo arcus DA, BG, et hoc est quod voluimus
300 declarare.

    [5.47] Et si linea HBZ fuerit contingens circulum [FIGURE 7.5.5a, p. 442], tunc angulus EBZ erit equalis angulo cadenti in portione BAD, et sic arcus BGD respicit angulum apud circumferentiam equalem angulo EBZ. Et angulus EBZ est equalis angulo BEA. Ergo angulus
5 BEA est equalis angulo qui est apud circumferentiam quem respicit arcus BGD, et arcus BG est equalis arcui BA, quia diameter qui exit ex B est perpendicularis super lineam AG, quare dividit ipsum in duo

282 in inter. L3; om. O; apud R/circumferentia: circumferentiam R/quem: quam ER/GD: GB O; CD R   283 et . . . GB (284) mg. a. m. E/BEG: BEH O; BEC R/post angulo scr. et del. qui est O/quem: quam R   284 AD GB: DGA BZC R; AGB E   285 probatio huius om. R/ extrahemus: extrahamus L3OER/post extrahemus add. enim R; scr. et del. B P1/ex B corr. ex E B O/HBZ: BZ R   286 AG: AC R/GZ: CZ R/est¹ om. S/GD: CD R   287 DZ: DGZ O/duobus arcubus transp. L3; om. P1/GD: CD R   288 DZ¹: DGZ C1/DBZ inter. O/DZ²: BZ P1C1O (alter. ex DZ a. m. C1)/arcus om. L3/duobus arcubus (289): duorum arcuum L3OE (alter. ex duobus arcubus a. m. E)   289 GD: CD R/post GD scr. et del. a O/AEB corr. ex AEEB O/AEB . . . angulo (290) inter. a. m. L3   290 post AEB add. est P1L3R/post equalis add. est C1E/quem: quam R   291 GD: CD R   292 item continuemus inter. O/post DZ add. et producamus ZB in H R/HBE: HBD R   293 angulis: arcubus O; corr. ex arcubus a. m. L3/BZD² inter. a. m. L3/post BZD² add. et FP1/BZD, BDZ: BDZ ZD O/respiciuntur: respiciunt F/a duobus om. S   294 DB BZ: BGD BFZ R/HBE: HBD R   295 DB: DA SP1L3C1O (alter. ex DB L3)/DB BZ: DBZ R/ZG: ZC R/et: ergo R/DABZ: DBZ R/post DABZ add. et arcus AB C1   296 post equalis¹ scr. et del. a C1/DA: AD C1; DGA R/BG: BZC R/angulus corr. ex arcus O   297 DA BG: DGA BZC R   298 BEG¹,²: BEC R/ergo . . . BEG² om. P1   299 quem: quam R/DA BG: DGA BZC R   1 et om. C1/fuerit om. SR/contingens: contingat R   2 erit: fuerit P1/portione: proportione L3O/BAD: HAD P1   3 BGD: BCD R/respicit corr. ex respiciet E; corr. ex respiciit F   4 EBZ¹: ABZ O/et . . . EBZ² om. S/ergo . . . BEA (5) om. F   5 BEA inter. E   6 BGC: BCD R/BG: BGE S; BC R; corr. ex G C1/qui: que R   7 AG: AC R/ipsum: ipsam SC1R

equalia. Ergo arcus AB erit equalis arcui BG; arcus ergo BGD erit equalis duobus arcubus BA, GD. Ergo angulus BEA est equalis angu-
10 lo qui est apud circumferentiam quem respiciunt duo arcus AB, GD. Et similiter declarabitur quod angulus BEG est equalis angulo qui est apud circumferentiam quem respiciunt duo arcus BG, AD, et hoc est quod voluimus.

[5.48] Item sit E extra circulum ABGD [FIGURE 7.5.5b, p. 443], et
15 extrahamus ex E duas lineas secantes circulum ABGD, et sint EAD, EBG. Dico ergo quod angulus GED est equalis angulo qui est apud circumferentiam quem respicit excessus arcus DG super arcum AB.

[5.49] Huius demonstratio: extrahamus lineam equidistantem linee BG. Erit ergo arcus ZG equalis arcui AB; erit ergo arcus DZ
20 excessus arcus DG super arcum AB. Sed arcus DZ respicit angulum DAZ, et angulus DAZ est equalis angulo GED. Ergo angulus GED est equalis angulo qui est apud circumferentiam DAZ, et hoc est quod voluimus.

[5.50] **[PROPOSITIO 6]** Hiis declaratis, sit visus punctum A [FIG-URE 7.5.6, p. 443], et sit punctum B in aliquo viso, et sit ultra corpus
25 diaffonum grossius corpore quod est in parte visus. Et sit superficies corporis diaffoni quod est ex parte B superficies circularis convexa ex parte visus. Ergo per duo puncta A, B transit superficies perpendicularis super superficiem corporis diaffoni, et non transit per illa superficies perpendicularis super superficiem corporis diaffoni in
30 qua reflectitur forma B ad A nisi una tantum. Hanc ergo superficiem corporis diaffoni signet circulus GED, cuius centrum quidem sit Z, et continuemus AGD. Linea ergo GZD erit perpendicularis super superficiem corporis diaffoni; punctum autem B aut erit extra lineam GD aut in ipsa.

35

8 erit equalis *transp. R*/BG: BC *R*/BGD: BCD *R*/erit²: est *R*   9 BA GD: AB CD *R*   10 AB . . . arcus (12) *mg. a. m.* E; *rep. in mg. a. m.* O/GD: CD *R*   11 BEG: BEC *R*   12 BG: BC *R*/BG AD *transp.* L3/*post* AD *rep. et del.* et (11) . . . AD (12) L3   13 *post* quod *add.* declarare *C1*   14 E *inter.* F/ABGD: ABCD *R*/et . . . ABGD (15) *mg. F; om.* S   15 extrahamus: extrahemus *C1OE*/ ABGD: ABCD *R*/sint: sunt L3   16 EBG: ABG *S*; EBC *R*/ergo *om. P1*/GED: CED *R*   17 *post* circumferentiam *add.* circuli *R*/excessus arcus *transp.* ER/DG: DC *R*   18 huius demonstratio *om. R*/extrahamus: extrahemus L3C1OE/*post* extrahamus *add.* enim *R*   19 BG: BC *R*/ZG: FC *R*/DZ: DF *R*   20 DG: DC *R*/super: supra *R*/AB: AD *FP1C1*/DZ: DF *R*   21 DAZ¹²: DAF *R*/et . . . DAZ² *mg. C1*/GED¹²: CED *R*/GED¹ . . . angulo (22) *inter. a. m.* L3/angulus² *om. R*   22 *post* est² *scr. et del.* non *P1*/DAZ: DAG *SFP1*/DE *O*; DF *R*/et *om.* S   24 *post* hiis *add.* ergo *FP1E*/A *inter.* O   26 est *rep. C1*/sit *om. C1*   29 *post* diaffoni *add.* nam L3O/et: si L3O (*deinde del.* L3)/illa: illas E   30 diaffoni *om. R*   31 qua: quo *O*/reflectitur: refringitur *R*/nisi . . . ergo *inter. a. m.* L3; et *inter* O   32 signet *inter. a. m.* L3; *om. O*/GED: CED *R*/quidem: quod *O; om. R*   33 AGD: ACZD *R*/GZD: CZD *R*   34 autem: ergo L3/lineam *om.* L3O   35 GD: GB *S*; CD *R*/aut in ipsa *om. O*/aut . . . ergo (36) *mg. a. m.* L3

[5.51] Si ergo B fuerit in linea GD, tunc visus A comprehendet B recte et sine reflexione, nam forma que extenditur per lineam GD extenditur recte in corpore diaffono quod est ex parte visus A, quia linea GD est perpendicularis super superficiem corporis diaffoni quod
40 est ex parte visus. Visus ergo A comprehendit B in suo loco et recte. Dico ergo quod forma B quod est in linea GD numquam reflectitur ad A.

[5.52] Huius demonstratio, quoniam punctum B aut erit in centro aut extra centrum. Si ergo fuerit in centro, tunc omnis linea per
45 quam extenditur forma B ad circumferentiam GED in rectitudine eius extenditur in corpore diaffono quod est ex parte eius, nam omnis linea exiens a centro circuli GED est perpendicularis super superficiem corporis diaffoni, et non exit a centro circuli GED ad visum A linea recta nisi linea ZA. Ergo forma B que est in centro non reflectitur ad
50 A ex circumferentia GED; ergo forma B numquam reflectitur ad A, si B fuerit in centro.

[5.53] Si vero fuerit extra centrum, aut erit in linea ZG aut in ZD. Sit ergo primo in linea ZG. Dico quod forma B non reflectitur ad A, quod si fuerit possibile, reflectatur ex puncto E. Et continuemus BE,
55 et extrahamus illud ad H, et continuemus ZE, et extrahamus ipsum ad T. Erit ergo linea ZET perpendicularis super superficiem corporis diaffoni quod est ex parte visus. Forma ergo B, quando extenditur ad lineam BE et reflectitur in puncto E, transit a perpendiculari TE ad partem H ad partem contrariam illi in qua est perpendicularis.
60 Forma ergo B non perveniet ad A secundum reflexionem, si B fuerit in linea ZG.

[5.54] Item sit B in linea DZ [FIGURE 7.5.6a. p. 443]. Dico ergo

36 ergo B transp. SC1OE/B: KB L3/in inter. S/GD: CD R/comprehendet: comprehendit FP1O
37 reflexione: refractione R/GD: GB F; CD R   38 A om. R   39 GD: CD R/super om. P1
40 visus^2 inter. a. m. L3/comprehendit: comprehendet SC1E   41 post forma add. puncti R/
quod: que L3/est om. E/linea GD: CD linea R/reflectitur: refringitur R   43 huius demonstratio om. R/post demonstratio add. est FP1   44 centrum: centra S; om. L3/post fuerit add.
B FP1 (mg. F)/in . . . forma (45) mg. a. m. E/post linea scr. et del. que C1/per . . . extenditur (45)
corr. ex extenditur per quam C1   45 post B scr. et del. ad L3/GED: CED R/in . . . extenditur
(46): extenditur in rectitudine eius R   46 eius: visus R   47 exiens: extensa FP1/GED:
CED R   48 diaffoni inter. a. m. E/GED: CED R   49 que: quod R/in rep. C1/reflectitur:
refringitur R   50 GED: CED R/reflectitur: refringitur R/A^2 om. L3   52 in^1 inter. a. m.
E/ZG: ZC R   53 sit: si O/ZG: ZC R/reflectitur: refringitur R/ad . . . reflectatur (54) mg.
a. m. E   54 reflectatur: refringatur R/puncto: ipso ER/post BE scr. et del. et extrahamus ip-
sum F   55 extrahamus: extrahemus S/ipsum: ipsam R; corr. ex ipsam F   56 T: P R; TG
S/ZET: ZEP R   57 est om. S/ex: in FP1/quando: quoniam S; corr. ex que C1   58 BE: AO
O/et inter. L3; om. SO/reflectitur: reflectetur L3; refringitur R/E: GED O; corr. ex ED L3/TE: PE
R; T L3O; DG S; TG P1; alter. ex DE in TG F   59 ad partem^2 om. R   60 forma ergo transp.
FP1/perveniet corr. ex pervenit F/secundum: per L3/reflexionem: refractionem R/si . . . dico
(62) mg. C1   61 ZG: GZ L3C1; ZC R

quod forma B non reflectitur ad A, quod si est possibile, reflectatur
ex E. Et continuemus BE, et extrahamus lineam ad R, et continuemus
65 ZE, et extrahamus lineam usque ad T. Et reflectatur forma B ad A
per lineam EA. Sic ergo angulus REA erit angulus reflexionis; angulus autem RET erit angulus quem continet linea per quam extenditur
forma et perpendicularis exiens a loco reflexionis. Angulus ergo REA
est minor angulo RET, et linea BZ aut minor linea ZE aut equalis ei,
70 nam B aut est inter duo puncta D, Z aut in puncto D. Ergo angulus
EBZ aut est maior angulo BEZ aut equalis ei. Sed angulus AER est
maior angulo EBZ; ergo angulus AER est maior angulo BEZ. Ergo
angulus AER est maior angulo RET, quo prius erat minor, quod est
impossibile.

75 [5.55] Ergo forma B non reflectetur ad A ex E, nec ex alio puncto
circumferentie GED, neque ex alia circumferentia circulorum qui fuerit in superficie corporis diaffoni in quo est B. B ergo existente in linea
GD non comprehenditur a visu reflexive, quare non comprehenditur
nisi unum punctum solum.

80 [5.56] Item sit B extra lineam GD [FIGURE 7.5.6b, p. 444], et extrahemus superficiem in qua est perpendicularis AD et punctum B.
Hec ergo superficies erit perpendicularis super superficiem corporis
diaffoni, et punctum B non reflectitur ad A nisi in hac superficie, non
enim transit per duo puncta A, B superficies perpendicularis super
85 superficiem corporis diaffoni nisi illa que transit per lineam AD, et
non exit ex linea AD superficies que transit per B nisi una tantum.
Hec ergo superficies signet in superficie corporis diaffoni circulum
GED. Forma ergo B non reflectetur ad A nisi ex circumferentia GED;
reflectatur ergo ex E. Dico ergo quod non reflectitur ex alio puncto

62 item . . .DZ *inter. a. m.* L3/DZ: BZ L3/ergo *om.* FP1   63 *ante* B *scr. et del.* puncti P1/reflectitur: refringetur R/reflectatur: refringatur R; reflectitur SL3C1O; *alter. ex* reflectitur *in* reflectetur E   64 E: K
S/*post* E *scr. et del.* ZE C1/extrahamus: extrahemus C1O/*post* extrahamus *add.* BE R/ad . . . lineam (65)
mg. a. m. E/R: T C1   65 ZE: ZD O/*corr. ex* ZD *a. m.* L3/extrahamus: extrahemus C1L3OE/lineam
om. L3/T: P R/reflectatur: reflectetur S; refringatur R   66 sic *corr. ex* sit F/REA: KEA P1/reflexionis:
refractionis R   67 RET: KET P1; REP R/extenditur forma (68) *transp.* L3   68 et *om.* P1/reflexio-
nis: refractionis R   69 RET: KET P1; REP R/*post* aut[1] *add.* est R   71 aut[1]: autem F/est[1] *om.* F/est[1]
. . . aut[2] *inter. a. m.* L3/*post* aut[2] *scr. et del.* est F/AER: ER F   72 *post* angulo [1] *add.* BEZ O/EBZ: BEZ E;
corr. ex BEZ L3/BEZ . . . angulo (73) *inter. a. m.* L3; *mg.* O; *om.* ER   73 RET: REP R   75 reflectetur:
reflectitur E; refringetur R   76 GED: CED R/circumferentia *corr. ex* circumferentie L3/fuerit: fiunt
FP1R; fuerint C1   77 in[2] *inter.* E/B ergo *transp.* R/ergo *om.* E   78 GD: CD R/comprehenditur[1]:
comprehendetur R/*post* comprehenditur[1] *add.* ipsum R/a . . . comprehenditur[2] (79) *inter. a. m.* L3/re-
flexive: per refractionem R; *corr. ex* reflexione *a. m.* E   79 punctum solum *transp.* C1R   80 item:
iterum S/GD: CD R/extrahemus: extrahemus R; *corr. ex* extrahamus P1C1   81 AD *om.* R/AD . .
. perpendicularis (82) *mg. a. m.* L3   83 reflectitur: refringetur R/*post* hac *add.* in S   84 B: D L3
85 diaffoni *om.* P1/*post* diaffoni *rep. et del.* non[2] (83) . . . superficies (84) O   86 ex linea: per lineam
C1/superficies: superficiem L3O   88 GED[1,2]: CED R/forma ergo *transp.* L3/reflectetur: reflectitur
L3OE; refringetur R/*post* ad *scr. et del.* nisi P1

90 quam E.

[5.57] Reflectatur ergo, si possibile est, ex alio puncto, qui, ut dictum est, erit in circumferentia GED. Sit ergo M. Et continuemus lineas BE, EA, BM, MA, ZE, ZM, et secent se linee BM, ZE in C. Et extrahamus BE usque ad H, et BM ad N, et EZ ad T, et ZM ad L.
95 Erit ergo angulus HET ille quem continet linea per quam extenditur forma et perpendicularis exiens a loco reflexionis, et angulus HEA erit angulus reflexionis, et NML angulus ille quem continet linea per quam extenditur forma et perpendicularis exiens a loco reflexionis, et angulus NMA erit angulus reflexionis.

100 [5.58] Angulus HET aut erit equalis angulo NML, aut erit minor, aut maior. Si equalis, angulus HEA, qui est angulus reflexionis, erit equalis angulo NMA, qui est angulus reflexionis. Angulus ergo AMB erit equalis angulo AEB, quod est impossibile. Si minor, erit angulus HEA minor angulo NMA; angulus ergo AMB erit minor angulo AEB,
105 quod est impossibile.

[5.59] Si maior, extrahemus lineam EB in partem B ad F, et extrahemus MB usque ad O. Angulus ergo EBM erit equalis angulo qui est apud circumferentiam quem respiciunt duo arcus EM, FO, et cum angulus HET fuerit maior angulo NML, erit angulus ZEB maior
110 angulo NML. Et cum angulus ZEB fuerit maior angulo NML, angulus MZT erit maior angulo MBE, et excessus anguli MZE super angulum MBE erit equalis excessui anguli ZEB super angulum ZMB, nam duo anguli apud C sunt equales. Arcus ergo qui respicit angulum MZE, cum fuerit apud circumferentiam, erit duplus ad arcum ME.

115 [5.60] Si ergo angulus MZE fuerit maior angulo MBE, tunc arcus

---

89 reflectatur: reflectetur *FP1*; reflectitur *L3O*; refringatur *R*/reflectitur: reflectetur *SO*; refringetur *R*/ex²: ab *C1*   91 reflectatur: refringatur *R*/ergo: enim *R*/qui: quod *R*   92 in *om. S*/GED: CED *R*   93 *post* MA *add.* et *L3*/ZE¹: ZEK *O*/ ZE ZM *alter. ex* EK ZM *in* EZ *L3*/ZM *om. SFP1*/BM² *corr. ex* GM E/ZE² *inter. a. m. L3; om. O*/*post* in *add.* puncto *R*/C: G *R*   94 extrahamus: extrahemus *FP1C1E*/EZ *om. S*/T: P *R*/L *corr. ex* B *O*   95 ergo *inter. E*/ HET: HEP *R*   96 reflexionis: refractionis *R*   97 reflexionis: refractionis *R*/et *inter. a. m. L3; om. O*/et . . . NMA (99) *om. F*/NML: unus *S*/NML angulus *transp. P1*/*ante* ille *add.* erit *P1* 98 et¹ *inter. C1*/reflexionis: refractionis *R*   99 NMA: MNA *S*/reflexionis: refractionis *R* 100 *post* angulus *add.* igitur *R*/HET: HEP *R*/NML: MNL *SE*   101 qui . . . equalis (103): que est angulus reflexionis angulus ergo AMB erit equalis qui est angulus reflexionis erit equalis angulo NMA *O*/reflexionis: refractionis *R*   102 reflexionis: refractionis *R*   104 angulo¹: aut *L3*/NMA: MNA *S*   106 extrahemus¹: extrahamus *SR*/extrahemus²: extrahamus *SL3C1OR*   107 O: A *SO*/EBM: EMB *SFP1ER*   108 quem: quam *S*; que *F*/duo: duos *L3* 109 angulus¹ *om. FP1*/HET: HEP *R*/fuerit: sit *R*/NML *corr. ex* NMB *O*   110 NML¹: MNL *S*; NMB *O*; *corr. ex* NMB *L3*/fuerit: sit *R*; *mg. C1*/NML²: MNL *SL3*; NMB *O*/*post* NML *add.* et *FP1*   111 MZT: MZP *R*/super: supra *R*/*post* angulum *rep. et del.* angulum *F*; *scr. et del.* reflexionis *P1*   112 anguli *om. S*/ZEB: ZBE *SL3OE*/super: supra *R*/ZMB: ZBM *F*; ZNB *L3O*   113 C: G *R*; *corr. ex* E *O*/ergo: vero *R*/*post* qui *add.* sunt *P1*/MZE *alter. in* ZME *a. m. E* 114 ME: MT *SO*; *corr. ex* MT *a. m. L3*

ME duplicatus erit maior duobus arcubus ME, FO. Et erit excessus arcus ME duplicati super duos arcus ME, FO equalis excessui arcus ME super arcum FO. Excessus ergo anguli MZE super angulum MBE est ille quem respicit apud circumferentiam excessus arcus ME super
120 arcum FO. Sed excessus arcus ME super arcum FO est minor duobus arcubus ME, FO; ergo excessus anguli MZE super angulum MBE est minor angulo MBE. Ergo excessus anguli ZEB super angulum ZMB est minor angulo MBE. Ergo excessus anguli HET super angulum NML est minor angulo MBE. Ergo excessus anguli HEA, qui est angu-
125 lus reflexionis, super angulum NMA, qui est angulus reflexionis, est multo minor angulo MBE.

[5.61] Sed excessus anguli HEA super angulum NMA est excessus anguli AMB super angulum AEB; ergo excessus anguli AMB super angulum AEB est minor angulo MBE. Sed excessus anguli AMB su-
130 per angulum AEB est duo anguli MAE, MBE. Ergo duo anguli MAE, MBE sunt minores angulo MBE, quod est impossibile. Forma ergo B non reflectetur ad A ex alio puncto preter quam ex E, et hoc est quod voluimus.

[5.62] **[PROPOSITIO 7]** Cum ergo forma B non reflectitur ad A nisi
135 ex uno puncto, non habebit nisi unam ymaginem. Sed locus ymaginis diversatur secundum diversitatem loci in quo est B. Continuemus ergo BZ [FIGURE 7.5.6b]. Linea ergo BZ aut concurret cum linea EA, aut erit ei equidistans, et concursus aut erit in parte EB, ut in K, aut in parte A, ut in R. Et cum BZ fuerit equidistans linee EA, erit ut
140 linea BZ, que est media inter duas lineas KBZ, BZR.

[5.63] Si ergo concursus harum duarum linearum fuerit in K, erit

116 duplicatus: duplatus *FP1*   117 duplicati: duplati *FP1*/super: supra *R*; *corr. ex* per *E*/*post* FO *scr. et del.* et erit excessus *S*/*ante* excessui *add* cum (?) *F*   118 super[1,2]: supra *R*/excessus *mg. C1*/MBE: MFE *O*   119 ille: iste *R*/ME *om. P1*/super: supra *R*   120 super: supra *R*   121 FO: FE *P1*/super: supra *R*/*post* super *scr. et del.* si *S*/*post* MBE *rep. et del.* est (119) . . . MBE (121) *L3*   122 ergo excessus *transp. R*/ergo . . . MBE (123) *om. E*/super: supra *R*   123 est *corr. ex* etiam *C1*/angulo *inter. O*/MBE: MBZ *FP1*/HET: BOT *OE*; HEP *R*/super: supra *R*/*post* angulum *scr. et del.* rectum *C1*   124 *post* HEA *scr. et del.* super angulum ?? est minor angulo MBE ergo excessus anguli HEA *E*   125 reflexionis[1,2]: refractionis *R*/super: supra *R*/super . . . reflexionis[2] *rep. FP1*; *inter. a. m. L3*; *om. O*/est[2] *om. FP1*   126 MBE *inter. O*   127 super: supra *R*   128 super[1,2]: supra *R*/ergo . . . AEB (129) *rep. L3*   129 est . . . AEB (130) *om. S*; *scr. et del. L3*/minor . . . est (130) *mg. a. m. E*/MBE: MDE *E*/super: supra *R*   130 est: sunt *R*/ergo . . . MBE[1] (131) *om. S*   131 B *inter. L3*   132 reflectetur: reflectitur *FP1L3E*; refringetur *R*/ex[1] . . . A (134) *om. L3*   134 forma *om. R*/reflectitur: refringatur *R*   135 non: nec *R*/habebit: habebis *SFP1C1O*/nisi *om. E*   136 quo: qua *S*   137 ergo[1]: enim *R*; *inter. E*/linea ergo BZ *mg. F*/BZ *om. P1*   138 erit[1] *om. P1*/ei equidistans *transp. P1*   139 ut[1] *corr. ex* aut *L3O*/ut[2]: aut *L3O*; *deinde corr. L3*   140 que *om. ER*; *scr. et del. L3*/est: sit *L3ER* (*alter. ex* est *a. m. L3*)

ymago ante visum, et erit forma manifesta et comprehensa a visu in
K. Si vero concursus fuerit in R, erit ymago punctum R, et tunc forma
comprehendetur a visu in eius oppositione, sed non manifeste, ta-
145   men quia comprehenditur a visu extra suum locum. Hoc autem declaratum est in loco in quo locuti sumus de reflexione. Et si linea BZ
fuerit equidistans linee EA, tunc ymago erit indeterminata, et forma
comprehendetur in loco reflexionis. Huius autem causa similis est
illi quam diximus in loco reflexionis, cum fuerit reflexio per lineam
150   equidistantem perpendiculari.

    [5.64] Ex predictis ergo patet quod res que comprehenditur a visu
ultra corpus diaffonum grossius corpore quod est ex parte visus non
habet nisi unam ymaginem, neque comprehendetur nisi unum tantum. Hec vero reflexio est a concavitate corporis diaffoni quod est
155   ex parte visus contingentis convexum corporis diaffoni quod est ex
parte rei vise, et hoc est quod voluimus.

    [5.65] **[PROPOSITIO 8]** Et si corpus diaffonum grossius fuerit ex
parte visus et subtilius ex parte rei vise, tunc visus non habebit nisi
unam solam ymaginem, nam tunc visus erit ut B [FIGURE 7.5.6b] et
160   res visa ut A, et cum forma A reflectetur ad B, reflexio erit in superficie perpendiculari super superficiem corporis diaffoni, et erit differentia communis inter illam superficiem et superficiem corporis diaffoni circulus ut circulus GED. Et erit punctus reflexionis ut E, et erit
linea reflexa ut EH.
165   [5.66] Sequitur ergo ut forma que extenditur per lineam AE et reflectetur per BE, quando extenditur ex B per lineam BE quod reflectitur per lineam AE. Si ergo forma A reflectatur ad B ex alio puncto
quam ex E, sequitur quod forma B reflectatur ad A ex illo puncto.
Sed iam declaratum est quod, cum forma extensa fuerit per lineam

141 ergo: vero R / concursus: concusus L3 / duarum *inter.* E / duarum linearum *transp.* E / fuerit: fuit S
142 et² *om.* C1   143 *post* K *add.* et C1 / *post* concursus *scr. et del.* harum S / in *inter* E; *mg. a. m.* L3; *om.*
SFC1O / *post* erit *add.* R O   144 *post* non *add.* tam P1L3C1R (*inter. a. m.* L3) / manifeste tamen *transp.*
E / tamen *corr. ex* tam E ; *om.* P1L3C1   145 quia *inter. a. m.* L3 / comprehenditur *alter. in* comprehendetur E / extra: ex S / hoc autem: et hoc C1   146 reflexione: ratione S / et *om.* ER   147 indeterminata: indeminata S   148 reflexionis: refractionis R / est *inter. a. m.* L3   149 reflexionis: refractionis
R / reflexio: ratio S   151 ex: et C1   153 comprehendetur: comprehenditur R / unum: una P1
154 reflexio: refractio R / quod *om.* ER / est² *om.* R   156 et . . . voluimus *mg. a. m.* F.   157 grossius
fuerit *transp.* ER   158 subtilius . . . vise: ex parte rei vise subtilius C1 / habebit: videbit R   160 A
reflectetur *inter. a. m.* E / reflectetur: refringetur R / reflexio: refractio R   161 perpendiculari: perpendicularis OE   162 et superficiem *mg. a. m.* E   163 GED: CED R / punctus reflexionis: punctus
refractionis R   164 reflexa: refracta R / EH: AEK SFP1L3ER   165 sequitur: sequetur FP1 / extenditur: extendetur R / reflectetur: refringetur R   166 quando: quam F; quoniam OE; *corr. ex* quoniam
L3; *om.* R / extenditur: extendetur C1E; extandatur R / *post* extenditur *scr. et del.* per B P1 / per lineam BE
om. FP1 / quod: et R / reflectitur: reflectetur C1E; refringatur R   167 A *inter.* E / reflectatur: reflectitur
E; refringitur R   168 sequitur: sequetur R / quod *inter. a. m.* E / reflectatur: refringetur R

170   BE et reflexa per lineam AE, numquam reflectetur ex B alia forma ad
       A, quare A non reflectetur ad B nisi ex uno puncto, nec habebit nisi
       unam ymaginem. Et si A fuerit in perpendiculari exeunti ex B ad cen-
       trum spere, tunc B comprehendet A in rectitudine perpendicularis, et
       patet quod forma A non reflectetur ad B, ex quo patuit quod forma B,
175   cum fuerit in perpendiculari, non reflectetur ad A. Cum ergo gros-
       sius corpus fuerit ex parte visus et subtilius ex parte rei vise, tunc res
       visa non habebit nisi unam ymaginem et unam formam tantum, et
       hoc voluimus.

       [5.67] **[PROPOSITIO 9]** Item, iteremus figuram, ponentes in cir-
180   cumferentia GED punctum ex parte G [FIGURE 7.5.7, p. 444]. Et sit
       E, ex quo extrahemus lineam equidistantem linee AD, et sit linea
       ET. Et continuemus ZE, et extrahamus illam usque ad H. Et sit pro-
       portio anguli ZEK ad angulum KET duplicatum maxima proportio
       quam angulus quem continet linea per quam extenditur forma cum
185   perpendiculari possit habere ad angulum reflexionis quem exigit ille
       angulus, quoad sensum. Anguli enim reflexionis qui fuerint inter
       duo corpora diaffona diversa in diaffonitate a luce transeunte per illa
       diversantur, quorum diversitas quoad sensum habet finem quem, si
       excesserint, sensus non comprehendet quantitatem reflexionis, com-
190   prehendet enim centrum lucis transeuntis per duo corpora in recti-
       tudine linee per quam lux extenditur, cum videlicet expertum fuerit
       hoc per instrumentum.
       [5.68] Et ponamus angulum DZT equalem angulo KET. Erit ergo
       angulus ZKE duplus ad angulum KET, et sic proportio anguli ZEK
195   ad angulum ZKE erit maxima proportio inter angulum quem conti-

170 BE . . . lineam *mg. a. m. L3* / reflexa: refracta *R* / reflectetur: refringetur *R* / ex . . . B (171) *om.*
R / alia . . . B (171) *scr. et del. E*   171 A¹: Q *O* / quare: quia *F* / quare A *inter. a. m. L3; om. O* / uno
puncto *transp. ER*   172 exeunti: exeunte *R* / ad: et *O*   173 tunc *corr. ex* nunc *O*   174 reflectetur: reflectitur *FP1L3*; refringetur *R* / *post* patuit *rep. et del.* quod (174) . . . patuit *C1*   175 non
inter. a. m. L3; om. O / reflectetur: refringetur *R* / grossius corpus (176) *transp. C1*   176 vise *corr. ex*
visus *F*   177 habebit: habet *P1* / unam formam *om. FP1* / formam *corr. ex* ymaginem *L3* / et
hoc voluimus (178) *om. R*   178 *post* hoc *add.* est *E; add.* est quod (est *inter.*) *C1*   180 GED *om.*
S / et sit E (181) *inter. a. m. E*   181 E: T *P1* / ex quo *mg. F* / extrahemus: extrahamus *R* / AD:
ad T *C1*; AB *ER* (*alter. ex* AD *a. m. E*) / sit: si *O*   182 et¹ *om. L3O* / ZE: ZET *O; corr. ex* ZET
L3 / illam usque *transp. deinde corr. L3* / H *corr. ex* A *C1*   183 ad: et *S* / KET: K et *L3O* / du-
plicatum: duplicata *O; corr. ex* duplicatam *L3* / proportio: proportione *E*   185 reflexionis:
refractionis *R* / *post* reflexionis *scr. et del.* qui fuerit inter duo corpora diaffona *C1* / ille angulus
(186) *transp. FP1*   186 reflexionis: refractionis *R*   187 diaffona *om. R* / diaffona diversa
transp. FP1   188 diversantur *corr. ex* diversitantur *S* / quoad sensum *mg. a. m. C1* / quoad . .
. finem: habet finem quoad sensum *C1* / quem: quoniam *SL3O*   189 excesserint: excesserit
ER / reflexionis: refractionis *R* / comprehendet: comprehendit *P1*   190 transeuntis . . . corpora
om. R / *post* duo *scr. et del.* puncta *P1*   191 expertum: experimentatus *R* / *post* fuerit *add.*
et *C1*   193 angulum *corr. ex* cangulum *C1* / angulum DZT *transp. C1*   194 ZKE: ZK *O* / sic:
si *O* / ZEK . . . angulum¹ (195) *om. FP1*

net prima linea et perpendicularis et inter angulum reflexionis. Sed
linea EK concurret cum linea AD; concurrant ergo in B. Et extrahamus ex E lineam equidistantem TZ. Concurret ergo cum ZG extra
circulum ex parte G. Sit concursus in A. Extrahamus BE usque ad L.
200 Erit ergo angulus LEA equalis angulo ZKE, et angulus LEH equalis
angulo ZEK. Erit ergo angulus LEA angulus reflexionis quem exigit
angulus LEH. Si ergo B fuerit in aliquo viso, et corpus diaffonum
cuius convexum est ex parte A fuerit continuatum ex E usque ad B,
et non fuerit distinctum apud circumferentiam GED ex parte B, tunc
205 forma B extendetur per lineam BE, et reflectetur per lineam EA, et
comprehenditur a visu A per verticationem AE.

[5.69] Et angulus AEH potest dividi pluribus proportionibus
earum que fuerint inter angulos reflexionis et angulos quos continent
perpendiculares cum primis lineis qui fuerint inter duo corpora diaf-
210 fona. Sic ergo in linea DB erunt plura puncta quorum forme extendentur ad arcum GE et reflectentur ad A, et forma totius linee in qua
est ille punctus reflectetur ad A ex arcu GE.

[5.70] Cum ergo visus fuerit in corpore diaffono et res visa fuerit in alio diaffono grossiori, et fuerit superficies corporis diaffoni
215 grossioris que est ex parte visus sperica convexa, et visus fuerit extra circulum cuius convexum est ex parte visus, et fuerit remotior a
visu quam punctum remotius ex duobus punctis sectionis facte inter
perpendicularem et circumferentiam, et corpus diaffonum grossum
quod est ex parte visus fuerit continuum usque ad locum in quo est
220 res visa, et non fuerit decisum apud circulationem que est ex parte rei
vise, tunc visus poterit comprehendere illam rem visam et reflexe et
recte, et ymago huius rei vise erit centrum visus.

[5.71] Item, si fixerimus lineam AGB et revolverimus figuram

195 quem: quam S / continet: continent SFP1L3O   196 post perpendicularis add. exiens a puncto
refractionis R / reflexionis: refractionis R / sed: si C1   197 EK: ET O / concurrant: concurrat FP1;
alter. in concurrat L3   198 E corr. ex B L3 / concurret: concurrit FP1; concurrat L3   199 post A
add. et R   200 et . . . ZEK (201) rep. O   201 ZEK: ZK O; corr. ex ZK L3 / post ZEK rep. et del. et
(200) . . . ZEK (201) L3 / reflexionis: refractionis R / exigit: excetat O   202 si: cum L3 / B fuerit transp.
L3 / viso: visu SFP1L3OE; corr. ex visu C1   203 A alter. ex B L3   204 GED: GD L3   205 post
B scr. et del. et non fuerit S / extendetur: extenditur FP1C1 / reflectetur: reflectitur O; refringetur R / per
lineam transp. deinde corr. S   206 comprehenditur: comprehendetur L3R / A om. P1   207 post
et add. quia R / angulus AEH transp. C1 / post AEH add. quos continent O; scr. et del. continent L3 / post
dividi add. in FP1 / proportionibus earum (208) transp. FP1   208 earum: eorum L3 / reflexionis: refractionis R   209 post lineis rep. et del. lineis F / qui: que R   210 extendentur: extenduntur C1ER
211 GE: EG R / reflectentur: reflectetur FP1; refringuntur R / A: ?? L3   212 est . . . reflectetur: sunt
illa puncta refringetur R / ille om. E / post punctus scr. et del. et L3   213 et om. O   214 post in
add. corpore E / alio diaffono transp. E / post diaffono add. viso S; add. corpore FP1 / grossiori: grossiore
R / corporis om. R   215 post fuerit scr. et del. ex parte P1   216 cuius: eius S / post fuerit add. ille
circulus R   217 facte: facto C1   220 res visa transp. L3 / decisum: derisum S / circulationem
que: circulum qui ER   221 rem om. F / reflexe: reflexive C1; refracte R   222 ymago . . . vise:
huius rei vise ymago ER

AEB in circuitu AB, et pars superficiei corporis diaffoni quod est ex
225 parte rei vise fuerit sperica, tunc punctum E signabit circumferentiam
in superficie circulari convexa que est ex parte visus, ex qua circumferentia reflectetur B ad A. Sed ymago in tota circumferentia reflexionis erit una, scilicet centrum visus. Ymago ergo huiusmodi rei vise
etiam est una. Sed ex hac positione accidit quod visus comprehendat
230 formam rei vise apud locum reflexionis, ea causa quam diximus in
conversione ex speculis, cum fuerit conversio a circumferentia in aliqua spera, et fuerit ymago centrum visus.

[5.72] Ergo huius rei vise forma comprehenditur a visu circularis
apud circulum reflexionis et in rectitudine perpendicularis transeun-
235 tis per visum et rem visam simul, et hoc est quod voluimus.

[5.73] **[PROPOSITIO 10]** Item sit A visus [FIGURE 7.5.8, p. 444],
et sit B in aliquo visu et ultra corpus diaffonum grossius illo in quo
est visus. Et sit superficies corporis quod est ex parte visus circularis
concava, cuius concavitas sit ex parte visus. Dico ergo quod B unam
240 solam habebit ymaginem et unam tantum formam apud A.

[5.74] Et sit centrum concavitatis G, et continuemus AG, et extrahemus ipsam recte usque ad Z. Erit ergo AZ perpendicularis super
superficiem concavam, et B aut erit in AZ, aut extra. Sit ergo primo
in linea AZ. A ergo comprehendet B in rectitudine AB, cum AB sit
245 perpendicularis super superficiem concavam, et numquam ipsam reflexe; quod si est possibile, reflectatur forma B ad A ex E, et continuemus BE, GE. Extrahamus BE usque ad T.

[5.75] Angulus ergo TEG est ille quem continet linea per quam
extenditur forma et perpendicularis exiens a loco reflexionis, et quia
250 corpus quod est ex parte A est subtilius illo quod est ex parte B, erit
reflexio ad partem contrariam illi in qua est EG. Linea ergo ET, quando reflectetur, removetur a linea EG, et linea ET non concurrit cum

223 fixerimus: fuerimus *F* / AGB: ABG *C1*   224 AB *corr. ex* AEB *P1* / est *om. S*   225 fuerit *om. C1* / signabit *corr. ex* significabit *C1*   227 reflectetur: reflectitur *L3O*; refringetur
R / reflexionis: refractionis *R*; *mg. F*   228 huiusmodi: huius *C1*; *om. ER*   229 est: erit
R / quod: ut *R* / comprehendat: comprehendit *L3E*   230 reflexionis: refractionis *R* / post ea
add. de R   231 conversione: reflexione *R* / conversio: reflexio *R* / a circumferentia *corr. ex*
ad circumferentiam *C1* / a . . . aliqua: in aliqua a circumferentia *C1*   232 *ante* spera *scr. et*
del. in *C1* / spera: sperica *O*; *corr. ex* sperica *L3*   233 huius: huiusmodi *L3* / vise *om. FP1* / comprehenditur . . . circularis: a visu circularis comprehenditur *R*   234 reflexionis: refractionis
R / post et *add.* et punctum eius superius circa D videtur *R*   235 est *om. O*   237 visu:
viso *R*   240 habebit: habet *SL3O*   241 extrahemus: extrahamus *ER*   242 erit ergo AZ
inter. a. m. E   243 aut² . . . AZ (244) *mg. a. m. E*   244 *post* in¹ *add.* A E   245 ipsam
om. R / reflexe: reflexere *S*; reflexam *L3O*; refracte *R*   246 possibile: impossibile *FP1* / reflectatur: reflectetur *L3*; refringatur *R*   247 *post* GE *add.* et *R* / extrahamus: extrahemus *L3*
248 continet: continuet *P1*; continent *SC1O*   249 reflexionis: refractionis *R*   250 est²
om. E / est subtilius *transp. R*   251 reflexio: ratio *S*; refractio *R*

linea BA aliquo modo. Forma ergo B non reflectetur ad A; non ergo comprehendetur reflexe, sed comprehendetur recte. Ergo non habebit apud visum nisi unam formam tantum, et hoc est quod voluimus.

[5.76] [**PROPOSITIO 11**] Item iteremus figuram, et sit B extra lineam AZ [FIGURE 7.5.9, p. 445], et extrahemus superficiem in qua est AZ et B. Hec ergo superficies erit perpendicularis super superficiem concavam, et non reflectetur forma B ad A nisi in hac superficie, non enim erigitur perpendicularis super superficiem concavam aliqua superficies equalis que transit per A nisi illa que transit per AZ. Sed per AZ et per B non transit nisi una sola tantum. Forma ergo B non reflectetur ad A nisi in superficie transeunte per lineam AZ et per B. Et sit differentia communis inter hanc superficiem et superficiem concavam arcus HDE, et reflectatur forma B ad A ex H.

[5.77] Dico ergo quod non reflectetur ex alio puncto; quod si fuerit possibile, reflectatur ex M. Et continuemus lineas AH, BH, GH, AM, BM, GM, et extrahamus HB recte usque ad T, et BM recte usque ad N, et GH recte usque ad L, et GM recte usque ad O. Et perficiamus circumferentiam HDE, et secet lineam AG in K. A ergo aut erit in linea KD aut extra in partem K. Si ergo A fuerit in linea KD, aut erit in G aut in altera duarum linearum GD, GK.

[5.78] Si ergo fuerit A in G, tunc forma B non reflectetur ad A, linee enim que continuant corpus circulare cum G sunt perpendiculares super superficiem corporis quod est ex parte A. Reflexio autem non

252 reflectetur: reflectitur *L3E;* refringitur *R/post* EG *scr. et del.* linea *S/*linea ET *om. R/*ET *om. L3/*concurrit: concurret *P1ER*   253 *post* BA *add.* linea ergo ET quando reflectitur non concurret cum linea BA *SC1E/*B: AB *L3/*reflectetur: reflectitur *E;* refringetur *R*   254 comprehendetur reflexe *transp. L3/*reflexe: refracte *R/*comprehendetur² *om. R/*comprehendetur recte *transp. L3*   255 apud *om. L3*   257 iteremus: reiteremus *FC1;* revertemus *S/post* B *scr. et del.* et *P1/*extra lineam *scr. et del. P1*   258 AZ: ZA *F;* AE *O; om. P1/*et . . . AZ (259) *om. L3O/*extrahemus: extrahamus *R/post* extrahemus *scr. et del.* in *F*   259 et *om. R*   260 reflectetur: refringetur *R/post* nisi *scr. et del.* in circu *P1*   261 aliqua: alia *R*   262 *post* transit² *scr. et del.* per A *O/*AZ *corr. ex* Z *L3*   264 reflectetur: refringetur *R*   266 HDE *corr. ex* HED *F/*reflectatur: reflectetur *S;* refringatur *R/*B *om. F/*H: HG *O; corr. ex* HG *L3*   267 ergo *mg. F/*reflectetur: refringetur *R/*fuerit possibile (268) *transp. ER*   268 reflectatur: refringatur *R/*BH *inter.* P1L3 (*a. m. L3)/*AM . . . GH (270) *mg. C1*   269 GM: GB *L3;* NB *O/*extrahamus: extrahemus *L3O/*usque *om. L3/*T: C *R/*recte² *om. L3O/*ad² *om. E/*N . . . usque¹ (270) *mg. a. m.* E   270 recte¹ *om. S/*O: P *R; alter. ex* BC *in* C *S/post* perficiamus *add.* usque ad *L3O (deinde del. L3)*   271 HDE: HED *ER/*aut *om. L3/*erit: fuerit *L3*   272 KD¹: KZ *SFP1L3O/*partem: parte *R/*A *om. L3/*linea *om. R*   273 GD: ZD *SFP1O; corr. ex* ZD *a. m. L3/*GK: DR *O;* DK *SFP1; alter. ex* DF *in* DK *a. m. L3*   274 fuerit A *transp. FP1/*B *inter. a. m. L3/*reflectetur: reflectitur *FP1;* refringetur *R/*A² *corr. ex* ea *L3*   276 *post* corporis *scr. et del.* ex *P1/*reflexio: refractio *R*

erit per ipsam perpendicularem sed ab ipsa; forma B ergo non reflectitur ad A, si A fuerit in G.

280 [5.79] Et si A fuerit in GD, tunc linea HT erit inter duas lineas HA, HG, et ideo linea MN erit inter duas lineas MA, MG, nam reflexio est ad partem contrariam parti perpendicularis, nam corpus diaffonum quod est ex parte visus est subtilius illo quod est ex parte rei vise. Et si linea HT fuerit inter duas lineas HA, HG, et A fuerit in linea GD,

285 tunc angulus BHA erit ex parte D, et similiter angulus BMA erit ex parte D, et erit B ultra lineam GHL, videlicet ex parte K a linea GHL. Et erit angulus THG ille quem continet linea per quam extenditur forma cum perpendiculari, et similiter angulus NMG, et erit angulus THA angulus reflexionis, et similiter angulus NMA.

290 [5.80] Angulus autem NMG aut erit equalis angulo THG, aut maior, aut minor. Si equalis, AMN erit equalis angulo AHT; ergo angulus BHA erit equalis angulo BMA, quod est impossibile. Si maior, tunc angulus AMN erit maior angulo AHT, et sic angulus BMA erit minor angulo BHA, quod est impossibile.

295 [5.81] Si minor, tunc angulus AMN erit minor angulo AHT, et sic totus angulus AMG erit minor toto angulo AHG. Et erit diminutio anguli AMN ab angulo AHT minor quam diminutio anguli AMG ab angulo AHG. Sed diminutio anguli AMG ab angulo AHG est equalis diminutioni HGM ab angulo HAM, duo enim anguli qui sunt in sec-

300 tione linearum AH, MG sunt equales. Ergo diminutio anguli AMN a diminutione de angulo AHT est minor quam diminutio anguli HGM ab angulo HAM.

[5.82] Et extrahamus duas lineas HA, MA ad duo puncta E, C. Erit ergo angulus HAM ille quem respiciunt in circumferentia duo arcus

277 erit: fit *R* / ab ipsa: extra ipsam *R* / B *om. SF* / B ergo *transp. R*   278 reflectitur: refringetur *R*   279 A *inter. E* / A fuerit *transp. C1* / HT: HC *R* / HA: AH *SFP1* / HA . . . lineas (280) *om. S*
280 MN: NM *R*; MZ *O* / reflexio: refractio *R* / post est *add. et C1*   282 *post* parte² *scr. et del.* vi *F*
283 HT: AT *P1*; HC *R* / A *om. FP1*   284 BHA *corr. ex* HBA *L3* / erit² *om. L3*   285 K: R *O* / GHL: HGL *FP1L3ER*; BGL *SO*   286 THG: CHG *R* / quam: quem *SO*   287 *post* perpendiculari *add.* exeunte a loco refractionis *R* / NMG: MNG *SL3*   288 THA: CHA *R* / reflexionis: refractionis *R* / NMA: MNA *SL3*   289 NMG: MNG *SL3E* / THG: TBG *L3*; CHG *R*   290 AMN: AM *S*; ANM *L3O* / AMN erit: erit NMA *R* / AHT: AHC *R* / AHT . . . angulo (291) *rep. SP1*   291 si: sit *S*
292 AMN: AM *E*; ANM *FP1L3*; NMA *R* / AHT: AHC *R*   293 angulo . . . minor (294) *om. FP1*
294 tunc *inter. C1* / AMN: ANM *L3E*; NMA *R* / sic: si *SO; corr. ex* si *L3*   295 AMG: ANG *O*
296 AMN: NMA *R; om. O* / AMN . . . minor *inter. a. m. L3* / post angulo *rep. et del.* AMN ab angulo *C1* / AHT: AHC *R* / AHT . . . angulo¹ (297) *om. SO* / quam . . . AHG¹ (297) *mg. a. m. L3* / AMG: NMG *L3C1E*   297 AHG¹: NHG *FP1*; THG *L3C1*; GHT *E* / AMG: AMN *S*; AMH *O* / post angulo² *add.* toto *C1* / AHG² *alter. in* AGH *a. m. E*   298 HGM: HMG *FP1*   299 AH: HA *C1* / MG: MN *S*; MH *L3O* / diminutio: duo *O* / AMN: ANM *SE*; NMA *R* / a diminutione (300) *scr. et del. C1; om. R*   300 de: ab *R; corr. ex* ab *a. m. E* / post angulo *add.* anguli *SL3C1O (deinde del. L3C1)* / AHT: AHC *R* / est minor *transp. R*   1 ab . . . HAM *inter. a. m. L3; om. O*   2 lineas *om. R* / HA: AH *C1ER* / C: O *R*   3 HAM: HTM *O; om. S* / duo . . . angulum (4) *om. P1*

5 HM, EC, et angulum HGM respicit in circumferentia arcus HM duplicatus, et cum angulus HGM est minor angulo HAM, erit arcus HM duplicatus minor duobus arcubus HM, EC. Et erit diminutio arcus HM
duplicati a duobus arcubus HM, EC sicut diminutio arcus HM ab arcu
EC. Ergo diminutio anguli AMN ab angulo AHT erit minor angulo
10 quem respicit apud circumferentiam diminutio arcus HM ab arcu EC;
est ergo minor angulo HAM. Excessus ergo anguli BMA super angulum BHA est minor quam angulus HAM. Sed excessus anguli BMA
super angulum BHA sunt duo anguli HAM, HBM. Ergo duo anguli
isti simul sunt minores angulo HAM, quod est impossibile.

15 [5.83] Et si A fuerit in linea GK, tunc linea HT erit inter duas lineas
HG, HA, et similiter linea MN erit inter duas lineas MG, MA. Erit
ergo angulus BHA ex parte K, et similiter angulus BMA erit ex parte
K, et erit B infra lineam GMO, scilicet ex parte D a linea GMO. Et
uterque angulus THG, NMG est ille quem continent linea per quam
20 extenditur forma et perpendicularis, et uterque angulus THA, NMA
erit angulus reflexionis.

[5.84] Si ergo angulus THG fuerit equalis angulo NMG, tunc angulus THA erit equalis angulo NMA, et sic angulus BHA erit equalis
angulo BMA, quod est impossibile. Si vero fuerit maior, tunc angulus
25 THA erit maior angulo NMA, et sic angulus BHA erit minor angulo
BMA, quod est impossibile.

[5.85] Et si fuerit minor, tunc angulus THA erit minor angulo
NMA, et sic totus angulus GHA erit minor angulo GMA. Ergo angulus HGM erit minor angulo HAM, et erit diminutio anguli HGM ab

---

4 EC: EO *R*; *inter. a. m. L3* / HGM: GHM *O*; *corr. ex* GHM *L3* / duplicatus *corr. ex* duplis *C1*
5 et . . . duplicatus (6) *scr. et del. E; rep. et del. primam C1; om. S* / HGM *om. P1* / est: sit *R* / angulo
om. P1   6 *post* HM *add.* et *FP1* / EC: EO *R*   7 *post* HM[1] *add.* duplicatus HM *P1* / duplicati:
duplicatus *S* / a: minor *S* / *post* HM[2] *add.* et *FP1* / EC: EO *R*; *corr ex* et *a. m. L3*   8 EC: EO *R* /
AMN: NMA *R* / AHT: AHC *R*   9 HM: GM *SO*   10 arcu *om. FP1* / EC: EO *R* / *ante* est *add.*
sed angulus quem respicit apud circumferentiam diminutio arcus HM ab arcu EO est minor
angulo HAM *R* / *post* ergo[1] *add.* diminutio anguli NMA ab angulo AHC *R*   11 super: supra
R / est: erit *L3* / minor: maior *O*; *corr. ex* maior *L3*   12 super: supra *R*   13 duo . . . isti: isti
duo anguli *R* / simul *inter. a. m. L3* / simul sunt *transp. C1* / HAM: HMA *SO*   15 HT: HC *R*
16 HG . . . lineas *mg. a. m. E* / MN: NM *C1*; M E / lineas *inter. a. m. L3*; *om. O*   17 ergo *inter.*
E / K: R *SO* / et . . . K (18) *om. S*   18 K: R *O* / GMO[1,2]: GMP *R* / et[2] . . . perpendicularis (20) *rep. O*
19 THG: CHG *R* / NMG: MNG *SL3O* / continent: continet *P1R*   20 *post* perpendicularis *add.*
exiens a loco refractionis *R*; *scr. et del.* uterque angulus THA MNA est ille quem continent linea
perpendicularis *L3* / THA: CHA *R* / NMA: MNA *SL3*   21 reflexionis: refractionis *R*   22 angulus[1] *om. ER* / THG: CHG *R* / angulo *om. ER* / NMG: MNG *SL3*   23 THA: CHA *R* / NMA:
MNA *SL3* / sic: si *O*; *corr. ex* si *L3* / equalis angulo (24) *transp. C1*   24 *post* impossibile *add.*
et *ER* / si *inter. L3* / vero *om. L3ER* / maior: minor *L3*   25 THA: CHA *R* / erit[1] *corr. ex* est *C1* /
maior: minor *L3* / NMA: MNA *S*   27 si *inter. L3*; *om. O* / si fuerit *transp. deinde corr. S* / THA:
CHA *R* / *post* minor *rep. et del.* minor *S*   28 NMA: ?MA *S* / NMA . . . angulo *mg. S* / NMA . . .
angulo (29) *om. O* / et . . . HAM (29) *mg. a. m. L3* / GHA: GH? *S* / *post* minor *add.* toto *R* / angulus .
. . erit (29): erit angulus HGM *R*

30 angulo HAM minor quam angulus GMA, ut prius declaravimus, et diminutio anguli THA ab angulo NMA est minor quam diminutio anguli GHA ab angulo GMA; est ergo minor quam diminutio anguli HGM ab angulo HAM. Ergo diminutio anguli THA ab angulo NMA est minor quam angulus GMA. Sed diminutio anguli THA ab angulo 35 NMA est excessus anguli BHA super angulum BMA. Sed excessus anguli BHA super angulum BMA sunt duo anguli HAM, HBM; ergo isti duo anguli simul sunt minores angulo HAM, quod est impossibile.

[5.86] Si vero A fuerit extra lineam KZ et ad partem K [FIGURE 40 7.5.9b, p. 445], et corpus in quo est A fuerit continuum usque ad A, continuabimus duas lineas AH, AM, et secabunt circumferentiam in R et in Q. Et si angulus THG fuerit equalis angulo NMG, tunc angulus BHA erit equalis angulo BMA, quod est impossibile. Et si fuerit maior, tunc angulus THA erit maior angulo NMA, et sic angulus 45 BHA erit minor angulo BMA, quod est impossibile.

[5.87] Si vero fuerit minor, tunc angulus THA erit minor angulo NMA, et totus angulus GHA erit minor toto angulo GMA; ergo angulus HGM erit minor angulo HAM. Sed angulus MGH est ille quem respicit apud circumferentiam arcus HM duplicatus, et angu-50 lus HAM est ille quem respicit in circumferentia excessus arcus HM super arcum RQ. Ergo arcus HM duplicatus est minor excessu arcus HM super arcum RQ, quod est impossibile.

[5.88] Ergo si punctum B fuerit extra lineam AKG, tunc forma eius non reflectetur ad A nisi ex uno puncto tantum, quapropter non habebit nisi unam solam ymaginem, que ymago aut erit ante visum, 55 aut retro, aut in loco reflexionis, ut in precedentibus declaravimus, et hoc est quod voluimus.

[5.89] Si vero corpus diaffonum grossius fuerit ex parte visus et

30 GMA *alter. ex* MHA *in* MAH *C1; alter. ex* HMA *in* MAH *a. m.* E / ut: aut O   31 anguli . . . diminutio² *om.* C1 / THA: CHA R / NMA: MNA SL3   32 GHA: THA O; *corr. ex* THA L3   33 HGM: HAM S / THA: CHA R / NMA: MNA SL3   34 GMA: HAM C1; *alter. in* HAM *a. m.* L3; *alter. ex* MAH *in* HAM *a. m.* E / THA: CHA R / *post* THA *scr. et del.* ab O   35 NMA: MNA SFL3 / sed . . . BMA (36) *mg. a. m.* L3; *om.* SOER   36 *post* BMA *add.* qui ER; *scr. et del.* sed excessus anguli P1 / HBM *om.* P1   37 duo anguli *transp.* L3 / *post* sunt *scr. et del.* N P1   39 KZ: R O; KD C1R; *alter. ex* K *in* KD E / et *om.* ER   41 secabunt: secabit S   42 R: K FP1; Q R / Q: R R / et *inter.* C1 / THG: CHG R / NMG: MNG SL3 / angulus *inter. a. m.* E   43 et *om.* FP1 / fuerit: fuerint S   44 THA: CHA R / maior: minor FP1 / NMA: MNA SL3 / sic: si O   46 vero *corr. ex* non L3 / THA: CHA R   47 NMA: MNA SL3 / angulus *om.* E   48 MGH: HGM R   49 apud circumferentiam: ad circumferentia S / et angulus *mg.* C1   50 ille *om.* E / circumferentia *corr. ex* circumferentiam S   51 super: supra R / RQ: TQ P1 / ergo . . . RQ (52) *mg.* C1 / *post* minor *scr. et del.* accessu S / arcus² *om.* P1   52 HM: BM S / super: supra R / RQ: TQ P1L3 / *post* RQ *scr. et del.* sed excessus anguli BHA super angulum BMA E / est *inter.* S   53 AKG: ABG S; AHG O   54 non¹ *inter.* E / reflectetur: refringetur R / puncto tantum *transp.* L3   55 solam *om.* ER / *post* que *add.* est L3   56 reflexionis: refractionis R 57 *post* voluimus *add.* declarare R

subtilius ex parte rei vise, eisdem permanentibus figuris, tunc etiam
60 res visa non habebit nisi unam solam ymaginem, et hoc demonstra-
bitur, ut in conversa septime figure. Et omnia que declaravimus in
reflexionibus a convexo et concavo circuli sequuntur in superficiebus
spericis et columpnalibus, preter reflexionem circularem a circumferentia circuli, que non fit nisi in superficiebus spericis tantum. Hec
65 autem que diximus sunt ymagines visibilium que comprehenduntur
a visu ultra corpora diaffona simplicia, que sunt unius substantie et
quarum figura que est ex parte visus est una figura.

[5.90] Si vero corpus diaffonum fuerit diversum aut non consimilis diaffonitatis, tunc ymagines rei vise diversantur, et si superfi-
70 cies corporis diaffoni que est ex parte visus fuerit diversa, tunc loca
ymaginum rei vise diversantur, cum forme reflexionum ex superficie corporis diversantur etiam. Et si aliquis respexerit ad parvam
speram, aut aliquod corpus rotundum parvum, aut columpnale vitri,
aut cristalli, ultra quod corpus fuerit aliquod visibile, inveniet ymagi-
75 nem illius alio modo quam sit res visa in se, et forte inveniet rei vise
ymaginem aliam, et sic ambigetur super hoc. Sed huiusmodi reflexio
non est una sed due reflexiones, forma enim rei vise extenditur a re
ad speram aut ad aliud corpus rotundum columpnale, et reflectitur a
convexo spere aut columpne ad interius corporis, et extenditur intra
80 corpus quousque pervenit ad superficiem eius, et deinde reflectitur a
spera aut columpna apud concavitatem aeris continentis speram aut
columpnam. Et sic comprehensio huiusmodi rerum erit duabus diversis reflexionibus, quapropter ymago eius erit diversa ab ymagine
eius quod comprehenditur una reflexione. Nos autem loquemur de
85 hoc parum, quando tractabimus de deceptionibus visus que fiunt per
reflexionem.

58 vero om. P1   59 eisdem permanentibus: iisdem manentibus R   60 visa: una FP1/solam
ymaginem transp. ER/demonstrabitur: declarabitur ER   61 septime: octave O; corr. ex octave
a. m. L3   62 reflexionibus: refractionibus R/convexo corr. ex concavo O/sequuntur: sequitur
FP1L3O.   63 columpnalibus: columpnaribus R/reflexionem: refractionem R   64 que: qui P1/
fit: sit SL3   65 que¹ inter. a. m. L3   67 quarum: quorum R; quare SO; alter. ex quare in quod L3
70 post visus add. rei vise SL3O (deinde del. L3)/fuerit om. S/post tunc scr. et del. et si superficies S/post
loca add. etiam ER   71 post rei scr. et del. in s E/reflexionum: refractionum R   72 post corporis
add. diaffoni FP1R (deinde del. FP1)/diversantur: diversentur R   73 columpnale: columpnare R;
columpne O; corr. ex columpne L3   74 post corpus scr. et del. cor P1/aliquod: apud P1   76 sic:
si S/ambigetur: dubitabitur R/hoc: ea R/huiusmodi: eius FP1/reflexio: refractio R   77 due:
duo O/reflexiones om. R/post enim rep. et del. enim F/post re add. visa R   78 ad² om. C1/post
rotundum add. et P1/columpnale: columpnare R; corr. ex columpne L3/et inter. O; om. S/reflectitur:
refringitur R   79 columpne: columpnale SC1/ad: aut P1/intra: extra S   80 pervenit: perveniat R/reflectitur: refringitur R   81 apud: aut O/continentis: contingentis R   82 columpnam:
columpna L3O/huiusmodi: huius O/rerum erit transp. O/post erit add. ex L3   83 reflexionibus:
refractionibus R/quapropter: quia propter S   84 reflexione: refractione R/loquemur: loqua-
mur L3   85 fiunt: sunt S/post per scr. et del. f C1   86 reflexionem: refractionem R

CAPITULUM SEXTUM
Qualiter visus comprehendit visibilia secundum reflexionem

[6.1] In precedentibus iam declaravimus quod, cum forma reflec-
90   titur ab aliquo corpore diaffono ad aliud corporis alterius diaffonitatis, extenditur per lineam rectam donec perveniat ad superficiem
corporis diaffoni in quo est; deinde reflectitur in illo alio corpore diaffono per lineam aliam rectam que continet cum prima linea angulum.
Et cum forma extenditur per hanc lineam aliam super quam reflecti-
95   tur forma in secundo corpore, alia quecumque sit forma in secundo
corpore usque ad punctum sectionis inter duas lineas rectas reflectetur per primam lineam rectam.

[6.2] Et est manifestum per experientiam quod, si aliquis inspexerit aliquod corpus diaffonum quod differt in sua diaffonitate a diaf-
100   fonitate aeris, comprehendet omnia que sunt ultra de illis que opponuntur visui, et si cooperuerit alterum visum et aspexerit reliquo,
comprehendet etiam quecumque sunt ultra, sive illud sit corpus sive
aer, aut aqua, aut vitrum. Et similiter, si homo posuerit visum intra
aut in aliquo corpore grossiori aere, aut vitro, aut cristallo, videbit
105   omnia que sunt ultra de illis que sunt in aere. Et si aspiciens moverit
visum suum dextrum, aut sinistrum, et in omnem partem, et non re-
moverit ipsum multum a suo primo loco, tunc comprehendet etiam
omnia que prius comprehendebat, sive motus visus fuerit in aere aut
in vitro.

110   [6.3] Sed iam declaravimus experientia et demonstratione quod
nichil comprehendit visus de illis que sunt ultra corpora diaffona que
differunt a diaffonitate ab aere nisi secundum reflexionem, preter
quam unum punctum quod est in perpendiculari exeunti a centro
visus super superficiem corporis diaffoni. Ergo omne punctum com-

---

87 capitulum sextum *om. R*/capitulum . . . reflexionem (88) *om. S*   88 qualiter: quomodo
R/comprehendit: comprehendat *R*/reflexionem: refractionem *R*/*post* reflexionem *add.* capitulum sextum *R*   89 cum *inter. C1*/reflectitur: refringitur *R*   90 aliquo: alio *C1*/ad: aut
P1/corporis: corpus *R*/alterius: diverse *ER*/alterius diaffonitatis *transp. L3*   91 perveniat:
proveniat *F*   92 reflectitur: refringitur *R*/diaffono *om. P1*   94 et *om. P1*/hanc *om. S*/li-
neam aliam *transp. SC1ER*/aliam *inter. a. m. L3; om. O*/super: supra *C1*/reflectitur: refringitur *R*
95 sit forma *transp. SER*   96 rectas *mg. a. m. C1*/reflectetur: refringetur *R*   97 primam
lineam *transp. FP1*   98 inspexerit *corr. ex* aspexerit *C1*   101 cooperuerit: cooperuit *S*/*post*
alterum *scr. et del.* a *S*   102 illud sit corpus *corr. ex* corpus sit illud *L3*/sit corpus *transp. R*/
sive *om. FP1R*   103 aut[1,2]: sive *R*/vitrum *corr. ex* vitum *a. m. L3*/visum: vitrum *P1O*/intra aut
(104) *om. R*   104 grossiori: grossiore *R*/aut[1]: ut *R*/cristallo *corr. ex* cristillo *F*   105 que[1]
. . . illis *om. S*/ultra . . . sunt[2] *inter. a. m. L3*/*post* illis *scr. et del.* corporibus *E*   106 dextrum:
dextrorsum *R*/sinistrum: sinistrorsum *R*   108 fuerit *inter. a. m. E*/aut: sive *R*   111 nichil
. . . visus: visus nichil comprehendit *R*/ultra *mg. C1; om. E*   112 reflexionem: refractionem *R*
113 exeunti: exeunte *R; corr. ex* exeuntia *F*

115 prehensum a visu ultra corpus diaffonum, preter illud punctum predictum, comprehenditur ex forma que extenditur ex illo puncto ad
superficiem corporis diaffoni ultra quod est, et reflectetur a superficie illius corporis ad visum. Et cum unus visus comprehendit omnia
que sunt ultra corpus diaffonum, omne punctum existens ultra illud
120 corpus diaffonum extenditur forma eius per lineam rectam ad superficiem illius corporis diaffoni, et reflectetur ad illum visum unum,
preter quam illud punctum predictum.

[6.4] Et cum forme omnium punctorum que sunt in omnibus visibilibus existentibus ultra corpus diaffonum reflectuntur in eodem
125 tempore ad centrum visus unius, forma puncti quod existit apud
centrum illius visus, cum fuerit in aliquo visibili, reflectetur ad omnia
puncta que sunt in omnibus visibilibus existentibus ultra corpus diaffonum oppositum visui in eodem tempore et eodem modo. Et similiter est de omni puncto propinquo puncto quod est apud centrum vi-
130 sus, nam si visus motus fuerit ad omnem partem et non fuerit remotus a suo situ, comprehendet visibilia. Ergo forma cuiuslibet puncti
cuiuslibet visi, cum fuerit ultra aliquod corpus diaffonum, extenditur
ad superficiem corporis diaffoni ultra quod est, et reflectitur ad universum eius quod opponitur ei ex corpore aeris. Et non est aliquod
135 tempus magis appropriatum huic quam aliud, sed hoc est proprium
nature lucis et coloris que sunt in visibilibus, scilicet quod semper
extendantur a quolibet puncto cuiuslibet corporis lucidi per lineam
rectam que extenditur ab illo puncto, et reflectuntur in omni corpore
diaffono diverso, preter quam punctum quod est in perpendiculari.

140 [6.5] Et omnis forma puncti cuiuslibet visibilis existentis in aliquo corpore diverso ab aere extenditur in illo corpore in quo existit
et reflectitur in universo corpore aeris sibi opposito. Et illa forma
exit apud quodlibet punctum aeris, quapropter forma totius rei vise

---

114 super superficiem *inter. a. m. L3; om. O* / post ergo *scr. et del.* est *E* / omne: est *O*   116 ex¹
. . . extenditur *inter. a. m. L3; om. O*   117 reflectetur: refringitur *R*   118 comprehendit:
comprehendat *R*   119 *post* diaffonum *add.* forma *R* / omne . . . existens: omnis puncti existentis *R* / punctum *corr. ex* corpus *P1* / illud corpus (120) *transp. ER*   120 forma eius *om.*
R / per: ad *C1*   121 reflectetur: refringitur *R*   122 illud: illum *L3*   123 punctorum
om. S / sunt: sciunt *C1*   124 reflectuntur: refringantur *R* / in *om. C1* / eodem: eorumdem *L3O*
125 tempore: corpore *P1O* / quod: que *L3*   126 illius visus *transp. L3C1ER* / reflectetur: refringetur *R*   127 puncta *mg. F* / sunt *mg. a. m. L3; om. O* / post omnibus *add.* et *C1*   128 tempore: corpore *O*   129 omni: eodem *FP1* / propinquo puncto *inter. a. m. E; om. SFP1*
132 cuiuslibet *om. FP1* / extenditur: extendetur *ER*   133 diaffoni *corr. ex* diaffi *C1* / post
diaffoni *scr. et del.* vel *E* / reflectitur: refringetur *R*   134 ei: e *O* / corpore *corr. ex* parte *a. m. L3*
135 tempus *corr. ex* corpus *O* / magis appropriatum *transp. deinde corr. S* / sed *corr. ex* et *a. m. E*
136 quod: ut *R* / semper *om. P1*   138 reflectuntur: reflectatur *S*; reflectantur *C1E*; refringantur *R*   140 *post* forma *add.* cuiuslibet *SE* / puncti cuiuslibet *transp. R*   141 extenditur:
extendetur *R*   142 reflectitur: refringetur *R* / post opposito *add.* aeris *L3O* (*deinde del. L3*) / et:
est *O; corr. ex* est *L3* / post forma *scr. et del.* q *C1*

coniungitur apud quodlibet punctum aeris, et forma totius cuiuslibet
145 visi existentis in aliquo corpore diverso ab aere existit apud unumquodque punctum aeris oppositi illius rei vise. Et illa forma extenditur a quolibet puncto rei vise in corpore in quo est, et reflectitur apud
superficiem illius corporis, et pervenit ad illud punctum aeris. Et
ideo, si visus aspexerit aliquod corpus diaffonum diversum ab aere
150 ultra quod fuerit aliqua res visibilis, visus comprehendit illam rem,
nam forma illius existit apud punctum apud quod existit centrum
visus, propter hoc quod etiam, si visus comprehenderit aliquam rem
visibilem ultra aliquod corpus diaffonum diversum ab aere, deinde
motus fuerit a suo loco dextro aut sinistro, dum in suo motu fuerit
155 oppositus corpori diaffono et rei que est ultra, semper comprehendet
illam rem. Unde etiam plures aspicientes comprehendent unam rem
in celo et in aqua, et in uno eodem tempore, et hoc etiam est in eodem
corpore diaffono, scilicet quod forma visi congregatur apud quodlibet punctum corporis in quo est, nam forma puncti cuiuslibet eius
160 extenditur per lineam rectam, et inter quodlibet punctum corporis in
quo est visus et quodlibet punctum rei vise est linea recta.

[6.6] Forma ergo cuiuslibet puncti rei vise extenditur ad quodlibet
punctum corporis diaffoni in quo est res visa, et forma cuiuslibet rei
vise lucide congregatur apud quodlibet punctum corporis in quo exis-
165 tit, et congregatur apud quodlibet punctum cuiuslibet corporis diaffoni diversi a corpore in quo existit, quando inter rem visam et illud
corpus diaffonum diversum non interfuerit aliquod impedimentum.
Et forma rei vise que est apud quodlibet punctum corporis diaffoni
in quo extenditur ad illud punctum recte et forma illius apud quodlibet punctum corporis diaffoni diversi extenditur ad illud punctum

143 exit: erit *FP1L3E* / apud: ad *R* / quapropter *inter. a. m. L3* / quapropter . . . aeris (144) *om. S* / forma .
. . aeris (144) *om. L3*   145 in aliquo *inter. a. m. E* / existit *om. L3* / apud: ad *P1* / unumquodque: unum
quodlibet *S*   146 oppositi *corr. ex* oppositum *L3* / illius: illi *R* / rei: re *O* / et . . . vise (147) *mg. a. m.*
E / illa forma *transp. ER*   147 quodlibet puncto *transp. C1* / in[1]: a *E* / reflectitur: refringitur *R* / apud
superficiem (148) *corr. ex* a superficie *a. m. L3*   148 illud: illum *F*   150 aliqua res *transp. L3* / *post*
rem *add.* visam *L3*   151 illius *corr. ex* visus *O* / *post* illius *add.* visus *L3*   152 hoc *scr. et del. C1*; *om.*
S / quod *om. L3O* / etiam: et *R*; *corr. ex* et *a. m. E* / *post* visus[2] *scr. et del.* fuerit *P1* / comprehenderit: comprehendit *L3O*   154 motus *rep. C1* / motus fuerit *transp. L3* / a: in *O*; *corr. ex* in *L3* / suo loco *transp.*
SR / dextro: dextrorsum *R* / aut: vel a *S* / sinistro: sinistrorsum *R*   155 *post* rei *add.* vise *R* / que *mg.*
F / comprehendet: comprehenditur *P1*; *corr. ex* comprehendit *O*   156 comprehendent: comprehendet *FP1*; comprehendunt *ER* / *post* comprehendent *scr. et del.* in *C1*   157 *post* uno *add.* et *L3R*;
scr. et del. ed *P1*   158 corpore: tempore *S* / *post* visi *scr. et del.* congra *P1* / apud: ad *S*   159 puncti
cuiuslibet *transp. C1*   160 et *rep. C1* / corporis . . . punctum (161) *rep. P1*   161 quodlibet
punctum *transp. C1* / punctum *rep. F*   162 forma *mg. a. m. L3*   163 rei: re *O*   164 vise *om.*
R / quodlibet . . . corporis (165) *om. O* / *post* punctum *add.* cuiuslibet *ER* / corporis . . . corporis (165)
mg. a. m. L3 / *post* corporis *scr. et del.* diaffoni diversi a corpore *E*   165 cuiuslibet corporis *transp.*
ER (cuiuslibet *mg. a. m. E*)   166 diversi *inter. a. m. E*   167 diversum *mg. P1* / aliquod: aliquo
O / *post* imperimentum *scr. et del.* diversum *P1*   168 quodlibet *corr. ex* quolibet *F*   169 in . . .
diaffoni (170) *mg. a. m. L3*; *om. SO* / extenditur *rep. L3C1E*

170   reflexive, quando inter quodlibet punctum aeris et quamlibet rem
visibilem existentem in aliquo corpore diaffono diverso ab aere, fit
piramis reflexa, cuius capud est in aere punctus, et cuius basis est
illa res visa, et erit reflexio eius ad superficiem corporis diaffoni ab
aere diversi. Omnis ergo res visa in corpore diaffono diverso ab aere,
175   quando comprehenditur a visu, comprehenditur a forma extensa in
piramide reflexa adunata apud punctum existens in centro visus.
Hoc igitur modo comprehendit visus ea que comprehendit reflexive.

[6.7] In capitulo autem ymaginis declaravimus quod omne visum
comprehenditur a visu ultra ymaginem, et locus ymaginis est punc-
180   tum in quo secant se linea radialis per quam forma extenditur ad
visum et perpendicularis exiens a puncto viso. Si ergo ymaginati
fuerimus quod ab unoquoque puncto rei vise exit perpendicularis
ad superficiem corporis diaffoni in quo est res visa, tunc habebimus
quoddam corpus exiens a viso ad superficiem corporis diaffoni, unde
185   sequitur quod istud corpus secet piramidem reflexam, et illa superficies in qua se secant est ymago illius rei vise.

[6.8] Si ergo superficies corporis diaffoni in quo est res visa fuerit
equalis, tunc corpus ymaginatum continens omnes perpendiculares
erit equalis superficiei, quare ymago addet parum super rem visam.
190   Et si corpus fuerit spericum, et convexum eius fuerit ex parte visus,
et centrum eius fuerit super illam rem visam, tunc corpus ymaginatum erit piramidale, cuius capud erit centrum spere, et quanto
magis extenditur ad superficiem corporis sperici, tanto magis ampli-
ficabitur. Et si sectio fuerit inter rem visam et superficiem spericam,
195   tunc ymago erit amplior ipsa re visa. Si autem sectio fuerit ultra rem
visam, tunc ymago erit strictior re visa. Si vero res visa fuerit ultra
superficiem spericam, tunc corpus ymaginatum erit due piramides

171 reflexive: reflexionis *L3O*; refracte *R* / quando: quia *R*   173 reflexa: refracta *R* / in . . . punctus:
punctum in aere *R* / cuius *om. R*   174 reflexio: refractio *R* / diaffoni *om. R*   176 quando *corr.*
ex quod non *L3* / a¹,² *om. E* / comprehenditur *corr. ex* comprehendetur *E*   177 reflexa: refracta *R*
178 comprehendit reflexive: refracte comprehendit *R* / reflexive: reflexius *L3*   180 comprehenditur *corr. ex* comprehendetur *P1*   181 forma *om. E* / forma extenditur *transp. R*   182 et *inter. a. m.*
L3; om. O / perpendicularis *corr. ex* perpendiculares *P1*   183 quod *inter. a. m. L3; om. O* / unoquoque: uno quodlibet *S*   184 habebimus: habemus *FP1*   185 exiens *corr. ex* existens *C1* / viso:
visu *SOER; corr. ex* visu *L3* / unde sequitur (186) *mg. C1*   186 reflexam: refractam *R*   187 *ante*
in *add.* est *L3O (deinde del. L3)* / in *inter. a. m. L3; om. O* / se *om. E* / se secant *transp. FP1R* / illius *om. P1*
189 ymaginatum: ymaginum *E*   190 erit: est *C1* / *post* erit *scr. et del.* q *F* / ymago *om. S* / addet:
addit *ER* / *post* visam *add.* et si convexum corporis diaffoni *L3O (deinde del. L3)*   191 et² *inter. a.*
m. E / eius: cuius *O* / fuerit² *inter. a. m. E; om. R* / visus: rei vise *SC1OE; corr. ex* rei vise *L3*   192 *post*
visam *scr. et del.* et si corpus fuerit spericum convexum eius ex parte rei vise et centrum eius fuerit
eius fuerit super illam rem visam *E*   193 *post* piramidale *scr. et del.* tunc corpus ymaginatum
erit *P1*   194 ad superficiem: a superficie *ER*   196 ipsa: illa *ER* / re visa *transp. L3*   197 *post*
erit *scr. et del.* amplior *S* / visa fuerit *transp. deinde corr. E*

opposite, quarum capud centrum spere. Quare loca sectionis non cadent
inter corpus ymaginatum et piramidem, forte locus sectionis in quo
200 est ymago erit maior viso, forte minor, forte equalis.

[6.9] Si vero corpus diaffonum fuerit spericum, et concavitas eius
fuerit ex parte visus, tunc corpus ymaginatum erit piramis cuius
capud est centrum spere; quanto ergo magis extenditur hoc corpus
in partem superficiei spere, tanto magis adunatur et constringitur, et
205 quanto magis extenditur in aliam partem, tanto magis amplificatur,
superficies enim continua parva erit media inter centrum eius et
speram. Si vero locus sectionis huius corporis cum piramide reflexa
fuerit propinquior centro concavitatis quam res visa, erit ymago
minor ipsa re visa. Si autem fuerit remotior a centro concavitatis
210 quam res visa, et ymago maior est quam res visa.

[6.10] Et cum una res visa comprehenditur a pluribus visibus in
uno momento, omnes ymagines quas comprehendunt illi visus erunt
illo tempore in uno ymaginato quod est perpendiculare super superficiem corporis diaffoni. Et una res visibilis comprehenditur ab uno
215 homine in uno tempore ultra corpus diaffonum diversum a diaffonitate corporis in quo est visus utroque visu, et tamen comprehendit
illam unam. Si enim homo comprehenderit aliquod de eis que sunt
in celo, aut in aqua, aut ultra vitrum, et cooperuerit alterum visum,
nichilominus comprehendet illud reliquo, ex quo patet quod una res
220 existens ultra corpus diaffonum diversum ab aere comprehendetur
utroque visu et altero visu.

[6.11] Causa autem huius est, ut in tertio huius libri diximus, quoniam in omni puncto cuiuslibet visi comprehensibilis recte et utroque
visu in quo coniuncti fuerint duo radii utriusque visus consimilis
225 positionis quantum ad duos axes visuum comprehendetur unum, et
si in ipso congregati fuerint radii diverse positionis quantum ad duos

198 erit: erunt *R*/piramides: piramide *O*   199 opposite: oppositione *O*/quarum: quare
OE; corr. ex quare *C1*/*post* centrum *rep. et del.* centrum *L3*/quare: sit *O*/*post* quare *add.* cum
C1ER (*inter. a. m. E; deinde del. C1*)/non cadent *om. C1R*/non . . . inter (200) *scr. et del. E*   200 inter *corr. ex* intra *P1*/*post* piramidem *add.* posset esse diversa *C1; add.* possint esse diversa *R*;
inter. possit esse diversa *a. m. E*   202 fuerit: fuit *L3*/fuerit spericum *transp. S*/concavitas:
cavitas *S*   205 partem: parte *P1*/constringitur: constringatur *L3O*   206 *post* magis²
scr. et del. adunatur *S*   208 speram: spericam *O; corr. ex* spericam *L3*/reflexa: refracta *R*
211 quam¹ . . . est *mg. C1*/et: erit *R*/est *om. R*   212 et . . . visa *mg. a. m. L3*/una res *transp.*
L3/in uno (213) *corr. ex* uno *L3*   213 comprehendunt . . . visus: illi visus comprehendunt
R/*post* erunt *add.* in *FP1ER*   214 in *inter. a. m. L3; om. O*   215 corporis *om. P1*/*post* res
scr. et del. visa *E*   217 visu *corr. ex* visui *F*/tamen: tantum *S*/comprehendit: comprehendat
FP1/*post* comprehendit *add.* rem *R*   218 *post* illam *add.* rem *FP1*/homo comprehenderit
transp. deinde corr. S/aliquod: aliquid *FP1R*/eis: hiis *C1*   219 ultra: intra *FP1*/cooperuerit:
cooperuit *C1*   220 *post* illud *add.* in *L3O* (*deinde del. L3*)/*post* res *add.* visa *R*   222 visu²:
vis *S*   223 est *om. P1L3*/huius *om. R*/libri: libro *R*/quoniam: quod *FP1*   225 in quo *inter.*
a. m. L3; om. O   226 visuum . . . axes (228) *om. S*/comprehendetur . . . visuum (228) *mg. C1*

axes visuum, comprehendentur duo. Sed in maiori parte earum que comprehenduntur positio est consimilis. Hec autem que sunt diverse positionis respectu utriusque visus sunt valde rara, ut in tertio diximus.

[6.12] Et illud quod comprehenditur reflexe comprehenditur in loco ymaginis. Forma autem que est in loco ymaginis comprehenditur a visu recte; est ergo quasi esset in aere et comprehenderetur a visu recte. Positio autem huius forme, que est ymago respectu visus, est sicut positio alicuius rei vise earum que videntur recte, unde positio harum ymaginum respectu visus est in maiori parte consimilis. Et in omni puncto ymaginis congregantur duo radii duorum visuum consimilis positionis, quare una res videbitur una utroque visu.

[6.13] Et ut hoc evidentius declaretur, dicamus quod iam diximus quod omne punctum eius quod comprehenditur reflexive comprehenditur in loco ymaginis, que est inter punctum sectionis ex perpendiculari exeunti ex illo puncto super superficiem corporis diaffoni in quo est res visa et inter lineam radialem per quam extenditur forma ad visum. Cum ergo aspiciens comprehenderit punctum alicuius rei utroque visu, ymago illius puncti respectu utriusque visus est in perpendiculari exeunti ex illo puncto, que est eadem linea. Et cum forma illius puncti pervenerit ad duo puncta superficierum visuum, quorum situs respectu axis visuum est consimilis, tunc due linee per quas forme extenduntur ad utrumque visum perveniunt ad duo centra duorum visuum. Sunt ergo axes aut habentes ex axibus positionem consimilem.

[6.14] Et duo axes visuum semper sunt in eadem superficie, et omnes linee exeuntes a centro duorum visuum habentes positionem consimilem ab axe communi erunt in eadem superficie, axis enim communis semper est in eadem superficie, nam, si aliquod comprehenditur utroque visu in eodem tempore vera comprehensione, tunc axes

227 congregati: agregati P1OR / ad duos om. P1   228 visuum: visum S / comprehendentur: comprehenduntur L3 / sed: et ER / maiori: maiore R / earum: eorum R   229 est corr. ex eius a. m. L3; om. P1 / post diverse scr. et del. se L3   230 positionis: rationis S; corr. ex positiones L3 / sunt om. P1   231 post diximus add. tractatu R   232 reflexe: refracte R / comprehenditur . . . recte (234) om. P1 / in om. S   233 est: comprehenditur L3   234 est . . . recte (235) scr. et del. E; om. R / comprehenderetur: comprehendetur FP1L3OE   235 autem inter. L3; om. O / post huius scr. et del. que est ymago P1   236 alicuius: alterius R   237 respectu corr. ex reflexio a. m. L3 / ante visus add. unum L3 / maiori: maiore R   238 post duorum scr. et del. diversorum C1   239 res videbitur transp. L3O / post res add. visa ER / videbitur: videtur ER   240 post hoc scr. et del. videamus P1   241 reflexive: refracte R   242 que: qui R / ex: inter C1; et OE; corr. ex et a. m. L3 / perpendiculari: perpendicularis E; perpendicularem C1O; corr. ex perpendicularem L3   243 exeunti: exeuntis E; exeunte R; exeuntem C1O; corr. ex exeuntem L3 / ex: ab ER   247 exeunti: exeunte R / ex: ab L3O   249 quorum situs inter. a. m. L3; quo scitus O / visuum: visus ER   251 ergo axes transp. FP1 / aut corr. ex ut C1 / habentes: abentes S / ex . . . positionem: positionem ex axibus S / ex . . . consimilem (252): consimilem ex axibus positionem P1   254 post duorum scr. et del. fora P1   255 axis . . . superficie (256) om. S   256 aliquod: aliquid R / comprehenditur: comprehendetur E

concurrunt in uno puncto illius rei, quare sunt in eadem superficie. Item positio visuum naturalis est consimilis, et non exit a naturali nisi per accidens aut violentiam, quare axes eorum sunt in eadem super-
260 ficie, principium enim axium est unum punctum quod est in medio concavitatis communis nervi, a qua exit communis axis.

[6.15] Existentibus ergo duobus visibus in sui naturali positione, semper axes erunt in eadem superficie, sive sint moti, sive quiescentes. Si autem positio alterius visuum mutata fuerit respectu reliqui
265 propter aliquod impedimentum, tunc res una videbitur due, ut in primo declaravimus. Duo ergo axes in maiori parte sunt in eadem superficie, quare omnes duo radii habentes positionem similem ex duobus axibus erunt in eadem superficie. Due ergo linee per quas extenduntur forme unius puncti ad duo loca consimilis positionis sunt
270 in eadem superficie. Sed ymagines illius puncti respectu duorum visuum sunt in illis duabus lineis; ergo sunt in eadem superficie, sed ymagines illius puncti sunt in perpendiculari exeunti ex illo puncto; ergo sunt in loco sectionis inter superficiem in qua sunt linee radiales, que est una superficies, et inter perpendicularem, que est una linea.

275 [6.16] Sectio autem unius superficiei cum una linea est unum punctum; ergo ymagines unius puncti respectu duorum visuum, quando perveniunt ad duo loca consimilis positionis, sunt unum punctum, ex quo patet quod ymago totius rei vise respectu duorum visuum erit una, si positio ymaginis fuerit consimilis, quare res com-
280 prehenditur una utroque visu. Si vero positio fuerit parum diversa et videbitur res una, sed non vere sed cavillose. Si autem diversitas positionis fuerit multa, tunc forma rei videbitur due, sed hoc fit rarissime. Hec est ergo qualitas comprehensionis visus de visibilibus secundum reflexionem.

285 [6.17] Hoc autem declarato, dicamus universaliter quod omnia que comprehenduntur a visu comprehenduntur reflexive, sive visus et visum sunt in eodem diaffono aut in diversis, sive visum sit in oppositione visus, sive comprehendatur ab ipso reflexive. Nichil enim

258 concurrunt: concurrent *FP1L3*   259 *post* naturali *add.* positione *FP1ER*   260 accidens: axidens *S* / *post* aut *add.* per *C1ER* / axes: acces *S*   262 qua: quo *R*   263 visibus: visibilibus *P1* / sui: suo *R* / positione: prepons *S*   264 *post* sive² *scr. et del.* e *O*   265 *post* fuerit *add.* a *L3O* / reliqui: aliquod *L3O* (*deinde del. L3*)   266 una: visa *SC1ER* / due: duplex *R*   267 *post* primo *add.* libro *R* / maiori: maiore *R*   270 positionis: positioni *S*   271 puncti *om. R* / duorum visuum (272) *transp. FP1*   272 illis … in (273) *om. SE*   273 exeunti: exeunte *R* / ex: ab *C1*   274 sectionis inter: secutionis intra *S*   276 sectio: secutio *S* / superficiei … unius (277) *mg. a. m. E*   278 quando … visuum (280) *om. S* / unum punctum (279) *transp. ER*   280 si … ymaginis: sed ymago primo *P1* / positio ymaginis *transp. F* / fuerit: erit *SL3* / *post* fuerit *scr. et del.* parum diversa et videbitur *F*   281 positio: primo *P1* / positio fuerit *transp. C1* / parum: piramis *L3* / parum diversa *transp. S*   282 et: etiam *S*; *om. ER*   283 fuerit *mg. F* / tunc: cum *S* / videbitur: videbuntur *R*   285 reflexionem: refractionem *R*   287 reflexive: refracte *R* / sive *inter. a. m. L3* / *post* sive *scr. et del.* que comprehenduntur *S*   288 et … sunt *inter. a. m. L3*; *om. O* / sunt: sint *S*; fuerit *ER* / aut: sive *R* / visum² *corr. ex* visus *L3*

comprehenditur sine reflexione facta apud superficiem visus, nam
290 tunice visus, que sunt cornea, albuginea, et glacialis, sunt etiam diaf-
fone et spissiores aere. Et iam declaratum est quod forme eorum que
sunt in aere et in aliis corporibus diaffonis extenduntur in illis corporibus, et si occurrerint corpori diverse diaffonitatis ei in quo sunt,
reflectuntur in illo corpore diaffono. Forma ergo eius quod est in aere
295 semper extenditur in aere; cum ergo aer tetigerit superficiem alicuius
visus, tunc illa forma que est in aere reflectitur in superficie visus, et
tunc reflectetur omnimodo in corpore cornee et albuginee, reflexio
enim proprie est de numero formarum. Recipere autem formas et
reflexiones est proprium corporibus diaffonis; forme ergo eorum que
300 visui opponuntur semper reflectuntur in tunicis visus.

[6.18] Et iam patuit quod, cum forme extenduntur super lineas
perpendiculares super secundum corpus, pertranseunt recte in secundo corpore; forme ergo eorum que opponuntur superficiei visus reflectuntur omnes in tunicis visus, et que fuerint ex eis in extremita-
5 tibus linearum radialium perpendicularium super superficiem visus
pertransibunt recte cum reflexione formarum earum in tunicis visus.
Parti enim superficiei visus que opponitur foramini uvee multa opponuntur visibilia, quorum alia sunt apud extremitates linearum
radialium, et alia extra.

10 [6.19] Omnes enim linee radiales que sunt perpendiculares super
superficies tunicarum visus continentur in piramide, cuius capud
est centrum visus et cuius basis est circumferentia foraminis uvee,
et quanto magis extenditur hec piramis et removetur a visu, tanto
magis amplificatur. Et omnes forme eorum que sunt intra pirami-
15 dem extenduntur in rectitudine linearum radialium, et pertranseunt
in tunicis visus recte, et hec piramis dicitur piramis radialis. Linee
autem que extenduntur in hac piramide quarum extremitates sunt

---

289 comprehendatur: comprehenditur P1/reflexive: reflexe R   290 reflexione: refractione R/apud: ad P1   291 albuginea: albugina FP1; albugena O; albiginea S; corr. ex
albigena a. m. L3/et om. R   292 est om. SO   294 occurrerint: occurrent E/corpori:
corporibus ER; corpora C1/post diaffonitatis add. ab R/ei: eo R   295 reflectuntur: refringuntur R; inter. a. m. L3; om. O/quod: que ER   296 tetigerit: tangerit S; contingerit L3;
tangit R   297 reflectitur: refringitur R; figure O; figure et L3   298 reflectetur: reflectitur FP1; refringuntur R/albuginee: albugine FE; albiginee S; albigine O; albgine L3/post
albuginee add. et L3/reflexio: refractio R   300 reflexiones: refractiones R/est proprium
transp. C1   1 visui mg. a. m. L3; om. O/visui opponuntur transp. L3C1ER/reflectuntur:
refringuntur R   2 iam om. P1   3 secundum inter. a. m. E/pertranseunt: pertransi-
unt S   4 corpore: tempore S/superficiei: superficie O; corr. ex superficie E/reflectuntur:
refringuntur R   5 in² om. E   6 linearum radialium transp. S   7 pertransibunt:
pertranseunt R/reflexione: refractione R/earum: eorum S   8 parti: parati O; corr. ex
parati C1; corr. ex pati L3   11 super inter. a. m. L3; om. O   12 superficies: superficiem E
13 foraminis uvee transp. ER   14 hec om. E/hec piramis inter. a. m. L3; om. O   15 amplificatur: applicatur O   17 dicitur inter. a. m. L3/dicitur piramis om. FP1O/post dicitur
scr. et del. pinin et pertranseunt S/piramis: piramidis inter. a. m. L3

108 ALHACEN'S *DE ASPECTIBUS*

apud centrum visus dicuntur linee radiales.

[6.20] Forme vero eorum que sunt extra hanc piramidem num-
20 quam extenduntur per aliquam linearum radialium, tamen extendentur per lineas rectas que sunt inter ipsam superficiem visus que
opponitur foramini uvee. Et forme que extenduntur per has lineas
reflectuntur a diaffonitate tunicarum visus, et forma cuiuslibet puncti
eorum que sunt intra piramidem radialem extenditur ad superficiem
25 visus que opponitur foramini uvee in piramide cuius capud est illud
punctum et cuius basis est superficies que opponitur foramine uvee.
Et una linea earum que ymaginatur in hac piramide est linea radialis;
cetere autem omnes que sunt in hac piramide non sunt radiales, et
nulla earum est perpendicularis super superficies tunicarum visus.

30 [6.21] Et forma cuiuslibet puncti eorum que sunt intra piramidem
radialem extenditur super lineam omnem que potest cadere in illa
piramide cuius capud est illud punctum et cuius basis est superficies rei vise que opponitur foramini uvee. Et per unam istarum linearum transit forma que extenditur per illam in tunicis visus secun-
35 dum rectitudinem, et omnes forme alie extense in residuo piramidis
reflectuntur in tunicis visus et non pertranseunt recte. Omnia ergo
que opponuntur parti superficiei visus que opponitur foramini uvee
ex illis que sunt in aere, aut in celo, aut in aqua, aut consimilibus, et
ex illis que convertuntur a tersis corporibus, que perveniunt ad hanc
40 partem superficiei visus omnes reflectuntur in tunicis visus. Et forme
eorum que sunt intra piramidem pertranseunt recte in tunicis visus
cum reflexione formarum earum que extenduntur super piramidem
que remanent ex universo huius partis superficiei visus. Restat ergo
declarare quod forme que reflectuntur in tunicis visus comprehen-
45 duntur a visu et sentiuntur a virtute sensibili.

[6.22] In primo autem declaravimus quod, si membrum sensibile

---

18 autem: ergo C1 / quarum: quare O; *corr. ex* quare L3 / *post* quarum *add.* autem L3   21 aliquam:
aliquod O / tamen: cum C1O; *corr. ex* tum L3 / extendentur: extenduntur ER   23 opponitur: opponuntur FP1R   24 reflectuntur: refringuntur R   25 radialem *corr. ex* radialium L3   26 que:
quem P1 / uvee: linee O / in . . . uvee (27) *om.* S / piramide: piramidem O; *corr. ex* piramidem L3 / capud
alter. in caput *a. m.* E   28 ymaginatur *corr. ex* opponitur C1   29 *post* que *add.* non L3ER (*mg. a.*
m. L3; *inter.* E)   30 earum *om.* S / super *rep.* P1 / superficies: lineas SC1O; *corr. ex* lineas *a. m.* L3; *alter.*
in lineas *a. m.* E   32 super: supra S / *post* super *scr. et del.* piramidem C1 / illa piramide (33): illam
piramidem R   34 per *inter. a. m.* L3; *om.* SO / unam: una SO   35 visus *mg.* C1   36 *post* rectitudinem *scr. et del.* et P1   37 reflectuntur: refringuntur R / non *corr. ex* nunc L3 / pertranseunt: pertranseant L3; pertransierunt C1 / *post* omnia *add.* enim O / ergo: enim SFP1   38 uvee: linee O / *post*
uvee *scr. et del.* est L3   39 sunt *corr. ex* fiunt *a. m.* E / in³ *om.* SFP1O / *post* aut³ *add.* in ER   40 illis:
hiis FP1 / convertuntur: reflectuntur R / a *corr. ex* ex E   41 *post* visus¹ *add.* et forme eorum que sunt
intra piramidem pertranseunt recte L3O / omnes *om.* R / omnes reflectuntur *scr. et del.* L3 / reflectuntur:
refringuntur R / *post* visus² *add.* et in tunicis visus L3O (*deinde del.* L3) / et . . . visus (42) *om.* SL3O   43 reflexione: refractione R / earum: eorum L3   44 remanent: remaneant F / ex: in R   45 declarare:
declare S / reflectuntur: reflectur S; refringuntur R   46 virtute: visu S

sentiret ex quolibet puncto sue superficiei omnem formam ad ipsam
pervenientem, tunc sentiret formas rerum mixtas, unde membrum
sensibile non sentit formas nisi ex rectitudine linearum perpendicu-
50 larium super superficiem ipsius tantum, quare transeunt forme visibilium nec admiscentur apud ipsum forme visibilium. In hoc vero
tractatu monstravimus quod forme reflexe numquam comprehenduntur nisi in perpendicularibus exeuntibus a visibilibus super superficies corporum diaffonorum. Ergo forme reflexe in tunicis visus
55 non comprehenduntur a visu nisi in perpendicularibus exeuntibus a
visibilibus super superficies tunicarum visus, et hee perpendiculares
linee sunt exeuntes a centro visus.

[6.23] Forme igitur omnes reflexe in tunicis visus comprehenduntur a visu in rectitudine linearum exeuntium a centro visus; forme
60 ergo omnium visibilium que opponuntur parti superficie visus que
opponitur foramini uvee existunt in hac parte superficiei visus, et
reflectuntur in diaffonitate tunicarum visus, et perveniunt ad membrum sensibile, quod est humor glacialis, et comprehenduntur a virtute sensibili per lineas rectas que continuant centrum visus cum illis
65 visibilibus. Sed quod forma cuiuslibet puncti visi oppositi superficiei visus que opponitur foramini uvee existit in universo superficiei
huius partis, et reflectitur a tota hac parte, et pervenit ad humorem glacialem, et tunc ille humor sentit formam ad se venientem. Et virtus
sensibilis comprehendit omnia que perveniunt ad glacialem ex forma
70 visus puncti super unam lineam continuantem centrum visus cum
illo puncto. Hoc igitur modo comprehendit visus omnia visibilia.

[6.24] In hoc capitulo diximus quod eorum que opponuntur superficiei visus alia sunt intra piramidem radialem et alia extra, et cum
dixero superficiem visus, intellige nunc et amodo partem oppositam
75 superficiei uvee. Visibilia ergo que sunt intra piramidem radialem

47 *post* autem *add.* tractatu *R*   48 ipsam: ipsum *SC1*; se *R*   49 pervenientem: venientem *R*/sentiret . . . rerum: formas rerum sentiret *C1*/formas rerum *transp. R*/rerum *mg. C1*
50 *post* sentit *scr. et del.* ex quolibet puncto sue superficiei *S*   51 ipsius: illius *C1*/tran-
seunt: transiunt *S*/forme *inter. a. m. L3*; *om. O*   52 admiscentur: miscentur *L3O*/ipsum *om.*
FP1/forme visibilium *om. R*   53 quod *om. S*/reflexe: refracte *R*   54 visibilibus *corr.*
ex visibus *L3*   55 diaffonorum: diaffonum *S*/reflexe: refracte *R*   56 non *scr. et del. S*
58 linee sunt *transp. C1*/sunt *inter. a. m. L3E*; *om. SO*   59 reflexe: refracte *R*; *corr. ex* reflexive *L3*
61 ergo *om. FP1*/omnium *corr. ex* eorum *P1*/que[1] *inter. a. m. E*/*post* que[1] *rep. et del.* que *F*/opponuntur .
. . que[2] *om. S*   62 foramini uvee *transp. L3O* (uvee: linee *O*)/*post* uvee *add. et L3ER* (*inter. a. m. L3*)/et
om. R   63 reflectuntur: refringuntur *R*/*post* diaffonitate *add.* aeris *L3*   64 humor glacialis *transp. L3*
65 que *corr. ex* non *L3*/illis: ipsis *ER*   66 sed: scilicet *R*; *corr. ex* scilicet *E*/*post* quod *scr. et del.* in *E*/
puncti *om. SFP1*/*post* puncti *add.* cuiuslibet *C1OER*/superficiei *corr. ex* superficie *O*   68 reflectitur:
refringitur *R*/hac *om. ER*/parte: superficie *ER* (*alter. ex* parte *a. m. E*)   69 venientem: pervenientem *C1*   72 hoc igitur *transp. L3*/comprehendit: comprehenderit *P1*   73 *post* hoc *add.* autem *R*/
que *om. S*   74 radialem *om. ER*/*post* radialem *scr. et del.* comprehenduntur a visu *C1*/et[2] . . . uvee
(76) *mg. C1*   75 dixero: dico *R*/intellige: intelligere *FP1R*/*post* intellige *add.* oportet *R*/nunc: tunc *S*

comprehenduntur a visu ex rectitudine linearum radialium recte ex
formis eorum que extenduntur ad visum in rectitudine harum linearum etiam, hee linee enim sunt perpendiculares que exeunt a punctis
visibilibus que sunt intra piramidem super superficiem tunicarum
80   visus. Illa autem que sunt extra piramidem radialem comprehenduntur a visu ex formis reflexis et in rectitudine linearum exeuntium
a centro visus existentium extra piramidem radialem, et hee linee que
sunt extra piramidem possunt etiam dici linee radiales transumptive,
assimulantur enim lineis radialibus in eo quod exeunt a centro visus.
85   Restat ergo declarare per experientiam quod visus comprehendit ea
que sunt extra piramidem radialem.

[6.25] Dicimus ergo quod manifestum est quod lacrimalia et ea
que continent oculum sunt extra piramidem cuius capud est centrum
visus et cuius basis est circumferentia foraminis uvee, quod est par-
90   vum foramen in medio nigredinis oculi. Et si aliquis sumpserit acum
subtilem gracilem et posuerit extremitatem eius in postremo oculi et
inter palpebras, et quieverit visus, tunc videbit extremitatem acus.
Et similiter, si posuerit extremitatem acus in lacrimali, et si miserit
illam in oculo, et applicaverit extremitatem in latere nigredinis oculi
95   aut prope, videbit extremitatem acus. Item omnia que equidistant
superficiei rei vise ex locis continentibus visum sunt extra piramidem
radialem, et cum dico loca continentia visum intelligo illa a quibus
linee exeuntes ad medium superficiei visus secant axem piramidis
radialis. Et si homo erexerit indicem suum in parte sue faciei et prope
100 palpebram, videbit indicem, et similiter, si applicaverit indicem cum
inferiori palpebra ita quod superior superficies eius indicis sit equidistans superficiei visus quantum ad sensum, videbit superficiem indicis.

[6.26] Sed omnia ista loca sunt extra piramidem radialem, et hoc

76 superficiei: superficie *S* / que sunt *transp. C1*   77 rectitudine: rectitudinee *F*   78 harum *om. C1*   79 *ante* etiam *add.* radialium *P1C1* / etiam: et *L3C1ER* / enim: igitur *FP1*; *om.*
ER / sunt *corr. ex* super *L3*; *corr. ex* cum sint *a. m. E* / *post* que *scr. et del.* sunt *C1*   80 superficiem; superficies *R*   81 illa: ille *FP1*   82 reflexis: refractis *R*   83 *post* centro *scr. et*
del. e *P1* / extra: intra *L3* / radialem *mg. C1* / radialem . . . piramidem (84) *mg. F* / et . . . piramidem
(84) *scr. et del. E*   84 *post* piramidem *add.* radialem *ER (deinde del. E)* / etiam: enim *S* / linee
om. P1   85 assimulantur: assimilantur *FR*   86 *post* declarare *scr. et del.* vel *L3* / visus
comprehendit *transp. E*   88 est *om. L3*   89 oculum: circulum *R* / cuius capud *transp.*
FP1 / est centrum *transp. FP1* / est . . . visus (90): centrum visus est *R*   90 *post* est¹ *scr. et del.*
in *L3* / parvum: parum *SO*   91 sumpserit *corr. ex* suprisserit *C1*   92 gracilem: gracialem
S; corr. ex gracialem *E* / posuerit: posuit *S* / in postremo *corr. ex* ipso in posxtremo *S* / oculi *om. E*
93 inter: intra *L3* / tunc: tamen *L3* / acus: eius *R*   94 si¹ *om. E* / acus: eius *FP1* / miserit *corr.*
ex mit *F*   95 *post* extremitatem *scr. et del.* oculi *C1*   96 *post* prope *scr. et del.* et *E* / *post*
extremitatem *scr. et del.* visus *FP1*   97 continentibus: contraheritibus *L3*   98 loca *om.*
S / a: in *O; corr. ex* in *L3*   100 in: ex *ER (alter. ex* in *E)* / parte: partem *P1* / faciei: superficiei *FP1*
101 *post* videbit *scr. et del.* in *S* / et . . . indicem *om. L3*   102 inferiori: inferiore *R* / quod: ut *R*
103 *post* ad *add.* sen *F* / *post* sensum *add.* non *P1*

105 patet, nam piramis radialis quam continet foramen uvee est valde
subtilis, et extenditur recte, et piramidalitas eius non est ampla, unde
nichil ex ipso pervenit ad loca que circumdant oculum, et appropin-
quant corpori oculi, et equidistant superficiei oculi. Et inter omnia
loca continentia oculum et equidistantia superficiei visus et inter su-
110 perficiem visus sunt linee recte propter reflexionem earum a corporibus densis; cum aer qui est inter ipsam et superficiem visus fuerit
continuus, tunc forma horum visibilium pervenit ad superficiem visus super has lineas que sunt extra piramidem. Et cum hec forma
pervenit ad visum non per lineas radiales, et tamen comprehendetur
115 a visu, patet quod visus comprehendit illam reflexive. Ex hac igitur
experientia patet quod visus comprehendit multa eorum que sunt
extra piramidem radialem reflexive.

[6.27] Inductione etiam possumus ostendere quod visus comprehendit illa que sunt intra piramidem reflexive cum hoc quod compre-
120 hendit illa recte hoc modo. Accipias acum subtilem, et sedeas in loco
opposito albo parieti, et cooperias alterum oculorum, et ponas acum
in oppositione alterius oculi, et facias acum appropinquare ita quod
applicetur palpebre. Et ponas acum in oppositione medii visus, et
aspicias parietem oppositum, tunc enim videbis acum quasi corpus
125 diaffonum in quo est aliquantula densitas, et videbis quicquid est ultra acum ex pariete et apud acum quasi corpus latum, cuius latitudo
est multiplex ad latitudinem acus.

[6.28] Causa autem huius in secundo tractatu declarata est, scilicet quod, si res visibilis multum fuerit propinqua visui, videbitur
130 maior quam sit, et quanto magis fuerit illa propinqua, tanto magis
videbitur maior. Diaffonitas autem eius est quia visus comprehendit

---

105 ista: illa *L3* / loca sunt *transp. S* / et *corr. ex* sed *a. m. E*   106 patet: patebit *ER*   108 ipso:
ipsa *R* / *post* loca *scr. et del.* ymaginum *P1* / oculum *mg. a. m. L3*   109 corpori *mg. F*   110 loca
mg. C1 / et² . . . visus (111) *mg. a. m. E*   111 reflexionem: refractionem *R*   112 ipsam: ipsa
L3ER (*alter. ex* ipsam *L3E*)   114 sunt *inter. a. m. L3* / hec *inter. a. m. L3*; *om. O*   115 pervenit: pervenerit *L3C1*; perveniat *R* / *post* visum *scr. et del.* ad *C1* / radiales *corr. ex* reflexas *E* / et: sed
P1 / tamen: cum *C1*; *corr. ex* non *L3* / comprehendetur: comprehendatur *R*; *alter. in* comprehenditur *E*   116 comprehendit: comprehendam *L3*; comprehendat *R* / illam . . . comprehendit (117)
om. S / reflexive: refracte *R* / ex: et *O*; *corr. ex* et *a. m. L3*   117 comprehendit: comprehendet *C1*
118 re-flexive: refracte *R*   119 etiam: autem *ER*   120 intra: contra *O*; circa *S*; *alter. ex* circa
in extra *a. m. L3*; *corr. ex* extra *E* / *post* piramidem *add.* radialem *R* / reflexive: refracte *R*   122 albo: arbo *F* / cooperias: cooperiat *L3O* / ponas: ponat *L3O*   123 *post* in *scr. et del.* p *P1* / oculi *om.*
P1 / facias: faciat *L3O* / appropinquare *alter. in* transire *a. m. E* / *post* appropinquare *add.* acum *L3O*
(*deinde del. L3*) / quod: ut *R*   124 applicetur: applicet *O* / ponas: ponat *L3O* / *post* oppositione *rep.*
et del. in oppositione *S*   125 aspicias: aspicies *FP1*; aspiciat *L3* / videbis: videbit *L3O* / quasi *corr.*
ex quando *E*   126 quo: qua *FP1E*   127 ex pariete *om. L3* / corpus latum *transp. E* / latum
inter. a. m. E   129 causa: cum *O*; est aut *E* / huius . . . tractatu: in secundo huius tractatus *E* / in
. . . est: declarata est in secundo tractatu *L3* / est *mg. F*; *corr. ex* erit *E* / scilicet *om. E*   130 si *inter.*
a. m. L3; *om. O* / visibilis *corr. ex* invisibilis *F* / multum fuerit *transp. R*   131 illa *om. ER*

quicquid est ultra. Acus autem est corpus densum cooperiens quod est ultra, et quia acus est valde propinqua visui, ideo cooperuit de pariete multiplex ad sui latitudinem, piramis enim, cuius capud est
135 centrum visus et basis est latitudo acus, basis eius erit multiplex ad latitudinem acus. Et cum hoc visus comprehendit quicquid est ultra acum, nec cooperuit a visu aliquid de pariete, sed comprehendit quod est ultra quasi ultra corpus diaffonum.

[6.29] Et cum acus fuerit opposita medio visui, tunc non cooperiet
140 totam superficiem visus propter subtilitatem eius sed aliquam partem quanta est latitudo eius, et remanet ex superficie visus aliquid a lateribus acus, et exit forma eius ad illud quod est a lateribus acus de superficie visus. Forma autem exiens ad acum numquam perveniet ad visum, nec comprehendetur ab ipso; forma autem que pervenit ad
145 latera superficiei visus reflectitur ad visum, cum non recte perveniat ad centrum visus. Si ergo visus non comprehenderet illud quod opponitur acui ex pariete nisi recte, tunc illud quod opponitur acui ex pariete esset coopertum a visu. Cum igitur comprehendatur et non recte, patet ipsum comprehendi reflexive per formam que reflectitur
150 a lateribus acus ex superficie visus. Et hoc manifestatur etiam, si experimentator posuerit loco acus aliquod corpus latum cuius latitudo sit maior latitudine foraminis uvee, tunc enim nichil videbit omnino de pariete, nec videbit illud corpus diaffonum sed densum.

[6.30] Ex hoc ergo quod paries comprehenditur ultra acum ex gra-
155 cilitate eius et non comprehenditur ultra corpus latum scimus quod illa comprehensio est ex forma que pervenit ad acum ex superficie visus et reflectitur in tunicis visus. Et quia quicquid a visu comprehenditur reflexive comprehenditur in rectitudine perpendicularium, ideo illud quod comprehendit comprehendit reflexive ex forma eius

132 videbitur *inter. a. m.* L3/quia: quod L3; si O/visus *corr. ex* visis O   133 autem: aut F   134 et *inter.* OE (*a. m.* E)/acus *corr. ex* arcus L3   135 sui: suam R/*post* latitudinem *add.* basis R/piramis enim: enim piramidis R/enim: eius FP1   136 latitudo: altitudo R/basis eius *om.* R/*post* basis *scr. et del.* est latitudo S/erit: est L3/*post* multiplex *add.* ad latitu O   137 cum *om.* P1   138 cooperuit: cooperitur R/*post* cooperuit *scr. et del.* vi F/de: a FP1; *om.* E/comprehendit quod (139) *transp.* L3O 139 est *om.* L3   140 fuerit: fuit FP1L3/cooperiet: comperiet S   141 totam superficiem *transp.* FP1/eius *inter.* C1   142 superficie *corr. ex* parte *a. m.* E   143 *post* acus[1] *add.* exit ad acum L3O (*deinde del.* L3)   144 acum *corr. ex* locum S   145 ab ipso *om.* S   146 reflectitur: refringitur R 147 *post* ergo *scr. et del.* co F/non *om.* FP1/illud *om.* C1/quod . . . illud (148) *om.* FP1   148 acui ex pariete: ex pariete acui R/ex *corr. ex* et *a. m.* C1/ex . . . recte *corr. ex* nisi recte ex pariete E/nisi: visui S; nec O; *corr. ex* nec L3/tunc . . . pariete (149) *mg. a. m.* E/acui[2] *rep.* L3   149 esset coopertum *transp.* S 150 comprehendi *corr. ex* comprehenditur C1/reflexive: refracte R/reflectitur: refringitur R   151 *post* superficie *scr. et del.* acus S/*post* hoc *add.* iam ER/*post* etiam *add.* quod ER   152 posuerit: posuit F/acus: aeris L3O   153 foraminis: foraminum FP1/*post* foraminis *scr. et del.* vitri C1/videbit omnino *transp.* L3   154 nec: nunc S/sed *alter. ex* sedens ad *in* se ad S   156 scimus: simus S 157 *post* forma *add.* que per P1/ad *inter.* L3; *om.* O/acum: acus O; *corr. ex* acus L3/*post* acum *add.* et L3   158 visus[1] *corr. ex* eius E/reflectitur: refringitur R   159 reflexive: refracte R

160   quod opponitur acui per rectitudinem linearum exeuntium a centro
visus que continuant centrum visus cum eo quod opponitur acui ex
pariete, et hee linee secantur acu. Et visus comprehendit illud quod
est ultra acum etiam in rectitudine harum linearum, et comprehendit
acum etiam in rectitudine illarum, quare totam quasi formam com-
165   prehendit ultra corpus diaffonum in quo est aliquantula densitas.

     [6.31] Et si experimentator scripserit in bombace subtiliter et ap-
plicaverit ipsum parieti, et remotus fuerit a pariete in quantum possit
legere scripturam, et posuerit acum in oppositione medii visus, ut
s   primo fecit, et aspexerit bombacem, tunc poterit legere scripturam,
170   sed tamen videbit eam quasi ultra vitrum aut ultra corpus diaffonum
in quo est aliqua densitudo. Si ergo visus non comprehenderet illud quod opponitur acui de bombace secundum reflexionem, tunc
aliquid lateret de scriptura, acus enim debet cooperire de scriptura
multo magis se in quantitate latitudinis diaffonitatis quam tunc com-
175   prehendit propter remotionem bombacis a visu. Sed quia visui non
latet aliquid de scriptura, patet ipsum comprehendere illud quod opponitur acui, sed hoc non potest fieri recte. Restat ergo quod fiat reflexive.

     [6.32] Et si experimentator abstulerit acum, non destruetur re-
180   flexio que prius erat, non enim propter acum erat reflexio, sed crescit
reflexio eo quod reflectitur ex loco acus. Et cum experimentator abstulerit acum, comprehendet illud quod opponitur visui manifestius, nam comprehendet illud recte quod cooperiebatur acu, cum hoc
quod comprehendit illud reflexive, sicut comprehendebat cum cooperie-
185   batur, et propter hanc additionem comprehendit illud manifestius
quam antequam auferabat acum, ex qua experientia patet quod illud quod opponitur visui de illis que sunt intra piramidem radialem
comprehenditur reflexive et recte.

     [6.33] Ex hiis ergo omnibus declaratur quod omnia que compre-

160 quod: quia O / comprehendit[2] inter. a. m. L3; om. SC1O / reflexive: refracte R   161 rectitudinem corr. ex reflexionem a. m. E   162 que . . . visus om. FP1R   164 etiam: est FP1 /
harum: illarum FP1 / linearum mg. F / et . . . illarum (165) om. FP1   165 post totam add. ultra
FP1 / quasi: quare E; scr. et del. L3 / quasi formam transp. R / comprehendit: comprehendet R;
corr. ex comprehendet a. m. E   166 ante ultra inter. quasi a. m. L3 / ultra om. FP1 / quo: qua S
167 post in scr. et del. idem F   168 a: in S   171 post tamen add. non FP1E   172 quo:
qua P1 / aliqua: aliquantula C1 / densitudo: densitas R / visus inter. u. m. E; om. P1 / comprehenderet: comprehendet FP1C1; comprehendit R; alter. ex comprehendet in comprehendit a. m. E
173 reflexionem: refractionem R   175 magis alter. in maius a. m. E   176 visui: visum FP1
177 latet: patet SL3R; corr. ex patet a. m. O   178 quod: ut R / reflexive: refracte R   180 non:
si S / destruetur corr. ex struetur a. m. L3 / reflexio: refractio R   181 reflexio: refractio R;
ratio S   182 reflexio: refractio R / reflectitur: refringitur R   183 visui: visu C1   184 cum
inter. a. m. L3; om. SC1O   185 reflexive: refracte R / cum corr. ex quod L3   186 illud:
aliud L3   187 quam inter. a. m. L3; om. SFP1O / auferabat: auferret R   188 illis: istis L3
189 comprehenditur: comprehenduntur SC1E; comprehendentur FP1 / reflexive: refracte R

190   henduntur a visu quorum forme perveniunt ad visum recte, aut con-
         versive, aut reflexe omnia comprehenduntur secundum reflexionem
         factam apud superficiem visus, et quod illa que comprehenduntur
         secundum reflexionem factam a superficie visus quedam compre-
         henduntur reflexe et recte simul. Et ideo illud quod opponitur me-
195   dio visus est manifestius illo quod est in circuitu medii, et cum visus
         comprehenderit aliquod latum, comprehendet illud quod est in me-
         dio manifestius illo quod est in lateribus. Hoc autem declaratum est
         in secundo tractatu, in quo declaravimus qualiter hoc posset experiri
         et diximus quod causa huius est propter lineas radiales, et hoc est in
200   illis que sunt intra piramidem radialem. In illis autem que sunt extra
         causa est reflexio. Causa ergo universalis in hoc quod illud quod op-
         ponitur medio visus est manifestius quam illud quod est in circuitu
         est quoniam illud quod opponitur medio visui comprehenditur recte
         et reflexe simul. Hoc autem quod quicquid comprehenditur a visu
205   comprehenditur reflexe a nullo antiquorum dictum est.

CAPITULUM SEPTIMUM
In deceptionibus visus que fiunt secundum reflexionem

         [7.1] Fallacie que accidunt secundum reflexionem similes sunt eis
         que accidunt per conversionem, quod enim comprehenditur re-
210   flexive comprehenditur non in suo loco, cum comprehendatur in loco
         ymaginis, quapropter positio forme comprehense erit alia a positione
         rei vise, et similiter remotio in eis. Item reflexio debilitat formam
         reflexam, scilicet formam lucis et coloris que sunt in re visa. Et hoc
         potest intelligi quoniam, si aspexeris aliquid existens in aqua, et tu sis

191 conversive: reflexe *R*   192 reflexe: refracte *R*/omnia *om. R*/*post* comprehenduntur *rep. et del.*
a (191) . . . recte (191) *S*/reflexionem: refractionem *R*   193 illa: illorum *R*   194 reflexio-nem:
refractionem *R*/quedam: que *C1*   195 *ante* reflexe *scr. et del.* et *L3*/reflexe: reflexive *E*; refracte *R*/
illud: istud *L3*/medio visus (196) *transp. C1*   197 aliquod: aliquid *L3R*   199 *post* secundo *scr.
et del.* con *C1*/qualiter: quomodo *R*/hoc posset *transp. FP1*/experiri: experimentari *R*   200 *post*
causa *scr. et del.* eius *S*/huius *om. P1*/est¹ *inter. a. m. L3*; *om. O*   201 illis²: aliis *R*/autem *inter. a.
m. E*   202 causa est: cause *FP1*; *transp. ER*/reflexio: refractio *R*/ergo: autem *ER*   203 quod
om. S/*post* est² *scr. et del.* circumferentia *P1*/in . . . est (204) *inter. a. m. L3*   204 est *om. E*/quod
mg. a. m. C1/visui: visu *L3*; visus *R*/*post* visui *add.* et *FP1*/*post* comprehenditur *add.* et *SFC1*
205 et *om. O*/reflexe: refracte *R*   206 comprehenditur: comprehendatur *R*/reflexe: refracte
R/dictum est *mg. a. m. L3*   207 capitulum septimum *om. R*/capitulum . . . reflexionem (208)
om. S   208 in deceptionibus: de fallaciis *R*/que fiunt *om. E*/fiunt . . . reflexionem: accidunt
ex refractione *R*/*post* reflexionem *add.* capitulum septimum *R*   209 *post* fallacie *add.* autem
S/reflexionem: refractionem *R*/eis: iis *R*   210 conversionem: reflexionem *R*/*post* conversionem *scr. et del.* quod enim *C1*/reflexive: refracte *R*   211 comprehendatur: comprehenditur *L3*
212 positio *om. S*/a positione: positio *S*   213 et . . . eis *om. R*/reflexio: refractio *R*/formam:
forma *S*   214 reflexam: refractam *R*

215 obliquus a perpendicularibus exeuntibus in re visa super superficiem
aque multa obliquatione, et intuearis illud vere, deinde movearis et
moveas visum donec ponas ipsum in aliqua perpendiculari exeunte a
re visa super superficiem aque, et aspexeris, tunc videbis illud manifestius quam cum eras obliquus. Et nulla est differentia inter duos
220 situs nisi quia in primo forma que exit ad visum est reflexa et multum
obliqua; in secundo autem forma exit recte, aut quedam pars ipsius
exit recte, et quedam modicum oblique, aut fere recte. Ex hac igitur
experimentatione declaratur quod reflexio debilitat formas reflexas.

[7.2] Item ea que sunt in aqua, et ultra vitrum, et consimilia, quan-
225 do reflectuntur ad visum, deferunt secum colorem corporis in quo
existunt. In illis ergo que comprehenduntur reflexe ultra corpora diaffona accidunt propter reflexionem fallacie que non accidunt in eis
que videntur recte, scilicet diversitas positionis et distantie et debilitas lucis et coloris. Preterea accidunt eis ea que accidunt illis que recte
230 videntur, forme enim eorum que comprehenduntur reflexive comprehenduntur in oppositione visus et in rectitudine linearum radialium. Quicquid ergo accidit eis que videntur in rectitudine linearum
radialium accidit istis. Et in tertio declaravimus omnes illas fallacias
et causas earum, et que sunt etiam cause istorum. Sed in hiis accidit
235 magis et citius propter debilitatem huiusmodi formarum.

[7.3] Particulares autem deceptiones que accidunt propter figuras
superficierum corporum diaffonorum sunt multimode, sed accidunt
raro visui, ea enim que comprehenduntur ultra corpora diaffona
diversa ab aere sunt stelle et ea que sunt in aqua; illa autem que sunt
240 ultra vitrum et lapides diaffonos diversarum figurarum raro comprehenduntur a visu. Et non est ita de istis corporibus diaffonis ut
de speculis, specula enim sepius aspiciuntur ab hominibus, ut vide-
ant in eis suas formas, et habentur in domibus. Et similiter, quando
homo aspicit in quolibet corpore terso, etiam videbit formam eorum

215 aspexeris: aspexerit S / aliquid: aliquod E   216 in: a R   217 post et^1 scr. et del. intuetur illud P1 / et^2 inter. a. m. L3   218 post moveas scr. et del. illud P1 / visum corr. ex visus
C1 / exeunte: exeunti C1E   219 super om. E / videbis: videbit S / illud: illum S / manifestius:
manifestium S   220 eras: eas O; corr. ex eas a. m. L3 / duos om. S   221 post primo add. situ
FP1 / reflexa: refracta R   222 ipsius: illius FP1   223 exit: erit L3 / exit recte transp. FP1
224 reflexio: refractio R / reflexas: refractas R   226 reflectuntur: refringuntur R / secum:
suum S   227 reflexe: refracte R   228 accidunt: accidit OE / reflexionem: refractionem R
229 positionis . . . debilitas rep. P1   230 preterea: propterea FP1L3 / accidunt: accidit
SC1OE; corr. ex accidit L3 / ea: ista R / om. FP1   231 comprehenduntur: extenduntur
C1 / reflexive: refracte R   232 visus: illius O   233 quicquid . . . radialium (234) mg.
a. m. L3; om. O / accidit . . . radialium (234) om. S / eis inter. L3 / eis que transp. L3 / que: qui FP1
234 et om. L3O / post tertio add. libro R   235 etiam: et L3 / istorum: istarum R; earum FP1
236 citius corr. ex cin F / huiusmodi: harum ER   239 raro corr. ex torto a. m. L3   240 diversa corr. ex diaversa S / post sunt^2 scr. et del. stelle F   241 diversarum . . . raro mg. a. m. E
243 hominibus: omnibus FP1   244 in eis om. FP1 / suas formas transp. FP1

245 que sunt in oppositione, et similiter, si aspexerit aquam, videbit formam sui in ea, et videbit ea que sunt in oppositione, et non est ita
illud quod videtur ultra vitrum et lapides diaffonos, quia homines
raro aspiciunt ad illud quod est ultra vitrum et lapides diaffonos. Et
quia ita est, dicamus de deceptionibus reflexionis particularibus que
250 semper accidunt et sine difficultate, scilicet que accidunt in eis que
videntur in celo et aqua, et dicemus parum de hiis que videntur ultra
vitrum et lapides.

[7.4] Dicimus ergo quod semper visus fallitur in eis que comprehenduntur ultra corpus diaffonum diversum ab aere preter quam in
255 positione, et remotione, et coloribus, et lucibus eorum, ut in magnitudine eorum, et figuris quorumdam, ea enim que videntur in aqua
et ultra vitrum et lapides diaffonos videntur maiora. Stelle autem et
distantie inter stellas quandoque videntur maiores, quandoque minores.

260

[7.5] **[PROPOSITIO 12]** Sit ergo visus A [FIGURE 7.7.10, p. 446],
et sit BG ultra corpus diaffonum grossius aere. Dico quod BG videtur
maior quam sit.

[7.6] Sit ergo in primo superficies corporis diaffoni plana. A autem
aut est in perpendiculari exeunti a medio BG super superficiem cor-
265 poris, aut extra. Sit ergo in primis in ipsa, et sit illa perpendicularis
AMZ. Et extrahamus superficiem in qua sunt linee AZ, BG, et faciat
in superficie corporis diaffoni lineam DME. Linea ergo AM est perpendicularis super lineam DME, et superficies in qua sunt due linee
AZ, BG erit perpendicularis super superficiem corporis diaffoni.

270 [7.7] Et non transit per A et per aliquod punctum linee BG super-
fices que sit perpendicularis super superficiem corporis diaffoni nisi
illa in qua sunt linee AZ, BG, non enim transit per A superficies per-

245 aspicit: inspexerit *R*/quolibet . . . terso: quodlibet corpus tersum *R*   247 sui: suam *R*/ea²
om. ER/post ea² *scr. et del.* super *O*   248 videtur: videbit *ER*/ultra: ultrum *S*/diaffonos: diaffonos *C1*/quia *corr. ex* quare *L3*/quia . . . diaffonos (249) *om. S*   249 diaffonos: diaffones *C1*
250 ita *corr. ex* ista *F*/reflexionis: refractionis *R*/particularibus: perpendicularibus *O*; *corr. ex* perpendicularibus *L3*/post particularibus *add.* ea *L3O*   251 semper *inter. a. m. L3*/sine: si *O*/scilicet
inter. F   252 *post* et¹ *add.* in *L3C1ER*/parum *corr. ex* pararum *F*   254 dicimus: dicamus *R*/post
ergo *scr. et del.* omnis *F*   255 corpus diaffonum: corpora diaffona *L3O*/preter quam: presertim *R*
256 positione *corr. ex* oppositione *L3*/et²: in *R*/et lucibus *mg. F*/ut: et *R; inter. a. m. L3; om. O*
257 figuris *corr. ex* figuras *L3*/ea enim: earum *P1*   258 et¹ *inter. a. m. E*/diaffonos: diaffones
C1O/videntur: viduntur *O*/maiora . . . videntur (259) *mg. a. m. E*   261 visus A *transp. C1*
262 BG¹·²: BC *R*   264 in *om. R*/post plana *scr. et del.* est *S*/autem *om. R*   265 aut *om. P1*/aut
est *inter. a. m. L3*/est *om. O*/perpendiculari *corr. ex* perpendicularibus *L3*/exeunti: exeunte *P1R*;
corr. ex exeuntibus *L3*/BG: BC *R*   266 sit¹: si *O*/post perpendicularis *add.* exiens a centro *P1*
267 AMZ: AMBZ *O*; *corr. ex* AMBZ *L3*/sunt *corr. ex* sint *E*/post sunt *add.* due *L3*/linee: linea
F; *corr. ex* linea *P1*/post linea *scr. et del.* ABG *P1*/BG: BC *R*/faciat: faciet *R*   268 linea . . .
DME (269) *scr. et del. E; om. R*/ergo: ZGD *O*; *corr. ex* TGD *L3*   270 BG: BC *R*/diaffoni *inter.
a. m. L3; om. O*   271 BG: BC *R*/post BG *add.* super *S*   272 diaffoni: diaffani *R*

pendicularis super superficiem corporis diaffoni nisi illa que transit
per AZ, que linea est perpendicularis super superficiem corporis diaf-
275 foni, nec exit ex A perpendicularis super superficiem corporis diaffoni nisi linea AZ. Non ergo transit per A superficies que sit perpendicularis super superficiem corporis diaffoni nisi illa que transit per
lineam AZ, et non transit per aliquod punctum linee BG et per lineam
AZ nisi illa superficies in qua sunt due linee AZ, BG. Non ergo tran-
280 sit per A et per aliquod punctum linee BG superficies perpendicularis
super superficiem corporis diaffoni nisi illa in qua sunt linee AZ, BG;
non ergo reflectitur forma alicuius puncti eorum que sunt in BG nisi
ex linea DE.

[7.8] Et extrahamus ex B et G duas perpendiculares. Cadant ergo
285 in lineam DE in duobus punctis D, E, scilicet BD, GE. Et sit BG in primis equidistans linee DE, et reflectatur forma B ad A ex T et forma G
ad A ex H. Et continuemus lineas BT, TA, GH, HA, et extrahamus AT
ad L et AH ad K. Quia ergo Z positum fuit in medio linee BG, positio
B ex A erit equalis positioni G ex A, et sic distantia T ex A erit sicut
290 distantia H ex A, et sic angulus DTL erit equalis angulo EHK. Sed
duo anguli D, E sunt recti, et linea DT est equalis linee EH, quia TM
est equalis linee MH. Ergo DL est equalis EK.

[7.9] Et continuemus LK. Erit ergo equalis linee BG. Et continuemus AB, AG. Angulus ergo GAB erit minor angulo KAL, et linea
295 LK est diametrum ymaginis linee BG, nam omne punctum linee BG
reflectitur ab aliquo puncto linee TH, nam si forma B reflectitur ex
T, punctum quod est inter B et Z reflectitur ab aliquo puncto inter T
et M. Et ponamus super lineam BZ punctum N. Si ergo forma eius
reflecteretur ab aliquo puncto extra lineam MT ex parte D, tunc linea

273 BG: BC R / enim inter. L3   274 diaffoni: diaffani R / illa inter. a. m. E   275 linea est transp.
C1 / diaffoni om. R   276 diaffoni: diaffani R   277 linea . . . lineam¹ (279) rep. L3 (illa . . . lineam
mg. a. m. in prima) / non ergo transp. FP1 / transit per A: per A transit ER / post A add. super P1 / sit corr. ex
transit L3   278 diaffoni: diaffani R / post diaffoni rep. nec (276) . . . diaffoni (276) L3O (deinde del. L3) /
illa . . . lineam¹ (279) om. C1   279 linee BG transp. FP1 / BG: BC R   280 BG: BC R   281 et per
inter. a. m. E / BG: BC R / post BG add. super C1   282 diaffoni: diaffani R / linee inter. a. m. E / BG: BC R
283 reflectitur: refringetur R / BG: BC R   285 G: C R / cadant: cadent R   286 D E om. S / scilicet
inter. a. m. L3 / GE: CE R / BG: BC R   287 reflectatur: refringatur R; corr. ex reflectantur S / T: P R / G:
C R; inter. F   288 ad A inter. a. m. L3 / BT: BP R / TA: PA R; TLA O / GH: CH R / AT: AP R   289 et
om. E / AH: AB S / K: R O / Z: ZZ SO / in om. S / BG: BC R   290 G: C R / T: P R; corr. ex DT L3 / erit² . . .
A (291) mg. a. m. E   291 DTL: DIL S; DPL R / EHK: FBR O   292 DT: DP R / EH . . . linee (293) mg.
a. m. E / quia: quare P1 / TM: PM R   293 linee om. R / MH corr. ex TH P1 / post MH add. ergo DM est
equalis linee MH S / DL: DM S / EK alter. ex DE in ER O   294 LK: LR O / BG: BC R / et² . . . AG (295)
om. R   295 post AB add. et E / GAB: CAB R / KAL: RAL O / post KAL inter. suo toto a. m. E   296 LK:
MK S; LR O / diametrum: diameter R / linee¹ om. R / BG¹: BK S; BC R / BG²: BC R   297 reflectitur¹,²:
refringitur R / linee om. ER / TH: PH R   298 T¹: P R / post inter¹ add. punctum FP1 / reflectitur: refringitur R / post puncto add. extra lineam L3 / inter om. O / inter . . . lineam (300) mg. a. m. L3 / T: P R; inter. a.
m. E   299 et¹: Z O / super lineam corr. ex superficiem a. m. E / eius: enim O; N C1ER

300 per quam extenditur forma N secaret lineam BT, et sic forma puncti
     sectionis reflectetur ad A ex duobus punctis, quod est impossibile, ut
     diximus in capitulo de ymagine. N ergo non reflectitur ad A nisi ex
     aliquo puncto inter T, M, et similiter omne punctum in ZG non re-
     flectetur ad A nisi ex linea MH. Linea ergo LK est diameter ymaginis
5   linee BG; forma ergo BG videbitur in LK.
          [7.10] Item iam declaravimus quod forma reflexa est debilior rec-
     ta. Ergo forma BG, que comprehenditur reflexe, est debilior forma
     eius que comprehenditur recte, et propter debilitatem forme rei as-
     similat eam visus forme rei que videtur a maiori remotione, maior
10   enim distantia debilitat formam. Et iam declaravimus in secundo
     quod visus comprehendit ymaginem rei vise secundum quantitatem
     anguli respectu remotionis et positionis rei vise apud visum. Et angu-
     lus KAL est maior angulo GAB, et positio LK est sicut positio BG, et
     BG videtur in LK, et LK comprehenditur quasi in maiori distantia
15   BG propter debilitatem forme. Visus ergo comprehendit BG reflexive
     ex comparatione anguli maioris angulo GAB ad distantiam maiorem
     distantia BG et ad positionem equalem positioni BG, quapropter
     BG comprehenditur reflexive maior, et hoc duabus de causis, scilicet
     magnitudine anguli et debilitate forme. Causa autem magnitudinis
20   anguli est propinquitas anguli ex visu, et causa propinquitatis est re-
     flexio. Causa ergo qua BG comprehenditur maior est reflexio.
          [7.11] Item iteremus formam, et sit BG [FIGURE 7.7.10a, p. 446]
     non equidistans linee DE. Et extrahamus a remotiore extremitatum
     BG ex linea DE lineam equidistantem linee DE, et sit GQ. Et extra-
25   hamus AZ ad O. Erit ergo O in medio GQ, quare Z est in medio BG,

300 reflecteretur: reflectetur *FP1*; reflectitur *L3E*; refringeretur *R*/ab . . . puncto *om. FP1*/MT:
MP *R*   1 N: enim *L3*/BT: BP *R*/puncti *corr. ex* punctis *C1*/puncti sectionis (2) *transp. L3*
2 reflectetur: refringeretur *R*; *corr. ex* reflectitur *E*/*post* quod *scr. et del.* est *P1*/est impossibile
transp. FP1   3 *post* capitulo *add.* quinto huius libri *R*/N: non *P1*/non *mg. F*; *om. P1*/reflec-
titur: reflectetur *SFO*; refringitur *R*/*post* A *scr. et del.* nisi *E*/*post* ex *scr. et del.* inter *E*   4 aliquo puncto *mg. a. m. C1*; *om. SL3O*/inter *inter. a. m. E*/T: P *R*/ZG: GZ *FP1*; ZC *R*/reflectetur:
reflectitur *C1E*; refringetur *R*   5 LK: LR *O*/diameter . . . BG[1] *corr. ex* ymaginis linee BG
diameter *C1*   6 BG[1]: BC *R*/*inter. C1*/BG[2]: BC *R*/in *mg. F*/LK: AK *S*; LR *O*   7 item *corr.
ex* ia *F*/reflexa: refracta *R*   8 BG: BC *R*/reflexe: reflexive *P1*; refracte *R*   9 rei *om. E*/rei
. . . forme (10) *om. S*/assimilat . . . visus (10): visus assimilat eam *ER* (*mg. a. m. E*)   10 forme
rei *mg. a. m. E*/maiori: maiore *R*/maior *alter. in* minor *a. m. E*   11 *post* et *scr. et del.* im *F*/*post*
secundo *add.* libro *R*   12 *post* secundum *scr. et del.* formam *C1*   14 KAL: RAL *O*/GAB:
CAB *R*/LK: KL *P1*; LR *O*/est[2]: et *O*/BG: CB *R*; *inter. E*   15 BG: BC *R*/LK[1,2]: LR *O*/quasi in
maiori: in maiore quasi *R*/distantia *rep. SC1E* (*scr. et del.* BG *ante secundum C1*)   16 BG[1,2]: BC
R/reflexive: refracte *R*   17 GAB: CAB *R*   18 BG[1,2]: BC *R*/quapropter *corr. ex* qua *a. m. L3*/
quapropter BG (19) *mg. E*   19 BG: BC *R*/reflexive: refracte *R*/*post* hoc *add.* est *L3*   21 anguli[1]
om. S/ex visu: ad visum *R*/*post* propinquitatis[2] *add.* anguli *E*/reflexio: refractio *R*/*post* reflexio
rep. et. del. et (21) . . . reflexio *E*   22 BG: BC *R*/comprehenditur *mg. a. m. L3*; *om. O*/reflexio:
refractio *R*   23 formam: figuram *R*; *corr. ex* figuram *a. m. E*/BG: BC *R*   25 BG: BC *R*/ex
corr. ex et *C1*/ex linea DE *om. R*/*post* DE[1] *scr. et del.* et sit GQ *P1*/sit *inter. a. m. E*/GQ: CQ *R*

quia BQ est equidistans ZO. Ergo proportio QO ad OG est sicut BZ ad ZG, Et reflectatur forma Q ad A ex T, et forma G ad A ex H. Et continuemus AT, et pertranseat usque ad L, et continuemus AH, et pertranseat usque ad K, et continuemus LK. Erit ergo LK diameter
30 ymaginis QG. Et continuemus AQ, AG. Erit ergo angulus KAL maior angulo GAQ; A ergo comprehendit ymaginem QG maiorem quam QG. ut prius diximus.

[7.12] Linea autem QT secabit lineam BG in R. R ergo reflectetur ad A ex T; ergo B reflectetur ad A ex puncto inter duo puncta T, D, nam
35 si reflecteretur ex puncto inter T, M, accideret predictum impossibile. Reflectatur ergo B ad A ex F. Et continuemus AF, et pertranseat ad I. Et continuemus IK. Ergo IK erit diameter ymaginis BG, et positio IK in respectu A est similis positioni BG, quia IK aut erit equidistans ad BG, aut non erit inter illum et equidistantem diversitas que mutet
40 positionem, non enim est inter distantiam IK et distantiam BG a visu grandis diversitas, quare declinatio IK a linea equidistanti BG que exit ex K erit valde parva. Angulus ergo IAK est maior angulo BAG, et positio IK est similis positioni BG, et IK comprehenditur quasi remotior propter debilitatem forme eius. Linea ergo IK videtur maior
45 quam BG, ut in precedenti figura declaravimus. Sed IK est ymago BG; ergo BG videbitur maior quam sit, et hoc est quod voluimus.

[7.13] **[PROPOSITIO 13]** Item sit visus A [FIGURE 7.7.11, p. 447], et res visa BG, et extrahamus perpendiculares BD, GE, et continuemus DE. Et sit BG equidistans DE, et sit A extra superficiem BDGE,
50 cum eo quod continuatur cum ipsa. Et dividamus BG in duo equalia in Z, et extrahamus perpendicularem AH, et continuemus AZ, et sit

26 *post* AZ *add.* perpendicularem *C1* / GQ: CQ R / GQ . . . medio² *om. FP1* / quare *inter. a. m. L3* / *post* quare *add.* et C1 / Z: T *SL3;* et O / medio BG *transp.* E / BG: BC R   27 quia: et *SC1O; alter.* in quod E / ergo: et *L3ER (mg. a. m. L3); corr. ex* go F; *om. O* / QO: ZO *SO; inter. L3* / OG: OC R / est² *om. R*   28 ZG: ZC R / reflectatur: refringatur R / Q: quasi O; *corr. ex* AQ P1 / ad² *rep. P1* / T: P R / T . . . ex² *inter. a. m. E* / G: C R   29 AT: AP R / et² . . . QG (31) *om. O* / AH . . . continuemus (30) *mg. a. m. E* / et³ . . . AQ (31) *mg. a. m. L3*   31 QG: QC R / et . . . AG *om. R* / AQ: AD O / erit: eritque R / ergo *om.* E / KAL: FAL O   32 GAQ: CAQ R / comprehendit: comprehendet *SC1ER* / QG: AG O; QC R / quam *om. R*   33 QG: QC R; *om. L3O*   34 QT: QP R; que O; *corr. ex* QTE *L3* / BG: BC R / R¹: K *P1C1* / R²: RC O; *corr. ex* RC *L3; om. P1* / reflectetur: reflectitur E; refringetur R   35 T¹·²: P R / reflectetur: reflectitur E; refringetur R / puncto *corr. ex* positione *a. m. L3* / nam: cum *L3*   36 reflecteretur: reflectetur *P1O;* reflectentur F; refringeretur R / *post* reflecteretur *add.* ad A *L3* / puncto *inter. a. m. L3; om. SO* / T: P R   37 reflectatur: reflectetur *SC1;* refringatur R   38 I: L S / IK¹·²: IR O / BG: BC R   39 IK¹·²: IR O / BG: BC R   40 BG: BC R / illum: illam R / *post* que *scr. et del.* in L3   41 enim est *transp.* ER / distantiam IK et *om. P1* / IK: IR O / BG: BC R   42 IK: ZR O / equidistanti: equidistante R / BG: BC R   43 angulus ergo *transp.* ER / BAG: LAG F; BAC R   44 BG: BC R   45 IK: KI ER; KQ O; GQ *SFP1C1*   46 BG: BC R / precedenti: precedente R; *corr. ex* precen F   47 BG¹·²: BC R   49 BG: BC R / GE: CE R   50 BG: BC R / DE: DC *C1* / A *om. F* / extra: ex S / BDGE: BDCE R; BD GC *C1;* BG GT O   51 BG: BC R

AZ posita perpendicularis super BZG. Positio ergo B respectu A est similis positioni G respectu A, et distantia B ex A est equalis distantie G ex A. Et reflectatur B ad A ex T, et G ad A ex K. Positio ergo T re-
55 spectu A est similis positioni K respectu A, et distantia T ex A est sicut distantia K ex A.

[7.14] Et continuemus lineas BT, TA, GK, KA. Est ergo superficies in qua sunt due linee AT, BT perpendicularis super superficiem corporis diaffoni, quia est superficies reflexionis; perpendicularis ergo
60 BD erit in hac superficie, et perpendicularis que exit ex T. Linea ergo AT secat BD. Extrahatur ergo AT, et secet BD in L, et extrahatur AK, et secet GE in O. Erit ergo AL sicut AO, et erit BL sicut GO. Et continuemus LO, que est diameter ymaginis BG, et erit LO equalis BG. Et continuemus AB, AG. Utraque ergo superficies ALB, AOG est per-
65 pendicularis super superficiem corporis diaffoni, et tres superficies perpendiculares super superficiem corporis diaffoni que transeunt per puncta B, Z, G secant se in perpendiculari exeunti ex A super superficiem corporis diaffoni.

[7.15] Et erit angulus BTL angulus reflexionis, et linea BLD est
70 perpendicularis super superficiem corporis. Ergo linea AL est obliqua super ipsum. Linea ergo AT continet cum perpendiculari exeunti ex T super superficiem corporis angulum acutum ex parte L. Et extrahamus perpendicularem, et sit TC. TC ergo erit equidistans LD; angulus ergo TLD est acutus. Ergo angulus ALB est obtusus. Linea ergo
75 AL est minor quam linea AB, et similiter declaratur quod AO erit minor AG. Sed linee AL, AO sunt equales, et AB, AG sunt equales, et linea LO est equalis linee BG. Ergo angulus OAL est maior angulo GAB.

[7.16] Et positio LO est similis positioni BG, quia linea que exit ex
80 A ad medium LO est perpendicularis super lineam LO, quia LO est equidistans BG, et BG est perpendicularis in qua sunt AZ, DB. Li-

52 Z: IZ O; *corr. ex* IZ L3 / *post* AH *add.* super superficiem BCDE R / et sit AZ (53) *om.* P1   53 BZG: BZC R / B *inter.* L3; *om.* O   54 G: C R; *om.* FP1 / A¹: GA F / et . . . A² (55) *mg.* C1   55 G¹·²: C R / reflectatur: refringatur R / T¹·²: P R   56 T: P R / est² *om.* R   58 BT TA: BP PA R / GK: GH O; CK R / ergo *inter. a. m.* L3; *om.* O / ergo superficies *transp.* SC1E   59 AT BT: AP BP R   60 reflexionis: refractionis R / reflexionis perpendicularis *transp.* C1   61 et *om.* FC1 / *post* perpendicularis *inter.* erit in illa superficie *a. m.* E / T: P R / *post* T *add.* erit in illa superficie R   62 AT¹·²: AP R / secat: secabit ER / *post* BD¹ *add.* et F   63 GE: CE R / *post* sicut *scr. et del.* erit O C1 / et² *om.* C1 / GO: CO R; PO S; BO O; *corr. ex* BO L3   64 BG¹·²: BC R   65 AG: AC R / *post* superficies *add.* ergo L3O (*deinde del.* L3) / AOG: ARG S; ACG L3O; AOC R   66 et . . . diaffoni (67) *mg. a. m.* L3; *om.* O   67 que transeunt *om.* F   68 G: C R / exeunti: exeunte R   70 BTL: BPL R / reflexionis: refractionis R / est perpendicularis (71) *transp.* R   71 super *om.* F / *post* corporis *scr. et del.* diaffoni P1   72 *post* super *add.* superficiem C1 / ipsum: ipsam R / AT: AP R / exeunti: exeunte R / ex: a S   73 T: P R / *post* corporis *add.* diaffoni FP1L3 (*inter. a. m.* L3)   74 TC¹: PG R / TC² *om.* R   75 TLD: TLO FP1; PLD R / TLD . . . angulus *rep.* S   76 AL: AB O / est: erit C1   77 AG¹·²: AC R / AL *inter.* E / *post* AL *add.* et C1 / et . . . equales² *mg.* FE (*a. m.* E); *om.* S   78 BG: GB E; CB R / maior:minor F   79 GAB: CAB R   80 est *om.* L3 / similis: consimilis ER / BG: BC R

nea ergo LO est perpendicularis super lineam AZ. Linea ergo LO est perpendicularis super superficiem que continuat A cum medio LO; positio ergo LO respectu A est sicut positio BG respectu A. Sed LO
85   comprehenditur remotius propter debilitatem forme; ergo LO videbitur maior quam BG. Sed LO est ymago BG; ergo BG videbitur maior quam sit.

[7.17] Item iteremus formam [FIGURE 7.7.11a, p. 447], et sit BG non equidistans DE, et extrahamus GF equidistantem ad DE. Et con-
90   tinuemus AF, et sit T punctum ex quo reflectitur F ad A; B autem reflectitur ad A ex Q. Et continuemus AQ, et protrahamus illam ad C. Sic ergo erit C altius quam L, nam B est ultra lineam AF, unde linea AC est ultra lineam AL. Ergo C est altius quam L.

[7.18] Et continuemus CO. Erit ergo CO diameter ymaginis BG,
95   et erit CO maior LO, et AC minor AL. Et due linee AC, AO sunt in duabus superficiebus secantibus se, scilicet ACB, AOG, et differentia communis inter has duas superficies transit per A. Et due linee que exeunt ex A perpendiculariter super hanc differentiam communem inter has duas superficies sunt altiores duabus lineis AC, AO. Ergo
100   angulus CAO est maior angulo BAG, et remotiones CO, BG ex A non differunt multum, et linea CO aut erit equidistans BG, aut non erit ibi differentia sensibilis in positione. Positio ergo CO respectu A est sicut positio BG respectu A, et inter distantias CO, BG respectu A non est diversitas sensibilis, quapropter CO videbitur maior quam BG. Sed
105   CO est ymago BG; ergo BG videtur maior quam sit, et hoc voluimus.

[7.19] **[PROPOSITIO 14]** Item iteremus formam primam huius

82 BG^{1,2}: BC R/*post* BG^1 *add.* et BG est perpendicularis equidistans BG *deinde del.* perpendicularis P1/*post* perpendicularis *add.* super superficiem R/*post* AZ *add.* et L3/linea *om.* R   83 super . . . perpendicularis (84) *scr. et del.* E/lineam: superficiem L3O; *om.* S/lineam AZ: eandem superficiem R/AZ: AE S/*post* AZ *scr. et del.* DB F   84 super *mg. a. m.* E   85 est: et O/BG: BC R 86 remotius: remotior R   87 BG^{1,2,3}: BC R/sed . . . BG^2 *rep.* S/ergo *inter. a. m.* L3/ergo BG *transp.* L3   89 item *om.* FP1/iteremus: iteramus L3/formam: figuram R/BG: BC R   90 GF: CF R/equidistantem: equidistante FP1/ad: linee R/DE: GE L3   91 AF: F S; *corr. ex* F E/T: P R; *corr. ex* I O/reflectitur^{1,2}: refringatur R/F *corr. ex* B P1   92 ex Q *inter. a. m.* E/*post* et^2 *scr. et del.* sit T punctum L3/C: G R   93 erit C *transp.* C1/C: G R/altius: alicuius O; *corr. ex* alicuius L3; *corr. ex* alterius S/est *inter.* F/AF: B O; *corr. ex* FB L3   94 AC: AG R/C: G R; *inter. a. m.* L3/altius: alterius SO   95 CO^{1,2}: GO R/erit ergo CO *inter. a. m.* E/ergo *om.* F   96 et^1 *om.* L3O/CO: GO R/AC^{1,2}: AG R/*post* AC^1 *add.* est C1   97 secantibus se *transp.* L3/se *inter. a. m.* L3; *om.* O/ACB AOG: AGB AOC R   98 *post* communis *scr. et del.* sunt in duabus superficiebus C1/has duas *transp.* R/transit . . . superficies (100) *scr. et del.* E   99 *post* super *add.* illam superficiem corporis diaffani sunt extra R/differentiam communem *transp.* R   100 inter . . . superficies: in his duabus superficiebus et R/AC: AG R   101 CAO: GAO R/BAG: BAC R/et *inter. a. m.* C1/CO BG: GO BC R   102 differunt: different FP1/et: quia R/CO: GO R; C S/equidistans *corr. ex* equidistantes F/BG: BC R   103 CO: GO R   104 BG^{1,2}: BC R/*post* inter *scr. et del.* differentias C1/CO: GO R   105 CO: GO R/BG: BC R/BG . . . sit (106) *mg. a. m.* E   106 CO: GO R/ymago *corr. ex* ymaginatio E/BG^{1,2}: BC R/videtur: videbitur L3/*post* hoc *add.* est quod R

capituli [FIGURE 7.7.12, p. 448], et sit perpendicularis secans LK
AMOZ. Erit ergo LO medietas LK. Sed punctus Z videbitur in O,
quia videtur in perpendiculari ZM. Ergo BG videbitur in linea LK.
110 Et BZ est medietas BG, et LO est medietas LK, et LK videtur maior
quam BG; ergo LO videtur maior quam BZ.

[7.20] Causa autem magnitudinis BG est reflexio; ergo causa magnitudinis BZ est reflexio. A autem est in perpendiculari AZ, que exit
ab extremitate BZ super superficiem corporis diaffoni. Et hoc idem
115 sequitur in tribus figuris sequentibus primam, scilicet in secunda, et
tertia, et quarta huius capituli, scilicet quod visus comprehendit me-
dietates visibilium maiores quam sint. Et visus est in perpendiculari
exeunte ab extremitate medietatis aut super superficiem transeuntem
per extremitatem medietatis perpendicularis super superficiem
120 corporis diaffoni, nam punctus quod est medium ymaginis est in
perpendiculari exeunti a medio rei vise, sive res visa sit equidistans
superficiei corporis diaffoni, sive non.

[7.21] Item BN est quedam pars linee BZ. Et extrahamus perpendicularem NC. Ymago ergo N erit in linea NC. Sit ergo C ymago N.
125 C ergo aut erit in linea LC aut prope illam, quapropter LC aut erit
equalis linee BN aut fere. Sed in prima figura huius capituli declaravimus quod BG comprehenditur maior quam sit, et causa huius est
reflexio. Et reflexiones formarum que remotiores sunt a perpendiculari cadenti a centro visus super superficiem corporis diaffoni sunt
130 maiores reflexionibus formarum que sunt propinquiores perpendiculari. Reflexio ergo forme BN ad A est maior quam reflexio forme partis linee ZN ad A. Causa ergo que facit formam BZ videri maiorem facit
ut BN habeat maiorem proportionem ad ipsum quam illam quam
habet BZ ad BN. Ergo LC, que est ymago BN, comprehenditur maior

107 item *om.* L3 O / *post* iteremus *scr. et del.* continuemus O / formam: figuram R / primam *inter.
a. m.* L3; *om.* O   108 *post* secans *add.* lineam R / LK: AZ O / *post* LK *add.* ymaginem FP1
109 AMOZ: MOZ FP1; LRMO O / LK: LR O / sed punctus: et punctum R   110 BG: BC
R / LK: LR O   111 *post* et¹ *scr. et del.* B O / BG: BC R / medietas *corr. ex* medietata P1 / et³ . . .
autem (113) *mg. a. m.* E   112 BG: BC R / videtur: videbitur R   113 BG: BC R / reflexio:
refractio R / reflexio ergo *transp.* F   114 BZ *om.* E / reflexio: refractio R / *post* A *scr. et del.*
autem F / est² *inter. a. m.* L3; *om.* O   115 et . . . diaffoni (121) *mg. a. m.* L3; *om.* O   116 in²
om. L3 / et: in C1R; *om.* L3   117 et *om.* L3   119 *post* medietatis *add.* super superficiem
corporis diaffani R / aut: autem P1 / super *om.* S   120 perpendicularis: perpendicularem P1
121 diaffoni *om.* R / nam: iam O / punctus: punctum R / quod: qui E   122 exeunti: exeunte R
124 est: sit R / extrahamus: extramus S   125 NC¹⋅²: NG R / sit . . . C (126) *om.* S / ergo C
ymago *mg.* F / C: G R / *post* ymago *add.* ergo F   126 C: G R / *post* C *scr. et del.* ergo E / LC¹⋅²:
LG R   127 huius: huis S   128 BG: LG P1; BC R   129 reflexio: reflexiones P1; refractio
R / reflexiones: refractiones R   130 cadenti: cadente R   131 reflexionibus: refractionibus
R / perpendiculari: perpendicularibus C1   132 reflexio¹⋅²: refractio R / BN: BM O; *corr. ex* BM
L3 / est: erit L3   133 linee *om.* R / *post* facit¹ *add.* Z OE (*deinde del.* E); *scr. et del.* BZ L3 / formam:
ymaginem R / videri *om.* E   134 ipsum: ipsam R / quam¹: quod S / illam: illa P1

135 quam BN.

[7.22] Item si A non comprehenderit ymaginem BN maiorem quam BN, non comprehendet ymagines ceterarum partium linee BN que sunt propinquiores ad Z maiores ipsis partibus, nam forme ceterarum partium sunt minoris reflexionis quam forma BZ. Sed

140 reflexio est causa ymaginis; ergo A non comprehenderet LO maiorem quam BZ, nam comprehendit LO maiorem quam BZ. Ergo comprehendet BN maiorem quam sit. Et A est extra perpendiculares exeuntes ex BZ super superficiem corporis diaffoni, et linea que exit ex A ad medium BZ non est perpendicularis super BZ, et hoc idem

145 sequitur in tribus figuris, in secunda scilicet, et tertia, et quarta huius capituli.

[7.23] Omne ergo quod comprehenditur a visu ultra aliquod corpus diaffonum grossius aere, cuius superficies fuerit plana, comprehenditur maius quam sit, sive sit visus in aliqua perpendiculari exeunti

150 ex illo visu super superficiem corporis, sive sit extra, et indifferenter si diameter rei vise fuerit equidistans superficiei corporis, aut non equidistans.

[7.24] [**PROPOSITIO 15**] Item sit superficies corporis sperica, cuius convexum sit ex parte visus, et sit grossius aere. Et sit visus A

155 [FIGURE 7.7.13, p. 448] et res visa BG, et sit centrum spere ultra BG in respectu visus. Et sit centrum D, Z medium BG, et continuemus DB, DZ, DG, et extrahamus has lineas quousque concurrant cum superficie spere ad E, et M, et N. Et extrahamus ZM in parte M.

[7.25] In primo sit visus in linea ZM. Erit ergo AMZ linea recta.

160 Et in primo sit BD equalis GD. Sic ergo erit AZ perpendicularis super BG; positio ergo B respectu A erit similis positioni G respectu A. Et extrahamus superficiem in qua sunt DE, DN. Faciet ergo in super-

135 habet: habent *FP1*/LC: LG *R*/comprehenditur . . . BN (137) *mg. F*   138 *post* quam *add.* ipsam *R*/BN[1]: BM *OE*/comprehendet: comprehendit S; comprehenderit E   140 minoris: minores *SL3*/reflexionis: refractionis *R*/BZ: BN C1   141 reflexio: refractio *R*/*post* causa *add.* magnitudinis *R*/comprehenderet *corr. ex* comprehenderit E/LO: LC C1   142 BZ[1]: BN *SC1E*/nam . . . BZ[2] *rep.* P1; *inter.* L3; *om.* OR/LO *inter. a. m.* E/*ante* ergo *add.* A *R*/comprehendet: comprehendit *L3O*   143 BN maiorem *transp. R*/*post* et *add.* idem accidit si *R*/est . . . perpendiculares: extra perpendicularem est *R*/extra perpendiculares *transp. L3O* (*deinde corr.* L3)/perpendiculares: perpendicularem E/exeuntes: exeuntem *ER* (*alter. ex* exeuntes E)   144 super *om.* O   146 in[1] *inter. a. m.* L3/et[1] *om.* OR   148 omne ergo *transp.* FP1/aliquod *om.* E   149 fuerit *om.* C1   150 maius: magis *FC1*/exeunti: exeunte R   152 si: sive *R*/*post* si *scr. et del.* differenter sit E/rei vise *transp.* C1/aut: sive R   155 sit[2] *om. R*/*post* aere *scr. et del.* cuius superficies C1   156 BG[1]: BC R; *corr. ex* G *L3*/BG[2]: BC R   157 *post* centrum *scr. et del.* spere *S*/BG: BC R   158 DZ: DE *SC1OE*/DG: DC R   159 et[1] *om.* L3R/M[1] *inter.* L3/et[2] *om. R*/et N: ZN *P1C1L3OE* (*alter. ex* ZM L3)/in parte *rep.* P1/M[2]: MZ *SO*/*post* M *add.* et ER   160 in[1] *om. R*/ZM: MZ FP1   161 in *om. R*/BD *corr. ex* DB O/equalis *om.* *L3O*/GD: CD *R*/*post* AZ *scr. et del.* su C1   162 BG: BC *R*/G: C *R*/et *om.* S

ficie sperica arcum circuli magni. Sit ergo arcus EMN, et hec superficies est perpendicularis super superficiem spericam, nec fit reflexio
165   extra hanc superficiem, nam AZ est perpendicularis super superficiem corporis. Non ergo reflectitur forma alicuius partis BG ad A
nisi ex circumferentia EMN.

    [7.26] Reflectatur ergo B ad A ex H, et G ad A ex T. Positio ergo H
respectu A et distantia eius est equalis positioni et distantie T. Et con-
170   tinuemus BH, HA, GT, TA, et extrahamus AH ad K et AT ad L, et continuemus KL. Erit ergo AK equalis AL, et erit LK ymago BG, et erit
equidistans BG; erit ergo maior quam BG. Et continuemus AB, AG.
Erit ergo angulus KAL maior angulo BAG, et erit positio KL similis
positioni BG. Et inter KL et GB non est differentia in distantia, ut in
175   precedentibus diximus; ergo KL videbitur maior quam BG. Sed KL
est ymago BG; ergo BG videbitur maior quam sit, quia ymago eius est
maior se. Et hoc est quia forma eius est debilior quam vera forma, et
hoc est quod voluimus.

    [7.27] **[PROPOSITIO 16]** Si ergo BD, GD fuerint inequales, tunc
180   AK, AL erunt inequales, et sic BG, KL erunt oblique super lineam
AD. Erit ergo KL, ut in secunda figura huius capituli diximus, maior
quam BG in visu.

    [7.28] Item si A fuerit extra superficiem BZG, et BD, GD fuerint
equales aut inequales, declarabitur, ut in tertia et quarta figura huius
185   capituli, quod KL videbitur maior quam BG. Sed secet ante DM lineam KL in O. Erit ergo KO ymago BZ. Et erit angulus KAO maior
angulo BAZ, et positio KO est similis positioni BZ, et distantie KO,
BZ respectu A non differunt multum, quapropter KO videbitur maior

163 *post* extrahamus *scr. et del.* super S / superficiem *mg. a. m.* E / DN: DQ E; DZ O; *corr. ex* DZ
L3; *om.* P1 / *post* DN *add.* DM R   164 arcum: ateris O; *corr. ex* acuminis L3 / *post* ergo *scr. et del.*
superficies F / EMN: ENM S; EMD O; *corr. ex* EMD L3   165 est *om.* L3O / nec fit reflexio *inter.*
a. m. L3; *om.* O / reflexio: refractio R   166 extra: preter L3; pre O   167 *ante* corporis *add.*
spericam R / *post* corporis *add.* diaffoni FP1L3 (*inter. a. m.* L3) / reflectitur: refringetur R / BG: BC R
168 EMN: ENM SP1; MN O; *corr. ex* MN L3   169 reflectatur: refringatur R; reflectitur L3O;
corr. ex reflectitur *a. m.* E / ad¹ *rep.* R / G: C R / ad² *corr. ex* ex *a. m.* E / ex T *inter. a. m.* L3; *om.* O / T:
G R   170 A: E SFP1L3OE / est *om.* S / et² *om.* L3O / T: G R; *inter. a. m.* E   171 GT TA: CG GA
R / AH *corr. ex* A C1 / K: R O / AT: AG R   172 KL: RL O / AK: LK O / AL: AB SO; *corr. ex* AB
L3 / BG: BC R   173 BG¹˒²: BC R / AG: AC R   174 BAG: BAC R   175 BG: BC R / KL: LK
R / GB: BG C1; CB R / ut *om.* S   176 *post* quam *scr. et del.* sit quia ymago C1 / BG: BC R; *om.* S
177 BG¹: BC R; *mg.* F / ergo BG *inter. a. m.* L3 / BG²: BC R / sit: FO F / quia *inter.* E / *post* eius *scr. et*
del. quod P1   178 hoc *inter. a. m.* L3; *om.* O / est¹ *om.* L3O / *post* quia *add.* vero O; *scr. et del.* ver
L3 / *post* eius *scr. et del.* de P1 / quam: qua S   180 ergo: vero R / *post* ergo *scr. et del.* DB S / BD:
BG F / GD: BG E; BC R   181 et sic *om.* S / BG: BC R / oblique: obliqua E   182 secunda *inter.*
C1 / capituli: capitis R   183 BG: BC R; GD S   184 BZG: BZC R / GD: CD R   185 ut *om.*
E / tertia *corr. ex* tertio L3 / et . . . figura: figura et quarta ER   186 KL: LK P1 / BG: BC R; MBG
L3 / *post* secet *scr. et del.* an P1 / ante: ari P1; linea R / DM *corr. ex* DIM O   187 KL: BL O

quam BZ.

190   [7.29] Et A est in perpendiculari ZM, que exit ab extremitate BZ
super superficiem corporis; sit autem BC pars BZ, et sit KR ymago
BC. Ergo, ut in quinta figura huius capituli diximus, patet quod KR
videbitur maior quam BC. A autem est extra omnes perpendiculares
exeuntes ex BC super superficiem corporis, nam linea que exit ex A
195   ad medium BC non est perpendicularis super BC. Et quia BG, KL
sunt oblique super AZD aut super superficiem que transit per lineam
MD, et KO est ymago BZ, et LO est ymago ZG, et angulus quem
respicit KO apud centrum visus est maior angulo quem respicit BZ
apud centrum visus, et similiter angulus quem respicit OL est maior
200   angulo quem respicit ZG, ergo KO videbitur maior quam BZ, et similiter KR videbitur maior quam BC. Et omnia hec declarantur in
quinta figura huius capituli. Sed in hac positione est quedam additio,
scilicet quod KL, que est ymago BG, est maior in veritate quam BG, et
KO est maior quam BZ.

205   [7.30] In prima autem positione, scilicet in plana superficie, due
ymagines sunt equales duobus visis; ymago ergo KL et ymago KO
sunt in visu maiores ipsis rebus, et sic sunt in veritate. Et patet quod
angulus quem respicit KL apud centrum visus est maior angulo quem
respicit BG apud centrum visus, et angulus quem respicit KO apud
210   centrum visus est maior illo quem respicit BZ, cum visus fuerit extra
superficiem in qua sunt DE, DZ, ut in quarta huius capituli diximus.
Ergo si visus comprehenderit aliquid ultra corpus grossius aere,
cuius superficies fuerit sperica, et cuius convexum fuerit ex parte
visus, et cuius centrum fuerit ultra rem visam quantum ad visum,
215   comprehendet illud maius quam sit, sive fuerit visus in perpendicu-

191 *post* A *add.* et L3O (*deinde del.* L3)/que *om.* R/exit: erit O; exeunte R   192 sit[1]: si O/autem:
ergo ER/BC: BF R   193 BC: BF R/figura *om.* L3O   194 BC ... autem: BF si autem A R/A
inter. a. m. E/est *inter. a. m.* L3/extra: ex S/*post* extra *add.* superficiem in qua sunt R   195 *post*
corporis *add.* diaffani R   196 *post* BC[1] *add.* perpendiculariter R/non ... BC[2] *mg. a. m.* E/*post* est
add. idcirco R/*post* super *add.* ymaginem L3; *add.* superficiem linee R/BC[2] ... MD (198) *inter. a. m.*
L3/BC[2] ... est[1] (198) *om.* O/*post* BC[2] *add.* idem patebit R/et: nam R/BG: BC R   197 oblique:
erecte R/*post* super[1] *add.* lineam R/super[2] *inter. a. m.* E   198 *post* MD *scr. et del.* BZ L3/BZ ...
ymago[2] *om.* L3/ZG: C R; *corr. ex* GZ F   199 est ... visus (200) *inter. a. m.* E/BZ: LZ F/BZ ...
respicit (201) *om.* S   200 OL: LO C1   201 ZG: ZC R; ZHG O; *corr. ex* BG L3/KO *om.* S/BZ:
CZ R; *corr. ex* BGZ P1/et ... BC (202) *mg. a. m.* E   202 BC: LC S; BF R/hec: que L3O/*post* declarantur *add.* ut C1   203 *post* quedam *scr. et del.* positio C1   204 que *inter.* C1/BG[1,2]: BC
R/in ... [end of text] *om.* F   205 KO: KC S; KA O; KE E; KR R/*post* est *scr. et del.* m S/quam
om. R/BZ: BF R   206 autem *inter. a. m.* E/plana: plena E   207 visis: visibilibus R/*post* visis
add. apparent autem visui esse maiores R/ergo: vero R/*post* ergo *scr. et del.* G O/*post* KO *add.* in
superficie sperica a qua fit refractio R   208 in visu maiores: maiores in visu ER/sic: sicut C1
209 est ... visus (210) *mg. a. m.* E   210 BG: BC R/KO: BO O   211 illo: angulo ER/BZ:
DZ L3/visus[2] *om.* P1   212 DZ *corr. ex* DEZ S/*post* quarta *add.* figura ER   214 ex ... fuerit
(215) *om.* S   215 ultra *om.* E/*post* visum *add.* comprehendet illud magis ad visum P1

lari exeunti a re visa super superficiem spericam, sive extra, sive linea
que exit a centro visus ad medium rei vise fuerit perpendicularis
super rem visam aut obliqua, et hoc est quod voluimus declarare.

[7.31] Et hoc accidit in eis qui videntur in aqua, nam convexum
superficiei aque sperice est ex parte visus, et centrum superficiei aque
220 est ultra illa que comprehenduntur in aqua, et aqua est grossior aere.
Sed illud quod videtur in aqua, si aqua fuerit clara et pauca, forte
non comprehendet visus ipsum esse maius in aqua quam si esset in
aere, non enim differt quantitas eius tunc quantum ad sensum, scilicet quantitas eius in aqua et aere, tunc enim illa additio in aqua erit
225 parva, et ideo sensus non distinguet tunc illam additionem.

[7.32] Tamen experientia potest comprehendi hoc modo: accipe
corpus columpnale, cuius longitudo non sit minor uno cubito, et sit
aliquante grossitiei album, nam albedo in aqua manifestius distingui-
tur. Et sit superficies basis eius plana, ita quod per se stet equaliter
230 super faciem terre. Hoc preservato, accipe vas amplum, et sit superficies eius plana, et infunde in vas aquam claram in altitudine minori
longitudine corporis columpnalis. Deinde mitte illud corpus columpnale in aquam, et pone ipsum super suam basim in medio vasis. Erit
ergo aliqua pars huius corporis extra aquam, nam altitudo aque est
235 minor longitudine huius corporis. Tunc enim, cum quieverit aqua,
videbis partem corporis que est intra aquam grossiorem illa que est
extra aquam. Patet ex hac experientia quod omne visum comprehensum in aqua comprehenditur maius quam sit in veritate.

[7.33] Item sit corpus spericum, cuius convexum sit ex parte visus,
240 et res visa sit ultra centrum superficiei sperice, et sit illud corpus
grossius aere. Sed in assuetis visibilibus non est tale aliquid quod
videatur ultra corpus diaffonum spericum grossius aere ultra cen-

216 maius: magis *P1*/*post* sit *add.* secundum veritatem et etiam secundum apparentiam in visu *R*/*post* in *add.* A *L3O*   217 exeunti: exeunte *R*/*post* spericam *rep. et del.* spericam *S*/sive extra *inter. a. m. L3*; *om. O*   219 aut: sive *R*/obliqua: obliquam *O*; *corr. ex* obliquam *L3*   220 et: *ex S*; item *P1*/qui: que *L3OR*/aqua: aliqua *O*/nam ... aqua¹ (222) *om. S*   221 aque¹ *inter. a. m. L3*; *om. O*/sperice: spere *O*; spericum *R*   222 et aqua *rep. P1*/*post* est² *add.* grossius vel *E*/*post* grossior *add.* est *E*   223 si: sed *S*/si aqua *inter. a. m. L3*   224 comprehendet: comprehenditur *R*; *corr. ex* comprehendit *a. m. E*/visus: a visu *R*/ipsum: punctum *O*; *corr. ex* punctum *L3*; *om. R*/in² *corr. ex* maior *C1*   225 differt: differet *L3*/quantitas: quantitates *S*/tunc ... eius (226) *rep. S*/sensum: visum *L3O*   226 *post* et *add.* in *P1*/*post* additio *add.* et *L3O*/erit parva (227) *om. P1* 229 co-lumpnale: columpnare *R*   230 aliquante: aliquantum *O*; *corr. ex* aliquantulum *a. m. E*/grossitiei: grossiei *P1L3*; grossiciei *R*; *corr. ex* grossius *a. m. E*/distinguitur *om. E*   231 sit: si *O*; *inter. E*/basis *om. P1*/quod: ut *R*; *inter. a. m. C1*   232 faciem: superficiem *R*/preservato: observato *R*   233 *ante* eius *add.* basis *P1L3C1O* (*alter. ex* vasis *L3*)/altitudine: latitudine *P1*/minori: minore *R*   234 columpnalis: columpnaris *R*/*post* deinde *scr. et del.* n *S*/columpnale: columpnare *R* 235 pone *corr. ex* pene *C1*/*post* super *scr. et del.* superficiem *P1*/basim: vasim *SO*   236 aque: ante *S*   238 videbis: videbit *P1*/que¹ *corr. ex* quod *P1*/intra: ultra *P1*   239 extra *corr. ex* intra *P1*/*post* patet *add.* ergo *P1ER*/ex *inter. a. m. E*/omne *inter. a. m. E*/comprehensum: apprehensum *C1* 240 sit *inter. a. m. E*   241 sit¹ *corr. ex* si *O*

trum spere et res visa, cum hoc erit intra corpus spericum, hoc enim
non fit nisi corpus spericum fuerit vitreum aut lapideum, et fuerit
245   totum corpus spericum solidum, et res visa fuerit intra ipsum, aut
ut corpus spericum sit portio spere maior semispera, et res visa sit
applicata cum basi eius. Sed hii duo situs raro accidunt. Huiusmodi
ergo res non sunt de assuetis visibilibus; non ergo debemus negotiari
circa ea que accidunt huiusmodi visibilibus.

250   [7.34] Sed sunt quedam assueta que videntur ultra corpus diaffonum
spericum grossius aere, cuius convexum erit ex parte visus, cum res
visa fuerit ultra speram cristallinam aut vitream, et res visa fuerit in
aere non intra speram; positiones autem huiusmodi visibilium sunt
multimode. Sed hec raro comprehenduntur, et si comprehendantur,
255   raro videntur. Non ergo est conveniens distinguere omnes illas positiones; sumus ergo contenti una sola positione, scilicet quod visus
et res visa sunt in eadem perpendiculari super superficiem corporis
sperici.

260   [7.35] **[PROPOSITIO 17]** Sit ergo visus A [FIGURE 7.7.14, p. 449]
et corpus spericum BGDZ, et centrum eius sit E. Et continuemus AE,
et extrahamus eam recte, et secet superficiem spere in duobus punctis
B, D. Et extrahamus ipsam in parte D usque ad H. Et extrahamus ex linea
HAB superficiem equalem secantem speram. Faciet ergo in superficie
spere circulum BGDZ.

265   [7.36] In nona autem figura de capitulo ymaginis, diximus quod
in linea BD sunt plura puncta quorum forme reflectuntur ad A ex
circumferentia BGDZ, et quod forma totius illius linee reflectitur ad
A, si BGDZ fuerit continuum et non fractum in parte D. Reflectatur
HL ad A ex circumferentia BGDZ, et reflectatur H ad A ex G et L ad
270   A ex T. Forma ergo HL reflectetur ad A ex arcu GT. Et continuemus

243 sed . . . aere (244) *mg. C1* / est *om.* E   244 videatur *corr. ex* videtur *a. m.* E / *post* grossius
rep. et del. quod (243) . . . grossius E   245 erit: sit R / hoc² . . . spericum (246) *mg. L3; om.* P1
246 *post* nisi *add.* si SC1O   247 visa *om.* E / intra: inter S / ipsum *inter. a. m.* E   248 semispera
corr. ex semisperia O   249 applicata: amplificata S / basi: vasi O   250 assuetis: suetis O; *corr.
ex* suetis *a. m.* L3 / negotiari: occupari R   253 *post* res *scr. et del.* re O   254 cristallinam *inter.
a. m.* E / et . . . fuerit² *om.* R   255 huiusmodi: huius O   256 raro: rara P1 / comprehenduntur:
comprehendentur L3 / comprehendantur raro (257) *inter. a. m.* E   257 ergo est *transp.* R / illas
corr. ex illa S   258 sumus: simus P1R / contenti *corr. ex* contintenti S / quod: ut R   259 sunt:
sint R / eadem: eodem SO   262 BGDZ: BGZD R   263 *post* eam *scr. et del.* et L3   264 B
D *transp.* L3 / D¹: E O / et¹ *om.* O / D² *inter. a. m.* E   265 HAB: HBA R; HAG O; *alter. ex* BAG *in*
HAG L3 / faciet: facies SO; *corr. ex* facies *a. m.* L3   266 spere circulum *transp.* P1 / BGDZ: BGZD R
267 in *om.* R / nona: octava C1R / *post* nona *add.* vel octava P1L3 (*inter. a. m.* L3) / de: in R / *post* capitulo *add.* de ymagine R   268 in *inter. a. m.* L3; *om.* OE / *post* BD *add.* illius figura octave P1 / forme
inter. a. m. E / reflectuntur: refringuntur R / ex: et SC1O; *corr. ex* et L3   269 BGDZ: BGZD R; BG
C1; DG E; *alter. in* GD L3 / reflectitur: refringitur R   270 si . . . A¹ (271) *mg.* L3 / BGDZ: BGZD
L3R / fractum: fixus SP1O; fixum L3; factum E / D *correxi ex* B / reflectatur: refringatur R

lineas GMH, GA, LZT, TA. H ergo extenditur per GH et reflectitur per GA, et L extenditur per LT et reflectitur per TA. Et continuemus lineas EG, ET, EM, EZ, et extrahamus EM ad C et EZ ad F.

[7.37] Forma ergo que extenditur per AG reflectitur per GH, et
275 pervenit ad H, et forma que extenditur per AT reflectitur per TL et pervenit ad L. Hoc est si corpus diaffonum fuerit continuum usque ad G. Si ergo corpus spericum fuerit signatum apud superficiem spericam, tunc forma que extenditur per AG reflectitur per GM in partem perpendicularis que est EG, et cum forma pervenerit ad M,
280 reflectetur secundo in contrariam partem perpendicularis que est EMC. Reflectatur ergo ad K. Et ideo forma que extenditur per AT reflectitur per TZ, et cum fuerit reflexa ad Z, reflectetur secundo ad contrariam partem perpendicularis que est EZF. Sit ergo reflexio forme que pervenit ad Z per lineam ZO.

285 [7.38] Forma ergo K extenditur per KM, et reflectetur per MG; deinde reflectitur secundo per GA. Et similiter forma O extenditur per OZ, et reflectitur per ZT; deinde secundo reflectitur per TA. Forma ergo totius KO reflectitur ad A ex arcu GT. Et si linea AK fuerit fixa, et ymaginati fuerimus figuram AGMK circumvolvi circa AK,
290 tunc arcus GT faciet figuram circularem, ut armillam, a cuius universo reflectetur forma KO ad A, et erit ymago KO centrum visus, quod est A. Forma ergo KO videbitur in tota superficie circulari, que est locus reflexionis, que est in rectitudine linearum radialium, que

271 *ante* HL *add.* ergo *R*/ex[1]: et *SO*/BGDZ: BGZD *R*/reflectatur: refringatur *R*   272 T: P *R*/HL: HB *SO*/reflectetur: refringetur *R*/GT: GP *R*   273 GMH GA: GMG HA *SP1L3*/LZT: LZP *R; corr. ex* HZT *O*/TA: PA *R*/extenditur: ostenditur *SO; corr. ex* ostenditur *L3*/reflectitur: reflectetur *S*; refringitur *R*   274 GA *corr. ex* TA *P1*/L *inter. a. m.* E/LT: BT *O*; LP *R*/reflectitur: refringitur *R*/TA: PA *R*   275 EG: EH *SP1L3C1OE*/ET: et *SP1L3C1OE; om. R*/post EM[1] *add.* et *SP1C1O*/EZ[1]: ZE *L3*; Z *SO*/EM[2]: M *O; corr. ex* M *a. m. L3*/EZ[2]: EN *O*   276 *sequens R addidi* AG reflectitur per/reflectitur: refringitur *R*   277 *post* H *add.* diminutio H *L3O (deinde del. L3)*/que extenditur *mg. a. m.* E/AT: LT *SL3E*; ST *P1*; AP *R*; hec *O*/reflectitur: refringitur *R*/TL: LT *L3*; PL *R*; D E   278 *post* L *add.* et *L3*/est *inter.* E/diaffonum *inter. a. m.* E   279 ergo: vero *R*/post fuerit *add.* continuum *L3*   280 reflectitur: refringitur *R*/GM *corr. ex* EM *P1*   281 EG: EH *SP1L3OER*/EG . . . est (282) *mg. a. m. L3*/pervenerit: pervenit *P1*; perveniet *ER*   282 reflectetur: refringetur *R*/secundo: SO *P1*/contrariam partem *transp. R*/que est *rep.* C1/est: et *S*   283 reflectatur: refringatur *R*/ideo: iterum *C1*/post ideo *add.* etiam *R*/AT: AP *R*/reflectitur: refringitur *R*   284 TZ: TN *O*; PZ *R*/reflexa: refracta *R*/Z: N *O*/reflectetur: refringetur *R*; *inter.* L3E *(a. m.* E); *om. SO*/reflectetur secundo *transp.* L3   285 partem perpendicularis *transp.* P1/EZF: ENF *O*/reflexio: refractio *R*   286 Z: M *O; corr. ex* M *L3*/ZO: ZM *O*; ZA *SC1*   287 reflectetur: reflectitur *E*; refringitur *R*/MG: KM *O*/post MG *add.* et reflectetur per MH *L3O (deinde del. L3)*   288 reflectitur: refringitur *R*/GA: HA *O*/post similiter *scr. et del. ex* E   289 OZ: ZT *P1*; Z *S*; AZ *O; corr. ex* AZ *L3*/et . . . ZT *om.* P1/et . . . KO (290) *om. S*/reflectitur[1,2]: refringitur *R*/ZT: ZP *R*; GT *O; corr. ex* GT *L3*/secundo *om.* P1/secundo reflectitur *transp.* C1/TA: PA *R*   290 reflectitur: refringitur *R*/GT: GP *R*/si linea *corr. ex* similia *L3*/AK: AKO *R*   291 *post* figuram *scr. et del.* K *C1*/AGMK: KGMK *O*; AGPK *R*/AK: AKO *R*   292 GT: GP *R*/a cuius: acus *SO*   293 reflectetur: reflectitur *E*; refringetur *R*/KO[1]: KA *SO; corr. ex* KA *L3*/post KO[2] *add.* apud *R*/centrum . . . KO (294) *om.* P1

est figura armille. Forma ergo KO erit maior se, et erit figura forme
295 diversa a figura KO.

[7.39] Hoc autem potest experimentari sic: accipe speram cristallinam aut vitream rotundissimam, et accipe corpus parvum vel ceram parvam, ut granum ciceris, nam experientia per corpus parvum
erit manifestior. Et tingat ipsam colore nigro, et sit figura cere sperica. Deinde pones ipsam in capite acus, et pone speram cristallinam
300 in oppositione alterius oculorum, et claude alterum oculum. Et eleva
acum ultra speram, et aspice ad medium spere, et pone ceram in oppositione medii forme, ita quod sit opposita medio spere in una linea
recta quoad sensum. Et respice ad superficiem spere, tunc enim videbis in ipsa superficie nigredinem rotundam in figura armille. Si vero
5 non videbis ipsam, move ceram ante et post donec videas nigredinem
rotundam. Tunc aufer ceram, et abscindetur nigredo; deinde redeat
cera ad suum locum, et videbis illam nigredinem rotundam.

[7.40] Ex hac ergo experientia patebit quod, si res visa fuerit ultra
corpus diaffonum spericum grossius aere, et visus, et res visa, et cen-
10 trum corporis sperici fuerint in eadem linea recta, tunc visus comprehendet illam rem visam in figura armille.

[7.41] **[PROPOSITIO 18]** Si vero BGDZ fuerit in corpore columpnali, et corpus fuerit grossius aere, tunc forma KO videbitur apud
arcum GT et apud arcum sibi equalem et sibi respondentem ex arcu
15 BD, sed hec forma non erit circularis, quia figura AHMG, cum fuerit
circumvoluta circa AK, non transibit per illam lineam arcus GT
per totam superficiem columpnalem. Sed forte reflectetur forma ex
aliquibus portionibus columpne, sed erit continua recte, nam superficies ex LK que transit per axem columpne facit in superficie columpne

295 reflexionis: refractionis R / est² om. E   296 erit¹: videbitur R / se: SO P1 / post se add. ipsa R
297 diversa: diverse R   298 experimentari sic transp. S / cristallinam: cristallam O   299 vitream: auream SOE / corpus . . . vel om. L3O / vel . . . parvam (300) om. SE / vel . . . ciceris (300):
ut granum ciceris vel ceram parvam R   300 post ceram scr. et del. ex L3 / ut: aut C1 / post ciceris
add. vel ceram parvam mg. a. m. E / experientia corr. ex experimentia S   1 tingat: tingant SC1OE;
tingas P1R / post figura scr. et del. spere L3   2 pones: ponas SR / post in scr. et del. medio C1 / capite acus transp. C1 / pone: pones P1; ponas R   3 alterum oculum transp. L3   4 eleva: leva
L3O / eleva acum: elevatum C1 / ceram corr. ex speram P1   5 forme: spere C1   6 ad inter. a.
m. E / enim videbis (7) transp. E   7 ipsa: illa ER / post superficie add. spere ER (inter. a. m. E) / nigredinem: nigredine O / figura: figure S; figuram C1   8 non om. E / videbis: videas R / ipsam:
eam P1R   9 aufer: aufers C1 / abscindetur: abcidetur S; abscidetur O   10 et inter. L3 / post et
add. iterum R; scr. et del. tunc C1   11 ante ex add. et SL3C1O / ergo om. C1   12 et visus mg. a. m. E
13 fuerint: fuerit SC1O; fiunt P1 / comprehendet; comprehendit R   15 BGDZ: BEDZ O;
BGZD P1R / columpnali: columpnari R   16 ante et add. etiam SP1C1OE / apud om. P1   17 GT:
GP R / GT . . . arcum² mg. a. m. E / arcum² om. S / arcu corr. ex arcui O   18 BD: BZ SO / AHMG:
AHMK O; AHPG R / cum: et S; non O / post cum scr. et del. volverit P1 / fuerit: fuerint OE; corr. ex
fuerint L3   19 per om. S / illam lineam transp. P1R / GT: GP R

20 que est ex parte A lineam rectam que transit per B et extenditur in
longitudine columpne. Et non reflectitur forma KO ex illa linea recta,
nam KZ erit perpendicularis super illam lineam rectam. Non ergo
erit forma rotunda, si fuerit corpus columpnale, sed erunt due forme,
quarum altera reflectitur super alteram. Videbitur ergo KO esse duo,
25 quorum utrumque erit maior KO, et forma utriusque erit diversa a
forma KO, et tamen ille due forme erunt idem punctum, scilicet centrum visus.

[7.42] In visibilibus autem assuetis nichil est quod comprehenditur a visu ultra corpus diaffonum spericum grossius aere, cuius
30 concavum sit ex parte visus, nam si fuerit ex vitro aut aliquo lapide,
oportet quod sit portio spere concava et quod res visa sit intra illam
speram, aut quod superficies eius que est ultra concavitatem sit plana
et res visa adhereat illi. Et illi duo situs non inveniuntur, aut raro; non
ergo solicitemur circa huiusmodi.

[7.43] Item non invenitur aliquod corpus subtilius aere, cuius super-
35 ficies que est ex parte visus sit plana aut convexa, et non invenitur
aliquod corpus subtilius aere ultra quod comprehenditur aliquid
nisi corpus celi et ignis. Et non dividetur a corpore aeris superficies
que distinguat unam partem ab alia, sed quanto magis appropinquat
aer celo, tanto magis purificatur donec fiat ignis. Subtilitas ergo eius
40 sit ordinate secundum succesionem, non in differentia determinata.
Forme ergo eorum que sunt in celo, quando extenduntur ad visum,
non reflectuntur apud concavitatem spere ignis, cum non sit ibi superficies concava determinata. Nullum ergo invenitur corpus subtilius
aere in quo extenduntur forme visibilium et reflectuntur apud super-
45 ficiem eius ad visum nisi corpus celeste, et corpus celeste est speri-

20 columpnalem: columpnarem R / forte reflectetur: refringetur forte R / reflectetur *corr. ex*
reflectitur *a. m.* E   21 columpne: columpnarie O; columpnaribus R; *corr. ex* columpnaribus E; *corr. ex* columpna L3 / sed: et R / post continua *add.* in una parte et similiter in alia
R / recte *om.* R   22 LK: K O; *corr. ex* K L3 / que: et SP1O; *inter. a. m.* E / post que *add.* etiam ER;
scr. et del. et C1 / facit . . . columpne² *inter. a. m.* L3   23 que¹: non O / est *om.* P1 / post A *scr. et*
del. columpnam E   24 reflectitur: refringitur R; flectitur O; *corr. ex* flectitur *a. m.* L3 / forma
KO *transp.* C1   25 KZ: KB R / rectam *mg. a. m.* E   26 corpus columpnale *transp.* C1 /
columpnale: columpnare R   27 reflectitur: refringitur R / super: per L3 / esse: erit S; *om.* L3O
28 *post* erit¹ *scr. et del.* diversa P1 / maior: maius ER   29 tamen: cum hoc S; cum O / ille *om.*
L3O / post erunt *add.* apud R / punctum *corr. ex* punctus P1   31 est *om.* S / comprehenditur:
comprehendatur R   32 corpus diaffonum *transp.* ER   34 quod¹·²: ut R / portio: proportio
SL3OE / spere: spera SP1L3C1 / concava *inter. a. m.* L3; *om.* O / et *inter. a. m.* C1   35 quod: ut R
36 *post* res *scr. et del.* illa S / post visa *rep.* sit² (34) . . . concavitatem (35) L3O / et illi *inter. a. m.*
E / raro: rara O   37 ergo *om.* S / solicitemur: solicitamur L3C1   39 et *inter. a. m.* L3; *mg.*
a. m. E; *om.* SO   40 aliquod: aliquid R / corpus *om.* ER / comprehenditur; comprehendatur R
41 nisi *om.* P1 / post celi *add.* et aque SOE (*deinde del.* E) / et ignis *om.* O / dividetur: dividitur R; videtur O; *corr. ex* videtur *a. m.* L3   42 distinguat: distinguit R   43 eius *om.* P1   44 sit: fit
P1L3R / ordinate: ordinata P1 / secundum: sed S; per O; *corr. ex* per *a. m.* L3 / successionem: suc-
cessinem S / determinata: terminata ER   45 ergo *om.* L3O / extenduntur: extendentur P1

cum concavum ex parte visus. Ergo omnes stelle que sunt in celo
extenduntur in corpore celi, et reflectuntur apud concavitatem celi,
et extenduntur in corpore ignis et in corpore aeris recte donec perveniant ad visum, et centrum concavitatis celi est centrum terre.

50   [7.44] Dico ergo quod stelle in maiori parte comprehenduntur non
in suis locis et quod semper comprehenduntur non in suis magnitudinibus, et cum hoc diversatur magnitudo uniuscuiusque earum
secundum locorum diversitatem. Diversitas autem locorum est propter positionem radiorum reflexorum, ut prius diximus. Diversitas
55 autem quantitatum est propter remotionem, propter remotionem
enim comprehendentur minores quam sint in veritate, ut diximus in
tertio tractatu, scilicet quod illa que in maxima remotione sunt comprehenduntur minora. Diversitas autem quantitatum secundum
diversitatem locorum accidit propter reflexionem, cuius causam hic
60 declaravimus. Et in quarto capitulo declaravimus quod forme stellarum que comprehenduntur a visu sunt reflexe.

   [7.45] Dico ergo quod omnis stella comprehenditur ex omnibus
locis celi per quos movetur in minori quantitate quam sit in veritate,
secundum quod exigit remotio eius, scilicet minor, si visa fuerit recte,
65 cum non fuerit inter illam et visum aliqua nubes aut vapor grossus.
Et omnis stella in vertice capitis aspicientis existens videtur minor
quam in alio loco celi, et quanto magis removetur a vertice capitis,
tanto magis apparet maior, ita quod in orizonte apparet maior quam
in alio loco. Et hoc est commune omnibus stellis remotis et propin-
70 quis.

   [7.46] Item si in aere fuerit vapor grossus ultra quam fuerit aliqua
stella, tunc comprehendetur maior quam si esset sine illo vapore, et
multum accidit quod vapor grossus sit in orizontibus, unde stelle in
maiori parte videntur in orizonte maiores quam in medio celi. Et hoc

46 reflectuntur: refringuntur R / ibi: sibi L3O   48 post aere scr. et del. cum quo vel C1 / in:
cum L3O / extenduntur: extendentur P1; extendantur R / et om. P1 / reflectuntur: refringantur
R; flectuntur O; corr. ex flectuntur a. m. L3   49 ad visum om. R / et corpus celeste mg. C1;
om. P1E   51 celi¹ ... corpore¹ (52) inter. a. m. L3 / reflectuntur: refringuntur R / celi²: eius L3
52 perveniant: perveniunt L3; perveniat OE   .   53 terre corr. ex spere P1   54 maiori: maiore
R / post comprehenduntur addidi non   55 comprehenduntur non: non comprehendantur
L3 / magnitudinibus corr. ex ymaginibus a. m. L3   57 diversitatem ... locorum² om. S / autem
om. P1   58 positionem ... reflexorum: radiorum reflexorum positionem E; radiorum refrac-
torum positionem R   59 quantitatum: qualitatum P1 / post remotionem¹ add. nam R / propter
remotionem² mg. C1 / propter² ... enim (60) mg. a. m. L3; om. O   60 enim om. P1R / comprehendentur: comprehenduntur C1ER / minores: maiores O; corr. ex maiores L3   61 post que
add. est P1   62 autem om. E   63 reflexionem: refractionem R   65 reflexe: refracte R
66 omnis: omnes O / post stella add. que O   67 in¹ om. ER / minori: minore R   68 fuerit:
fuit P1   69 post fuerit add. recte L3O / grossus corr. ex grossius C1   70 vertice: veritate O;
virtutem inter. a. m. L3 / videtur minor transp. L3   71 removetur: movetur O / post vertice scr.
et del. celi C1   72 quod: ut R / apparet: appareat R   73 est om. C1

75   apparet in distantiis que sunt inter stellas magis quam in magnitudinibus ipsarum stellarum, nam quantitas stelle quoad visum est parva,
et excessus in diversitate distantie inter stellas, cum fuerit in orizonte,
est grandis manifestius sensui, et maxime in distantiis spatiosis, et
maxime si in orizonte fuerit vapor grossus.

80

[7.47] **[PROPOSTIO 19]** Sit ergo circulus meridiei in aliquo orizonte
BK [FIGURE 7.7.15, p. 450], et differentia communis inter hunc circulum et concavitatem celi circulus MEZ. Et sit centrum mundi G et
centrum visus T, et extrahamus GT in partem T. Et occurrat circulo
meridiei in B, et secet circulum qui est in concavitate orbis in E. Erit
85   ergo B vertex capitis quoad visum T. Sit KL diameter alicuius stelle
aut distantia inter aliquas duas stellas, et linea TB transeat per
medium KL, et secabit illam in C. Ergo erit arcus KB equalis arcui BL.
Et continuemus duas lineas TK, TL. Erit ergo angulus KTL ille a quo
T comprehendit KL si recte comprehenderetur.
90   [7.48] Et reflectatur K ad T ex M, et L ad T ex Z. Et continuemus
GM, GZ, et pertranseant ad F, O. Et continuemus lineas KM, MT, LZ,
ZT. Forma autem que extenditur ex K per KM reflectitur per MT, et
GM est perpendicularis exiens ex M, quod est punctum reflexionis,
super superficiem corporis quod est in parte T, et quia corpus ZM est
95   subtilius corpore GT, erit reflexio MT ad partem perpendicularis MG.
M ergo erit inter duas lineas TB, TK, nam si M esset ultra TK, tunc
perpendicularis que exit ex G esset ultra TK, et forma K, cum extenderetur ad illud punctum, reflecteretur ad partem perpendicularis, et
non perveniret ad perpendicularem, et non perveniret ad T. M ergo
100 est inter duas lineas TK, TB, et similiter declarabitur quod Z est inter
duas lineas TB, TL.
[7.49] Et extrahamus TM ad Q et TZ ad R. Erit ergo arcus QK

75 ultra: intra *E* / quam: quem *L3C1R* / *post* fuerit[2] *scr. et del.* vapor *S*   76 comprehendetur *corr. ex* comprehenditur *a. m. E* / illo: alio *P1*; ille *O*   77 multum: multoties *R* / quod: ut *R* / grossus: grossius *S* / *post* sit *add.* ut *P1* / orizontibus: orizonte *R*   78 maiori: maiore *R* / *post* maiori *scr. et del.* quando *P1* / hoc *inter. E*   79 in[1] *corr. ex* quod *L3* / stellas: illas *ER*   80 parva *corr. ex* maior *E*   81 et: sed *R* / distantie *corr. ex* distante *C1* / fuerit: fuerint *P1R*   82 *post* grandis *add.* et *R* / manifestius: manifesta *E*; manifestus *R* / et[1] *inter. L3; om. O* / spatiosis: spationis *S*   84 meridiei *om. S* / orizonte *corr. ex* oriozonte *S*   85 BK: ABG *O; corr. ex* AB *L3*   86 et[1] *inter. a. m. E* / circulus: circulum *S* / G: DG *O; corr. ex* D *L3*   88 in E: ME *SP1L3*   89 sit: ergo *O; inter. a. m. L3* / *post* sit *add.* ergo *P1L3* / KL: KT *P1*   90 transeat: transibit *O*   91 secabit: secet *R* / erit arcus *transp. deinde corr. E*   92 duas *om. L3O* / KTL *corr. ex* KT *E*   93 comprehendit: comprehenderit *S*; comprehenderet *P1* / comprehenderetur: comprehenderet *R* / *post* comprehenderetur *scr. et del.* de *E*   94 reflectatur: refringatur *R* / L: B *C1*   95 GM ... continuemus *mg. a. m. E; om. S* / lineas *corr. ex* linea *L3*   96 *post* KM *add.* et *C1* / reflectitur: refringitur *R*   97 punctum: punctus *E* / reflexionis: refractionis *R*   98 est in parte *rep. E* / quia: quod *SO; corr. ex* quod *a. m. E* / ZM: Z *S*   99 reflexio: refractio *R*   100 M[1] *inter. a. m. E* / erit: est *P1* / lineas *om. S* / TB TK: DB DK *P1* / *post* ultra *add.* corpus *P1* / TK[2]: DL *P1*   101 TK: D *S*; T *P1OER; corr. ex* TL *C1*   102 *ante* ad[1] *scr. et del.* cum *C1* / reflecteretur: reflectetur *P1E*; refringeretur *R* / *post* perpendicularis *add.* GM *R*

equalis arcui LR, et angulus QTR erit minor angulo KTL. Et angulus
QTR est ille per quem T comprehendit KL reflexive, et angulus KTL

105 est ille per quem T comprehenderet KL si recte comprehenderetur.
Sed remotio KL a visu est maxima, quapropter quantitas eius non
certificatur, quare T estimat remotionem KL, sicut in secundo huius
libri diximus. Sed estimatio eius, quando comprehendit reflexe, non
differt ab estimatione eius, quando comprehendit recte, nisi quod

110 putat se recte comprehendere, cum reflexe comprehendat. T ergo comprehendit KL reflexive ex angulo minori illo ex quo comprehendit
illam recte et secundum comparationem ad illam eandem remotionem ad quam comparet illam, si recte comprehenderet. Sed visus
comprehendit magnitudinem ex quantitate anguli respectu remotio-

115 nis; T ergo comprehendit quantitatem KL reflexe minorem quam si
comprehenderet illam recte.

[7.50] Et si circumvolvamus figuram KTL circa TB immobili, faciet
circulum, et erunt anguli qui sunt apud T quos continent due linee
KT, TL et suos compares equales. T ergo comprehendit KL reflexive

120 in omni situ in respectu circuli meridiei, cum fuerit in vertice capitis,
minorem quam si comprehenderet eam recte. Et si TB secuerit KL in
duo equalia, tunc duo puncta Q, R erunt etiam inter duo puncta K,
L, et erit angulus QTR minor angulo KTL, et erit omnis angulus eius
exiens a puncto secans stellam, et linea que exit ex T in superficie

125 illius circuli secabit circulum, et comprehendetur minor quam sit. Et
sic tota stella videbitur minor quam sit.

[7.51] Stella ergo in vertice capitis comprehenditur minor quam si
comprehenderetur recte, et similiter distantia inter duas stellas, cum

---

103 perveniret[1]: perveniet E/ad[1] ... ad[2] mg. a. m. E/post perpendicularem add. GE R/post et add.
sic R/T M: tantum S   104 TK TB transp. R/TK ... lineas (105) om. S/et ... TL (105) mg. a. m. E
106 TZ: TE O/QK: QTK E; UK O; corr. ex QT C1   107 LR: LK L3/QTR: QRL O/erit ... QTR
(108) om. ER/KTL: KTR P1   108 per quem T inter. a. m. L3/T om. O/reflexive: reflexione S;
refracte R/post reflexive scr. et del. et angulus KTL est ille per quem comprehendit KL reflexive
L3/KTL: QTL S/post KTL add. et angulus QTL S   109 ille om. E/T: R O/KL corr. ex QKL C1/
comprehenderetur: comprehenderet R   111 estimat: existimat R/huius om. R   112 libri:
libro R/eius om. L3/quando corr. ex qnm L3/reflexe: reflexive C1; refracte R   113 differt ab:
differet ad L3O/estimatione: estimationem L3/nisi quod inter. a. m. L3; om. O   114 reflexe:
reflective L3; refracte R/comprehendat: comprehendit L3   115 reflexive: refracte R/post reflexive add. et S/minori: minore R   116 illam[1]: eam P1/et om. E/ad om. S   117 comparet:
compararet R; parara S/comprehenderet: comprehenderit E   118 respectu inter. a. m. E/remo-
tionis: remotonis P1; remotioris O; corr. ex remotioris L3   119 reflexe: refracte R/post si add.
BA O   120 post illam scr. et del. reflexe C1   121 circumvolvamus: circumvolamus O/figuram
inter. a. m. L3; om. O/KTL: KEL SL3C1/circa corr. ex cir L3/TB rep. OE/post TB add. BT C1/immo-
bili: immobilem R   122 T corr. ex TE C1   123 KT: TK L3/suos: sue S; sui P1R/reflexive:
refracte R   124 in[2] om. L3/vertice: veritate O   125 quam rep. SO/si[1]: cum E; inter. a. m. L3;
om. SP1C1O/comprehenderet corr. ex comprehenderit a. m. E/TB: BTB O; corr. ex BT L3   126 Q
R mg. E/etiam om. R   127 KTL: TL P1; QTL SO; corr. ex QTL L3   128 exiens: exieiens
S/post puncto add. T R/T: TO L3O; ea S

vertex capitis fuerit inter duas extremitates distantie, comprehende-
130 tur in omnibus positionibus minor quam si recte comprehenderetur,
et hoc est quod voluimus.

[7.52] **[PROPOSITIO 20]** Item si stella sive distantia fuerit infra
verticem capitis et orizonta, aut in orizonte, aut inter orizonta et
135 verticem capitis.
[7.53] Et sit visus A [FIGURE 7.7.16, p. 450] et vertex capitis B, et
continuemus AB. Et sit diameter stelle aut distantia DE equidistans
orizonti, et sit circulus verticalis qui transit per alteram extremitatem
diametri vel distantie circulus BD, et ille qui transit per aliam extremitatem circulus BE. Et sint due differentie communes inter duos
140 circulos et inter concavitatem orbis duo circuli HG, GZ. Forma ergo
D reflectitur ad A in superficie circuli BD. Et continuemus AD, AE.
Arcus ergo BD erit equalis arcui BE, quia DE est equidistans orizonti,
et reflectitur D ad A ex H, et E ad A ex Z.
[7.54] Et continuemus lineas AH, HD, AZ, ZE. Et sit centrum
145 mundi M, et continuemus MH, MZ, et pertranseant ad F, N. Erit ergo
MH perpendicularis exiens ex H super superficiem corporis diaffoni,
et erit HA reflexa ad partem HM; erit ergo reflexa ad partem contrariam illi in qua est HF. H ergo est altius quam AD, et similiter declarabitur quod Z est altius quam AE. Duo ergo puncta F, N sunt inter
150 duo puncta D, E, et angulus reflexionis qui est apud H est equalis
angulo reflexionis qui est apud Z, positio enim duorum punctorum
D, E respectu A est consimilis; tantum ergo distat F ex D quantum N
ex E.
[7.55] Et extrahamus AH ad T et AZ ad K. Distabit ergo T ex D
155 tantum quantum K ex E. Et continuemus TK. Erit ergo equidistans

130 minor: maior E   131 capitis *corr. ex* montis *P1* / comprehenditur: comprehendetur *P1* /
si *corr. ex* sit *P1*   132 cum: si *L3; om. O*   133 capitis *om. ER*   134 si *corr. ex* sit *P1* / recte
comprehenderetur *transp. P1; mg. a. m. E*   136 si: sit *OR; corr. ex* sit *L3* / distantia *corr. ex* distantie *L3* / fuerit . . . aut¹ (137) *om. R*   137 *post* et² *scr. et del.* in *L3*   138 verticem: vertice
O / *post* capitis *add.* equidistans orizonti *R*   141 qui transit *inter. a. m. E* / alteram: aliam *P1*
142 circulus *corr. ex* circuli *O* / aliam: alteram *L3*   143 circulus: circuli *O* / BE: DB *O; corr. ex*
DB *a. m. L3* / *post* BE *add.* ?? *C1*   145 reflectitur: refringatur *R* / in . . . A¹ (147) *om. R* / BD: DB
C1; BE *O* / AE: DE *E*   146 erit equalis *transp. P1* / *post* est *scr. et del.* qi *P1*   147 D: DA *P1*;
inter. L3; om. O / E: C *P1*   148 lineas *inter. a. m. E* / *post* lineas *scr. et del.* HD *P1* / *post* ZE *add.*
AD AE *R*   149 M: E *O*; *corr. ex* E *L3* / *post* M *add.* et continuemus punctum M *SL3C1OE*
(M *inter. a. m. E*) / MH MZ *transp. C1* / pertranseant: pertranseat *C1* / F *om. O* / FN *corr. ex* FQ *P1*
150 super: ad *R*   151 HA: AH *C1* / reflexa¹,²: refracta *R* / ad¹ . . . reflexa² *om. P1*   152 HF:
FH *R* / est altius *transp. P1* / altius: alicuius *SO; corr. ex* alicuius *L3; corr. ex* alterius *C1* / *post* quam
rep. et del. quam *S* / AD: D *O* / declarabitur: delectabitur *O*   153 quod: quam *O; corr. ex* quam
L3 / Z: TB *O; alter. ex* TB *in* T *L3* / altius: alicuius *SO; corr. ex* alicuius *L3* / duo ergo *transp. P1OR*
154 *post* E *add.* et zenith capitis *R* / reflexionis: refractionis *R* / *post* equalis *scr. et del.* angulo reflexionis qui est apud Z est equalis *L3* (Z *alter. ex* H)   155 reflexionis: refractionis *R*

DE; est ergo minor. Et linee AT, AK, AD, AE sunt equales, quia A est quasi centrum duobus circulis BD, BE. Due ergo linee AT, AK sunt equales duabus lineis AD, AE. Et basis TK est minor quam basis DE; ergo angulus TAK est minor angulo DAE, et angulus TAK est ille quo
160 DE comprehenditur reflexe, et angulus DAE est ille quo DE comprehenditur recte.

[7.56] Si ergo stella fuerit in orizonte aut inter orizonta et circulum meridiei, et fuerit diameter eius equidistans orizonti, videbitur minor quam si videretur recte, et hoc idem de distantia inter duas stellas, si
165 distantia fuerit equidistans orizonti, et hoc est quod voluimus.

[7.57] **[PROPOSITIO 21]** Item iteremus formam [FIGURE 7.7.17, p. 451], et sit diameter aut distantia erecta, scilicet in eodem circulo verticali. Et sit ille diameter aut distantia linea DE in circulo verticali BDE, et sit differentia communis inter hunc circulum et inter concavi-
170 tatem orbis circulus GHZ. Et contineumus AD, AE, et reflectatur D ad A ex H, et E ad A ex Z. Patet ergo, ut in precedenti figura, quod H est altius quam AD et quod Z est altius quam AE. Et continuemus lineas AH, HD, AZ, ZE, MH, MZ, et extrahamus MH ad T et MZ ad K. Erit ergo angulus AZM valde parvus, et angulus reflexionis eius
175 erit pars illius; erit ergo angulus EZK acutus, et similiter DHT acutus, et uterque angulus AHD, AZE est obtusus.

[7.58] Z autem aut erit in orizonte aut altius; erit ergo in extremitate perpendicularis exeuntis ex A super AB, aut altius illa, et H est altius quam Z. Ergo angulus AHM est minor angulo AZM; ergo
180 angulus DHT est minor angulo EZK. Ergo angulus AHD est maior angulo AZE. Et due linee MT, MK sunt diametri circuli BDE et diametri circuli GHZ; ergo MT est equalis MK, et MH est equalis MZ. Ergo HT est equalis ZK, et angulus DHT est minor angulo EZK; ergo linea HD est minor quam EZ.

156 consimilis: similis *L3O*/ex: a *R*   157 ex: ab *R*   158 T ex D *corr. ex* D ex T *L3*/ex: a *R* 159 K *corr. ex* E *L3*/ex: ab *R*/TK: TH *SOE*   160 AD: AF *SP1L3OER*   161 quasi *om. S*/*post* centrum *add.* mundi et *R*/duobus circulis: duorum circulorum *R*/BD: BDB *S*; BA *C1*/BE: DE *S* 162 duabus: duobus *C1*/est: erit *E*   163 DAE: ZA *O*/*post* et *scr. et del.* minor *C1*   164 reflexe: reflexive *C1*; refracte *R*; *corr. ex* reflexexe *P1*/reflexe . . . comprehenditur² *inter. a. m. L3*/DE comprehenditur: deprehenditur *P1*   167 et fuerit: tunc *O*   168 *post* idem *add.* est *R*/stellas: lineas *E*   169 et . . . voluimus *om. R*   170 formam: figuram *R*   171 aut *mg. a. m. E*; *corr. ex* ad *P1*/erecta: erectus *O*; *corr. ex* erectus *L3*   172 linea: lina *P1*/DE *inter. a. m. L3* 173 *post* sit *add.* a *C1*/hunc *inter. a. m. L3*/inter² *om. L3*   174 circulus *inter. a. m. E*/GHZ: GBZ *E*; *corr. ex* BHZ *C1*/reflectatur: refringatur *R*/D: B *O*   175 ad² *om. S*/precedenti: precedente *R* 176 quod *corr. ex* quam *L3*   177 et² *inter. L3*; *om. O*   178 reflexionis: refractionis *R*/eius *om. R*   179 erit¹: sit *L3O*/EZK: ZZK *O*/DHT: ZHT *O*; *corr. ex* GHT *P1*   180 et *inter. E*/est *om. R*/obtusus *corr. ex* equidistans *P1*   181 *post* altius *add.* si in orizonte *R*/in² *inter. E*   182 super: semper *P1*/H: hoc *O*   183 est¹: erit *C1*/AHM *corr. ex* HAM *L3*/est²: erit *R*/angulo *om. E*   184 EZK: EZH *O*

185   [7.59] Et due linee AD, AE sunt equales, et A est quasi centrum circuli BDE; ergo circulus qui continet triangulum AHD est maior circulo qui continet triangulum AZE, quia angulus AHD est maior angulo AZE. Et linea HD est minor, ut declaratum est, quam ZE; ergo HD distinguit de circulo continenti triangulum AHD arcum
190   minorem arcu simili arcui quem dividit ZE ex circulo continenti triangulum AEZ. Angulus ergo HAD est minor angulo ZAE.

   [7.60] Sit ergo angulus ZAD communis. Ergo angulus HAZ est minor angulo DAE, et angulus HAZ est ille quo A comprehendit DE reflexive, et angulus DAE est ille quo comprehendit DE, si illud com-
195   prehenderet recte. A ergo comprehendit DE reflexe minorem quam recte, et hec demonstratio sequetur si circulus BDE fuerit circulus meridiei.

   [7.61] Diameter ergo stelle, cum fuerit directus et rectus, et distantia inter duas stellas recta comprehenditur reflexive minor quam
200   recte, et hoc est quod voluimus.

   [7.62] Et omnis stella in celo comprehenditur rotunda; ergo dia-
   metri eius comprehenduntur equales. Et cum sit manifestum quod uterque diameter eius rectus et transversus secundum latitudinem comprehenditur minor quam si comprehenderetur recte, ergo uterque diameter eius declinis comprehenditur minor quam si compre-
205   henderetur recte. Et similiter distantie inter stellas comprehenduntur in omnibus locis et in omnibus sitibus minores quam si comprehen-
   derentur recte. Item diximus quod omnis stella existens in vertice capitis comprehenditur minor quam in omnibus partibus celi, et quanto magis fuerit remotior a vertice capitis, tanto magis comprehendetur

185 MK *corr. ex* MD *a. m.* L3/diametri^{1,2}: semidiametri R/diametri^1 . . . et^2 *rep.* S/*post* et^2 *add.* due linee MH MZ sunt R   186 MH: MK S   187 DHT: ZHT O/EZK: EZH O/ergo^2 . . . EZ (188) *mg. a. m.* E   188 HD: DH R/quam *inter.* O/EZ: DE O; *corr. ex* DE L3   189 AD *inter.* L3; *om.* O/*post* equales *add.* similiter due AH AZ sunt equales R/et^2: Z SO; quia R; *inter. a. m.* L3/*post* A *scr. et del.* et L3   190 *post* BDE *add.* et circuli GHZ R/*post* circulus *add.* quam O; *add.* quem SL3 (*deinde del.* L3)/est maior *transp.* R/est . . . AZE (191) *mg. a. m.* E   191 AZE: AEZ P1ER; *corr. ex* AZO O   192 AZE *corr. ex* AEZ C1/HD: DH P1; HK O; *corr. ex* HK L3 193 distinguit: distinguet L3O; distingunt P1/continenti: continente R   194 arcu *corr. ex* arcui P1/arcu simili *om.* S/continenti: continente R   195 triangulum *om.* R/*post* triangulum *scr. et del.* AHD arcum minorem C1/HAD: AHD O; *corr. ex* AHD L3/est minor *transp.* R 196 angulus *om.* R/HAZ: GAZ O; *corr. ex* GAZ L3   197 *post* DAE *rep. et del.* ergo^2 (196) . . . DAE E/HAZ: GAZ O; *corr. ex* GAZ L3/*post* ille *add.* sub R/A *om.* P1/DE reflexive (198) *transp.* ER (reflexive: refracte R)   198 *post* ille *add.* sub R/comprehendit: comprehenderit P1C1; comprehenderat S/*post* DE *add.* recte R/si . . . DE (199) *scr. et del.* E/illud *om.* R   199 recte *om.* R/A ergo *transp.* E/reflexe: reflexive C1; refracte R   200 sequetur: sequitur R/BDE . . . stelle (202) *scr. et del.* E   202 directus: directa R/rectus: recta R   203 recta: recte O; *corr. ex* recte L3/reflexive: refracte R   204 et . . . voluimus *inter. a. m.* E   205 comprehendi-tur: comprehendetur P1/ergo: quia R   206 quod: et O   207 uterque: utraque R/rectus: recta R/transversus: transensus S; transversa R   208 comprehenderetur: comprehendetur P1O/uterque: utraque R   209 eius: est L3O/*post* eius *add.* est E; *scr. et del.* de P1/declinis: declivis R/*post* comprehenditur *add.* equaliter R

210 maior, et quod maxima comprehenditur quando comprehenditur in
orizonte. Et hoc quidem testatur esse, scilicet quod stella videtur in
medio celi minor quam si fuerit intra medium celi, et similiter de distantiis, et stella maxima in orizonte et distantie similiter. Restat ergo
declarare causam quare hoc sit.

215   [7.63] Dico quod in secundo huius libri declaravimus, cum tracta-
vimus de magnitudine, quod si visus comprehenderit magnitudines
visibilium, comprehendet illas ex quantitatibus angulorum quos
respiciunt visibilia apud centrum visus, et ex quantitatibus remotionum, et ex comparatione angulorum ad remotiones. Et declaravimus
220 quod numquam visus comprehendit visibilium quantitates nisi remotiones eorum sint in rectitudine corporum propinquorum contin-
uorum et quod, si visus non certificaverit de remotionibus visibilium,
non certificabit quantitates visibilium. Et declaravimus illic etiam
quod, si visus non certificaverit distantiam visi, potest perpendere
225 distantiam eius et assimulare eam distantiis visibilium assuetorum
quibus tale visibile comprehenditur in tali forma et in tali figura;
deinde comprehendit magnitudinem illius ex quantitate anguli quem
respicit illud visibile apud centrum visus respectu remotionis quam
perpendit.

230   [7.64] Et remotiones stellarum non sunt in rectitudine corporum
propinquorum, quare visus non comprehendit quantitates eorum,
nec visus certificat distantias stellarum; visus ergo perpendit distantias stellarum et assimulat illas distantiis eorum que sunt terrestria
que comprehenduntur ex distantia maxima, et perpendit quantitates
235 eorum. Corpus autem celi non videtur sensui quod sit spericum et
quod concavum eius sit ex parte visus, nec visus sentit corporeitatem
celi, nec visus sentit de celo nisi solum colorem glaucum solummodo.
Corporeitas vero, et extensio secundum tres dimensiones, et rotun-

210 stellas: lineas E / comprehenduntur: comprehendentur P1   211 comprehenderentur: comprehenderetur SE   212 existens: eius SC1; ens P1; exiens L3O (deinde del. L3); om. ER   213 post
omnibus add. aliis R / quanto corr. ex quando a. m. C1   214 magis[1] om. L3OR / magis[2] om. R / comprehendetur: comprehenderetur SL3C1O   215 maior inter. a. m. E / quod: quam R / quando comprehenditur inter. a. m. E / comprehenditur[2] mg. a. m. L3   216 et . . . similiter (218) om. R / esse inter. a.
m. L3 / in . . . minor (217): minor in medio celi C1   217 celi inter. a. m. E / intra: inter O   220 post
secundo add. libro C1; add. tractatu R   221 si inter. a. m. E / comprehenderit: comprehenderet S
222 comprehendet: comprehendit R   224 ad corr. ex et a. m. L3   225 numquam visus transp.
L3R / comprehendit: comprehenderit L3   226 sint: sunt P1; inter. a. m. E / in rectitudine om. S / propinquorum: propinquiorum P1 / propinquorum . . . certificaverit (227) mg. a. m. L3; om. O   227 et
om. P1 / quod inter. a. m. E; om. L3 / certificaverit: certificavit C1; certificarit R / de om. R / remotionibus
visibilium: visibilium remotiones R   228 illic: illis S   229 si visus transp. ER / perpendere:
comprehendere C1   230 assimulare: assimilare R   233 illud: illum visum O / visibile corr. ex
visum L3   235 non corr. ex que L3   236 propinquorum: propinquiorum P1 / post quare scr.
et del. autem O / eorum: earum R   237 nec: neque R / visus[1] om. R / stellarum: earum R / post stellarum rep. et del. non (235) . . . comprehendit (236) S / visus[2] . . . stellarum (238) mg. a. m. E / distantias[2]:
distantiam E   238 assimulat: assimilat R / illas: illis S / que . . . eorum (240) om. E

ditas, et concavitas nullo modo possunt comprehendi. Et cum visus
240 non certificaverit aliquid, tunc assimilabit ipsum illis quibus assimilatur de rebus assuetis, unde comprehendit solem et lunam planas,
et corpora convexa et concava a distantia maxima plana, et arcus
quorum convexum aut concavum est ex parte visus comprehendet
rectas, nam si non comprehenderit propinquitatem medii et remotio-
245 nem extremitatum, et remotionem in mediis concavis et propinquitatem extremitatum, tunc assimilabit superficies convexas et concavas
superficiebus planis, et assimilabit arcus lineis rectis, assueta enim
visibilia in maiori parte sunt plana et recta.

[7.65] Nec visus, cum forma stelle pervenit ad ipsum, sentit quod
250 illa forma sit reflexa aut quod reflectetur ex superficie concava, et
quod corpus in quo est stella sit subtilius corpore in quo est visus.
Sed forma stelle comprehenditur sicut forme aliarum rerum qui
comprehenduntur in aere recte, et forme visibilium non reflectuntur
quando occurrunt corpori diverso ab aere propter visum, nec visus
255 sentit reflexionem eorum, nec superficiem a qua reflectuntur forme in
corporibus diversis in diaffonitate in proprietate naturali forme lucis
et coloris que extenduntur in corporibus diaffonis. Forme ergo stellarum reflexarum perveniunt ad visum sicut perveniunt forme eorum
que sunt in aere ad visum, et comprehenduntur sicuti comprehen-
260 duntur in aere.

[7.66] Visus autem comprehendit colorem celi, nec tamen certificat formam eius nudo sensu. Et cum visus comprehenderit aliquem
colorem existens in longitudine et latitudine super hoc quod comprehendet figuram et formam, comprehendet ipsum planum, assimilabit
265 enim ipsum aliquibus superficiebus assuetis, ut parieti et aliis, et hoc
modo comprehendet superficies convexas et concavas a maxima
remotione. Visus etiam comprehendit planitiem terre planam omnino,
nec sentit convexitatem eius nisi fuerint ibi montes et valles. Visus

240 non *om. S*   241 quod *om. ER*/sit *om. P1*/nec: neque *R*/nec visus *om. S*   242 nec: neque
R/solum[1] *om. R*   244 nullo: in illo *O*/et[2] *inter. a. m. L3*/cum: tamen *S*; si *R*   245 non *inter.
a. m. L3; om. O*/certificaverit: certificavit *P1*/illis . . . assimilatur: alicui *R*   246 planas: planos
R; corr. ex planetas *C1*   247 a *om. P1*/distantia maxima *transp. R*   248 comprehendet: comprehendat *C1*/*post* comprehendet *add.* lineas *R*   249 si *inter. L3*/non *om. SP1C1*/remotionem
corr. ex remotionum *C1*   250 *post* extremitatum *add.* in convexis *R*/in *om. R*/mediis: medii
R/concavis *om. R*   251 *post* extremitatum *add.* in concavis *R*/convexas et concavas: concavas
et convexas *E*   253 maiori: maiore *R*   255 illa: ipsa *C1*/reflexa: refracta *R*/reflectetur: refringatur *R; corr. ex* reflecteretur *C1*   256 est stella *transp. ER*   257 comprehenditur . . . qui
om. S/*post* aliarum *add.* stellarum *O*/rerum: regularum *O*/qui: que *R*   258 comprehenduntur:
comprehendentur *P1L3*/recte *om. P1*/reflectuntur: refringuntur *R*   259 *post* quando *scr. et del.*
contra *E*/corpori diverso: corpora diversa *O*   260 reflexionem: refractionem *R*/reflectuntur:
refringuntur *R*   261 in[2]: nisi *R*   263 reflexarum: refractarum *R*   264 in aere *inter. a. m.*
E/*post* et *add.* non *ER*/sicuti: sicut *P1ER*   267 aliquem colorem (268) *transp. ER*   268 existens
om. R/hoc *om. S*/comprehendet: comprehendit *ER*

ergo comprehendit superficiem celi planam, et comprehendet stellas
270 sicut comprehendit visibilia assueta separata que sunt in locis spatiosis. Et cum visus comprehenderit aliqua visibilia separata in aliquo
loco spatioso, et comprehenderit illa angulis equalibus, et comprehenderit quantitates distantiarum illorum visibilium, tunc illud quod
est remotius comprehendetur maius, nam quantitates magnitudinis
275 remotionis comprehendentur ex comparatione anguli quem respicit
illa remotio apud centrum visus ad distantiam remotam, et comprehendet quantitatem magnitudinis propinque ex comparatione anguli
quem respicit illud propinquum qui est equalis angulo quem respicit
distantia ad distantiam propinquam.

280 [7.67] Et hoc patet et esse testatur ei, scilicet quod duorum visibilium
que a visu comprehenduntur duobus angulis equalibus, quorum
distantie sunt diverse sensibiliter, remotior videbitur maior, et remotius videbitur maius. Nam si homo opposuerit se spatioso parieti,
deinde elevaverit manum donec opponat illam visui, et cooperuerit
285 alterum visum, et aspexerit reliquo, et posuerit manum mediam inter
visum suum et illum parietem, tunc manus eius cooperiet portionem
et latitudinem illius parietis, et comprehendet parietem et manum
suam simul. Comprehendet ergo manum suam angulo acuto, et in
hoc statu comprehendet latitudinem parietis maiorem quam latitu-
290 dinem manus multiplicem. Deinde, si moverit manum ita quod dete-
gatur illud quod manus cooperuerat de pariete, et aspexerit illud
quod detectum fuerit de pariete, et aspexerit ad manum, videbit illud
quod est detectum de pariete maius quam sit sua manus multipliciter. Et ipse comprehendet manum suam, et comprehendet parietem
295 duobus angulis equalibus, ex quo patet quod visus comprehendit
magnitudinem ex comparatione anguli ad remotionem.

269 comprehendet: comprehendit E   270 enim inter. O/aliis: aeris O   271 comprehendet: comprehendit R/a: in ER/maxima remotione (272) transp. ER   272 etiam: ergo C1ER/comprehendit:
comprehendet L3   273 nec corr. ex non L3/nisi: qui P1/ibi corr. ex in a. m. C1   274 ergo: autem
L3O/comprehendet: comprehendit C1ER   276 post comprehenderit scr. et del. quantitates distantiarum C1/post visibilia add. assueta L3ER (inter. a. m. L3)/separata om. R/aliquo loco (277) transp. ER
277 angulis: angulus O/et² om. S   278 illorum: illarum E; om. R/illud om. L3   279 magnitudinis remotionis (280) transp. P1ER (deinde corr. P1)   280 comprehendentur: comprehendetur
L3; comprehenduntur R/post anguli add. q L3O   281 illa om. P1/comprehendet: comprehendit
R/comprehendet quantitatem (282) transp. C1   282 ante quantitatem add. visus R/quantitatem:
quantitates P1/propinque: propinqui C1O   283 illud: illum P1C1   285 et¹: ex S/post et² add.
hoc C1/ei scilicet: AS O   287 distantie corr. ex distantia P1/remotior . . . et om. R   288 se: si
S/spatioso: spatio P1/parieti: parti O; corr. ex parti a. m. L3   289 opponat: apponat R/illam: illa P1
290 visum: visuum P1C1O/reliquo: reliquum O/posuerit: posuit L3   291 eius om. C1/cooperiet:
cooperietur O; corr. ex cooperietur L3/portionem: portione O   292 latitudinem: latitudine O/et¹
. . . parietem om. L3O/parietem . . . suam (293): manum suam et parietem ER   293 suam² inter.
a. m. E   294 statu alter. in situ a. m. E/latitudinem¹ corr. ex latinudinem C1   295 manus: maius
P1/quod: ut R   296 et . . . pariete (297) scr. et del. E; om. R/aspexerit: aspexit S

[7.68] Visus ergo comprehendet superficiem celi planam, nec sentiet concavitatem eius, et comprehendet stellas separatas in ipso. Comprehendet ergo stellas equales separatas inequales, nam comparat angulum quem respicit stella extrema propinqua orizonti apud centrum visus ad distantiam remotam, et comparat angulum quem respicit stella que est in medio celi et propinqua medio remotioni propinque. Et similiter comprehendit stellam que est in orizonte aut prope maiorem ea que est in medio celi aut prope. Comprehendit ergo eandem stellam et distantiam in diversis locis celi diverse quantitatis; sic ergo comprehendit eandem stellam aut distantiam in orizonte aut prope maiorem quam in medio celi aut prope, nam comparat angulum quem respicit illa stella apud centrum visus, stella existente in orizonte, distantie remote, et comparat angulum quem respicit illa stella apud centrum visus, stella existente in medio celi, distantie propinque. Sed inter angulum quem respicit stella apud centrum visus, stella existente in medio celi, et inter angulum quem respicit apud centrum visus, stella existente in orizonte, non est maxima diversitas, sed duo anguli sunt propinqui, quamvis diversi, et similiter distantie inter stellas. Et cum sensus comparaverit duos angulos propinquos in magnitudine ad duas diversas distantias in magnitudine, tunc remotior comprehenditur maior.

[7.69] Et quod certificat hanc causam est quod illi anguli quos eadem stella respicit apud centrum visus ex omnibus partibus celi, cum linee que continent ipsos fuerint reflexe, quoniam locus visus est centrum celi, et reflexiones formarum stellarum non diminuunt ex istis angulis diminutione maxima, et cum iste diminutiones non sint maxime, tunc diversitas inter angulos reflexos quibus stella comprehenditur et inter remotionem inter stellas a locis diversis celi non erit maxima diversitas. Et cum diversitas istorum angulorum non est

297 detectum: decretum *P1* / fuerit: est *S* / de . . . pariete (298) *om. S*   298 maius: manus *S* / multipliciter: multiplices *O*   299 ipse comprehendet: tamen comprehendit *P1* / comprehendet²: comprehendit *SP1*; *om. R*   1 remotionem . . . comprehendet (2) *rep. P1*   2 comprehendet: comprehendit *SP1OR*   3 sentiet: sentit *ER* / comprehendet: comprehendit *R*; *corr. ex* comprehendit *a. m. C1*   4 comprehendet: comprehendat *S*; comprehendit *R* / equales *mg. a. m. L3* / equales . . . inequales: separatas inequales equales *P1*   6 *post* centrum *add.* et *L3*   7 que est *om. R* / et *om. P1* / medio²: medie *SO*; *alter. in* meridiei *a. m. L3*   9 que *corr. ex* quia *C1* / celi *inter. a. m. E* / comprehendit: comprehendet *E* / ergo *inter. a. m. L3*; *om. O*   11 aut¹: et *ER*   12 maiorem . . . prope² *om. P1R* / quam: que *L3*; quidem *inter. a. m. E* / comparat *corr. ex* comparait *S*   13 respicit illa stella: stella illa respicit *P1* / in . . . existente (15) *rep. S*   14 respicit: respit *S*   15 apud . . . stella² *mg. a. m. E* / stella² *corr. ex* stellarum *P1* / stella existente *transp. R* / in *inter. a. m. C1*; *om. O* / *post* in *scr. et del.* or *P1* / distantie propinque *inter. a. m. L3*; *om. O* / *post* distantie *scr. et del.* celi *P1*   16 stella *inter. a. m. L3*; *om. O* / apud . . . stella (18) *mg. C1* / visus: visa *P1* / visus . . . existente (17) *corr. ex* stella existente visus *O*   17 in . . . existente (18) *rep. S*; *mg. a. m. L3*; *om. O* / *post* respicit *add.* stella *R*   18 est *inter. E*   20 sensus: centrum *P1*   21 magnitudine: ymagine stelle *P1* (*deinde del.* stelle) / distantias: substantias *O*   23 quod *inter. P1* / illi *om. R* / illi anguli *transp. E*   24 stella *inter. a. m. E*   25 reflexe: refracte *R* / *post* reflexe *add.* sunt quasi anguli per quos comprehenderetur recte *C1ER* (*mg. a. m. E*; comprehenderetur: comprehendentur *E*)

maxima, tunc magnitudo stelle non comprehendetur diversa maxima diversitate. Et quod demonstrat diminutiones angulorum reflexionis ab angulis quos continent linee recte non sunt maxime magnitudinis et quod sunt valde parve est quod dictum est in predicta experientia de capitulo reflexionis, in quo declaravimus quod visus comprehendit stellam reflexive, et videt stellam fixam ex polo mundi, et remotio est eius ab ipso in una revolutione, nam hec diversitas invenitur parva, ex quo patet quod anguli reflexionis sunt parvi. Unde per illam diversitatem que est inter ipsos non diversantur anguli quibus stella comprehenditur in locis diversis celi maxima diversitate.

[7.70] Sed magnitudo stelle et distantia stellarum differunt multum cum sint in orizonte et in medio celi; ergo causa diversitatis stelle et distantie in magnitudine in locis diversis celi non est diversitas angulorum reflexionis. Et iam declaravimus quod visus comprehendit magnitudinem comparando angulos ad remotiones; ergo si diversitas inter angulos fuerit modica et inter distantias et remotiones multa, tunc res videbitur ex maiori distantia maior. Causa ergo per quam videntur distantie stellarum in orizonte maiores quam in medio celi aut prope est illud quod sensus estimat illas distare in orizonte magis quam in medio celi, et hoc quod visus comprehendit stellas in locis diversis celi diversas in magnitudine est error perpetualis, quia causa eius perpetua, et est quoniam visus comprehendit superficiem celi planam, nec sentit concavitatem eius et equalitatem distantiarum eius a visu. Et constat in anima quod in superficie plana, que extenditur ad omnem partem, differunt distantie eius in visu et quod illud quod est propinquius est illud quod est proximius capiti. Comprehendit ergo illud quod est in orizonte remotius quam illud quod est in medio celi, et quod anguli quos respicit eadem stella apud centrum

26 post centrum add. centro L3O (deinde del. L3) / et inter. C1 / reflexiones: refractiones R; reflexionis SL3O; corr. ex reflexionis a. m. E / diminuunt: diminuuntur ER   27 istis: illis ER / diminutiones corr. ex diminutionem P1   28 reflexos: refractos R / comprehenditur: comprehendit O   30 maxima diversitas transp. L3 / istorum angulorum transp. S   31 post non scr. et del. est maxima P1   32 post demonstrat inter. quod C1 / reflexionis: refractionis R; reflexorum L3O   33 ab angulis: ad angulos R; corr. ex ad angulos a. m. E / sunt: est P1L3ER (alter. ex sunt L3)   34 post quod² scr. et del. predic-tum E   35 reflexionis: refractionis R / post quod scr. et del. d O   36 reflexive: refracte R / reflexive ... stellam² om. P1 / ex: in L3O   37 est om. L3 / est eius transp. R   38 reflexionis: refractionis R   39 que corr. ex quod L3   41 distantia: distantie R   42 sint: sunt R   43 post magnitudine add. celi P1L3O   44 reflexionis: refractionis R   45 post angulos add. remotionis C1ER / si inter. a. m. L3   46 et¹ inter. a. m. L3 / distantias: distantes S / et² inter. a. m. C1   47 maiori: maiore R / post maiori scr. et del. d L3 / causa: cum O; inter. a. m. L3 / ergo: autem P1 / per: propter R / quam: quem O   48 videntur: videantur L3 / post videntur scr. et del. forme P1 / in² om. P1   49 estimat: estimant P1 / in ... magis (50): magis in orizonte R   50 post celi rep. aut (49) ... celi L3O (distare ... magis² mg. a. m. L3; om. O) / et hoc inter. a. m. L3 / quod inter. a. m. C1 / comprehendit: comprehendet P1   51 locis om. P1 / locis diversis transp. R / perpetualis: perpetuus R; perpendicularis O; corr. ex perpetua a. m. L3   52 eius: est ER   53 sentit: sentiet L3 / distantiarum: distantiam S; distantie L3ER; corr. ex distantie P1

visus ex omnibus partibus celi non maxime diversantur. Et quod visus
55 comprehendit magnitudinem rei ex comparatione anguli quem res
respicit ad remotionem illius rei a visu, comprehendit ergo quantitatem stelle et quantitatem distantie que est inter stellas, cum fuerit in
orizonte aut prope, comparatione anguli ad distantiam remotam, et
cum fuerint in medio celi aut prope, ex comparatione anguli equalis
60 primo aut fere ad distantiam propinquam, et inter ipsam et inter
distantiam orizontis videtur diversitas maxima.

[7.71] Hec igitur est causa propter quam errat visus in diversitate
magnitudinis stellarum et distantiarum, et hec causa est fixa, perpetua,
immutabilis. Et visus comprehendit stellas parvas propter remotio-
65 nem earum, respiciunt enim apud centrum visus angulos parvos.
Sed sensus non certificat quantitatem remotionis stelle, sed estimat et
comparat remotiones stellarum cum remotionibus visibilium assuetorum que sunt in terra ita quod opinatur quod remotio stelle est sicut
remotio alicuius maxime remoti in terra. Comparat ergo angulum
70 quem facit stella apud visum, qui est parvus, ad remotionem sicut
remotio est eorum que sunt super terram, et sic comprehendit stellam propter hanc comparationem parvam. Et si visus esset certus
de quantitate remotionis stelle, tunc comprehenderet eam magnam,
et similiter est de omnibus que sunt super terram maxime remotis,
75 si comprehendantur parva est quia non certificatur remotio eorum,
et iam declaravimus hoc perfecte in tertio huius libri. Et sicut visus
errat in quantitate remotionis stelle, quia non est certus de ipsa, et
assimulat eam remotionibus que sunt super terram, sic errat in hoc
quod distantie earum in locis diversis celi sunt diverse, cum sint
80 equales, quia assimulat eas etiam distantiis diversis que sunt super
terram dextram et sinistram et opposite de quibus non est dubium
ipsas esse diversas. Et sicut error in remotione et magnitudine stelle
est perpetuus, sic error in diversitate distantiarum stellarum in locis
diversis celi et in diversitate magnitudinis est perpetuus, nam forme
85 harum distantiarum non diversantur apud visum in diversis temporibus sed semper sunt eodem modo, et visus assimulat eas distantiis

54 eius *corr. ex* visus *C1; om. R*   55 quod *om. R*/illud: id *R*/illud quod (56) *mg. a. m. E*/illud . . .
est[2](56) *rep. S*   56 proximius: proximum *R*; proximus *SL3O*; *corr. ex* proximus *C1*   57 ergo illud *mg. a. m. E*   60 rei *inter. a. m. L3; om. O*/anguli . . . comparatione (63) *om. P1*/quem: quam
L3/res   *om. L3*   65 ad *inter. a. m. L3; om. O*/et[1] *inter. a. m. L3; om. O*   66 videtur *inter. a. m. L3;
om. O*/diversitas maxima *transp. ER*   67 igitur est *transp. ER*/propter: per *L3*/errat: erat *S*/diver-
sitate magnitudinis (68) *transp. C1*   68 causa est *transp. E*/est fixa *transp. R*/*post* fixa *add.* et *R*/*post*
perpetua *add.* et *R*   69 parvas *corr. ex* per quas *a. m. L3*   71 *post* sed[1] *add.* et *ER*   74 remoti:
remota *P1*   76 est *om. E*/que *corr. ex* qui *C1*/super terram: in terra *R*   80 *post* comprehendantur *add.* et *E*/est: sunt *R*/non: si *E*/eorum: earum *SO*   81 *post* tertio *add.* tractatu *R*/sicut: cum
P1   83 assimulat: assimilat *R*/eam: ipsam *ER*/errat: erat *S*   84 sunt: sint *R*/sunt . . . cum *mg. a.
m. E*   85 assimulat: assimilat *R*/eas *corr. ex* eam *L3*/etiam: in *P1*   86 dextram . . . opposite *om.*

assuetarum rerum que maxime distant a visu super faciem terre.

[7.72] Accidit etiam eis que sunt in celo alia causa ad hoc quod videantur maiora in orizonte in maiori parte, scilicet vapores grossi
90 qui sunt oppositi inter visum et stellas. Et cum vapor fuerit in orizonte aut prope, et non fuerit continuus usque ad medium celi, erit portio spere, cuius centrum erit centrum mundi, quod continet terram, et sic abscindetur ex parte medii celi, et erit superficies eius que est ex parte visus plana, quare forma aut distantie que sunt ultra illum vaporem
95 videbuntur maiores quam sine illo vapore, in illo enim loco concavitatis celi ex quo loco reflectitur forma stelle ad visum existit forma stelle, et ex ipso extenditur recte ad visum, si in orizonte non fuerit vapor grossus.

[7.73] Si vero affuerit vapor grossus, tunc hec forma extendetur
100 ad superficiem vaporis qui est ex parte celi et existet in illa superficie, et sic visus comprehendet illam sicut comprehendet ea que sunt in vapore, scilicet quod illa forma extenditur in vapore grosso recte, deinde reflectitur apud superficiem vaporis que est ex parte visus ad contrariam partem perpendicularis existentis super superficiem
105 vaporis (que est plana), nam aer qui est ex parte visus est subtilior illo vapore; ex quo sequitur quod forma videatur maior quam si videretur recte, ut in prima figura huius capituli diximus. Et est cum corpus subtilius fuerit ex parte visus et grossius ex parte rei vise, erit superficies corporis grossioris plana. Forma ergo que pervenit ad
110 superficiem vaporis que est ex parte celi est res visa, et corpus in quo extenditur hec forma est vapor grossus, et aer in quo est visus est subtilior illo.

[7.74] Causa ergo principalis quare stelle et distantie stellarum videntur in orizontibus maiores quam in medio celi est illa predicta,
115 et est fixa perpetua. Si vero acciderit quod sit vapor grossus, crescit

87 ipsas: eas ER/sicut: sic P1E   88 sic inter. a. m. E/distantarium stellarum transp. L3O
89 ante est add. eius L3/est inter. a. m. L3/forme harum (90): formarum O   90 harum: earum
P1ER/non om. L3O   91 post sunt add. in C1E/assimulat: assimilat R   92 assuetarum:
assuetorum P1L3O/rerum om. L3O/maxime corr. ex magne S/faciem: superficiem L3R   93 accidit: accedit R/causa: tam S; ei cum L3O   94 maiori: maiore R/scilicet: si L3O/grossi
inter. a. m. L3   95 qui: que E/stellas: stellam R/vapor mg. a. m. E   96 portio: proportio L3O   97 erit centrum inter. a. m. E/quod: qui R; et S   98 abscindetur: ascindetur
SO/que: quod C1   99 quare inter. a. m. E/forma: forme ER   101 reflectitur: refringitur
R/existit . . . stelle (102): forma stelle existit ER/forma stelle (102) transp. P1   104 si: sine
L3O/affuerit: fuerit ER/affuerit . . . grossus: vapor grossus affuerit C1/post tunc add. enim C1/
extendetur: extenditur S   105 qui: que R   106 comprehendet: comprehendit R/ea:
eam P1   107 scilicet . . . vapore² mg. a. m. L3   108 reflectitur: refringitur R/que . . . visus
om. R/parte corr. ex pate C1   109 existentis: exeuntis P1L3ER; et scientiis O/super inter. a.
m. L3; om. SOE   111 videatur: videtur ER   112 recte mg. a. m. E/in inter. C1/et inter. a.
m. E/cum: circa P1/cum corpus (113) transp. deinde corr. L3   113 subtilius fuerit transp. C1
114 grossioris corr. ex grossius P1

magnitudo earum. Sed hec causa est in quibusdam locis semper et in quibusdam quandoque. Omnia ergo que diximus in hoc capitulo de illis que accidunt visui propter reflexionem deceptiones sunt ille que semper accidunt, aut in maiori parte, et sufficiunt in hoc quo indige-

120   mus de deceptionibus quarum causa est reflexio.

[7.75] Nunc autem terminemus hunc tractatum, qui est finis libri, etc.

125

116 hec: et *L3O* (*deinde del. L3*); *om. ER*/grossus: grossius *O*; *corr. ex* grossius *L3*; *corr. ex* grossior *C1*/aer *inter. a. m. E*   117 subtilior: subtilius *L3O*   118 ergo: vero *ER*   119 videntur: videantur *R*/orizontibus: orizonte *R*   120 vero *inter. a. m. L3*; *om. O*/quod: ut *R*/sit: si *S*/crescit: cressit *SO*   121 sed *corr. ex* et *E*/post in[1] *scr. et del.* ma *P1*/semper et *om. P1*   122 *post* quibusdam *rep.* quod (120) . . . causa (121) *O*; *add.* non et in quibusdam *deinde rep. et del.* crescit (120) . . . causa (121) *L3*/ergo: vero *L3O*   123 reflexionem: refractionem *R*/deceptiones sunt *transp. R*   124 maiori: maiore *R*/quo: quod *SC1*   125 *ante* de *scr. et del.* reflexionibus *E*/de *inter. a. m. E*/est *rep. O*   126 *post* libri *add.* penna precor siste quoniam liber explicit iste *L3*; *add.* explicit liber halhacen filii halhaycen filii aycen de aspectibus *E*; *add.* alhazen filii alhayzen opticae finis *R*   127 etc. *om. L3ER*/post etc. *add.* deo gratias explicit liber hacen filii hucayn filii haycen de aspectibus *SC1* (deo gratias *om. S*); *add.* deo gratias explicit liber septimus alhacen *P1*; *add.* deo gratias explicit expliceat videre scriptor eat *O*

FIGURES FOR
TRANSLATION
AND
COMMENTARY

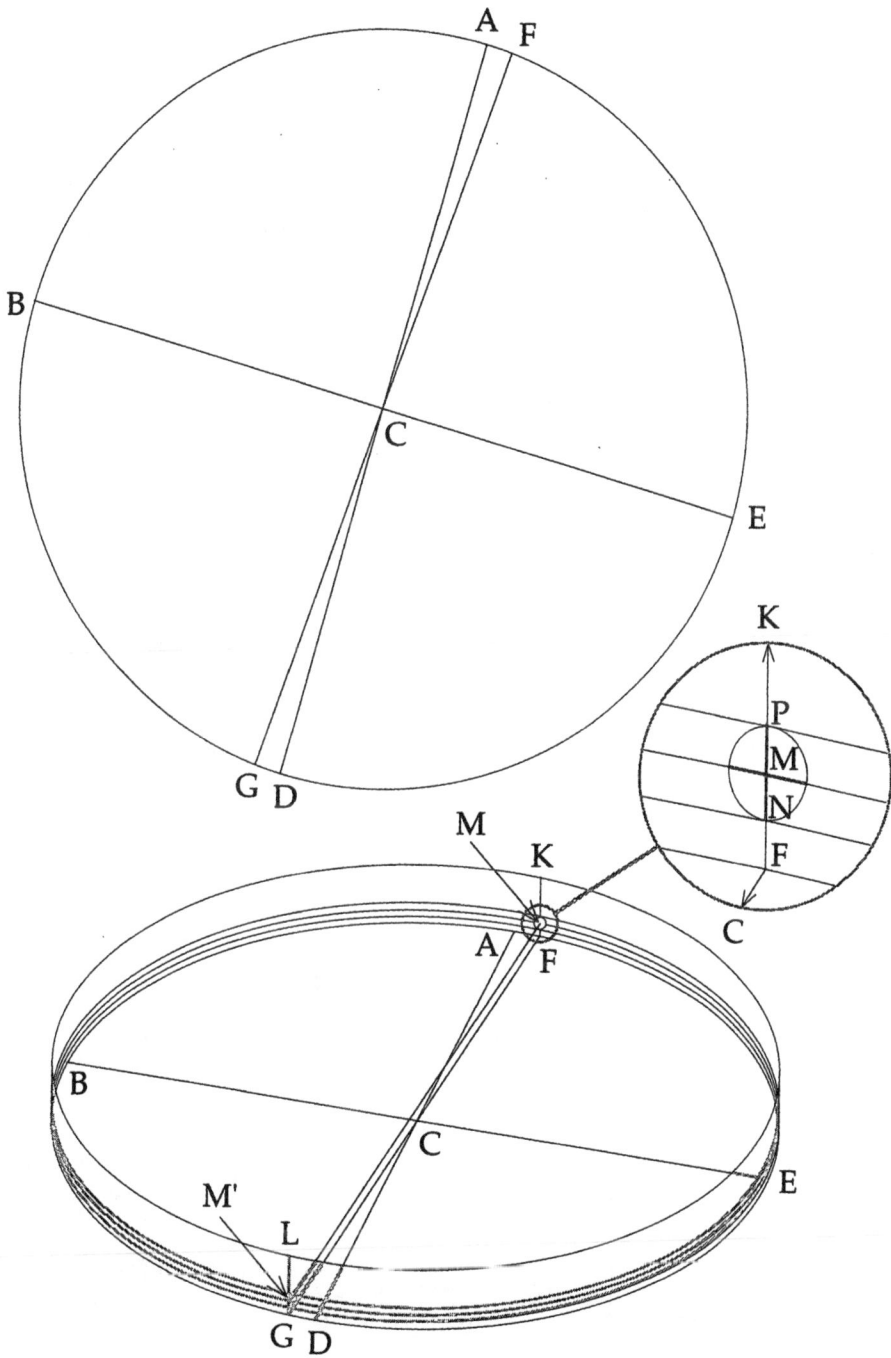

figure 7.2.1

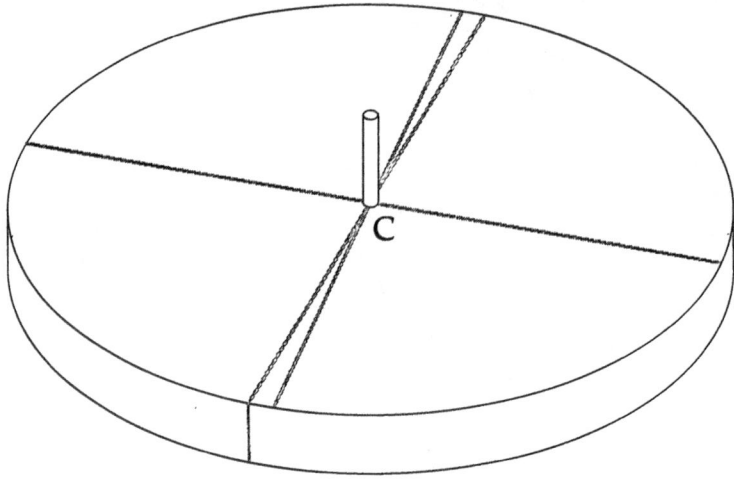

figure 7.2.2

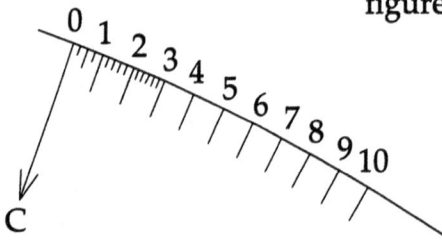

figure 7.2.3

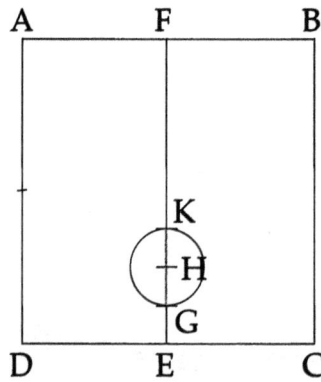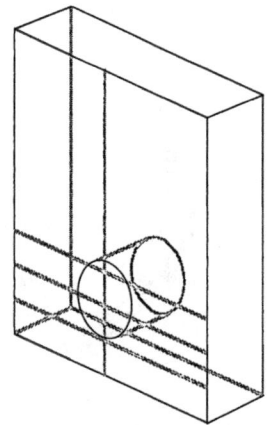

figure 7.2.4

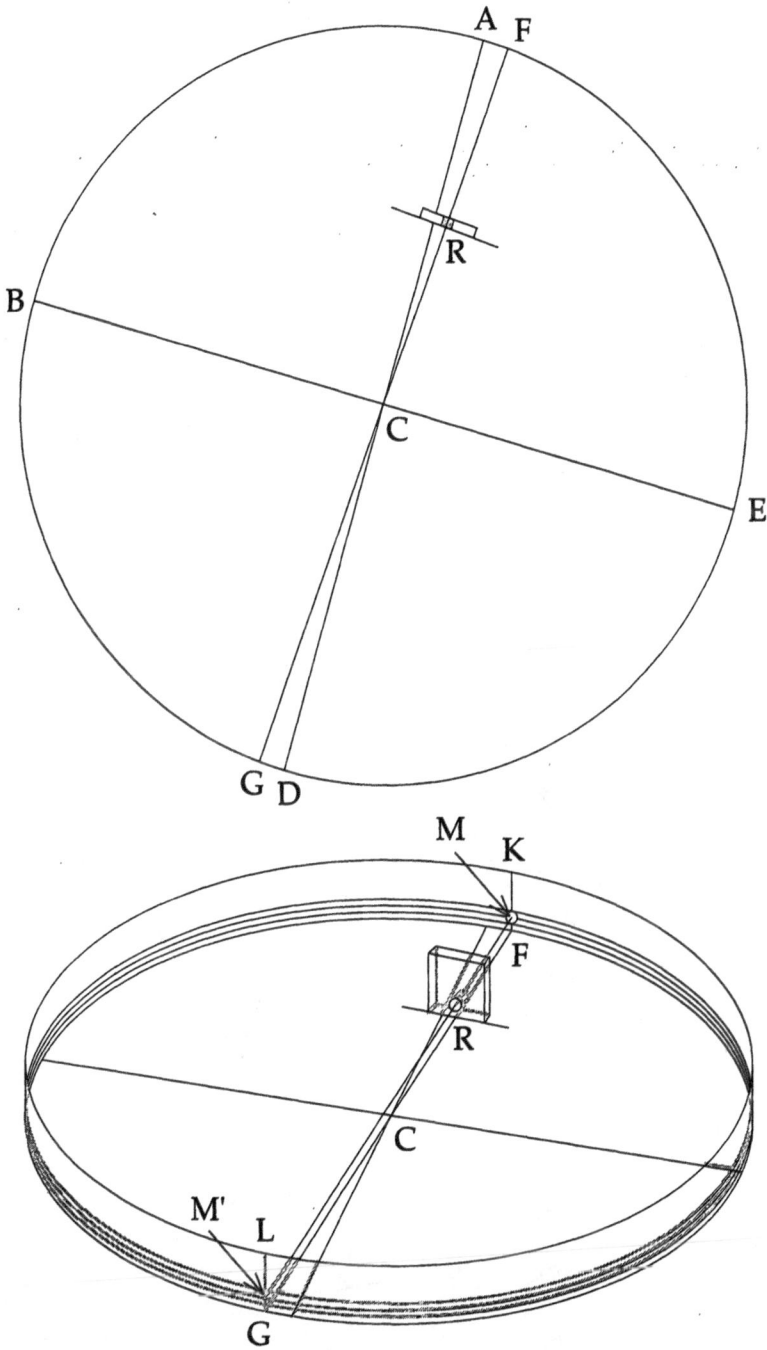

figure 7.2.5

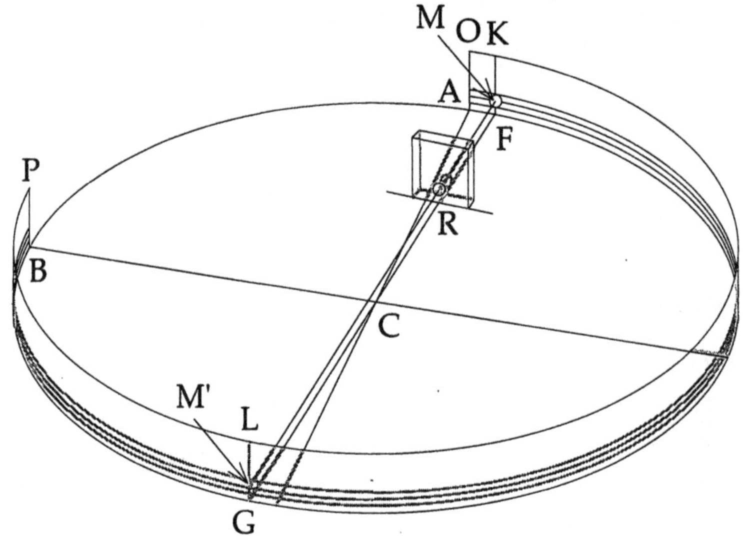

figure 7.2.6

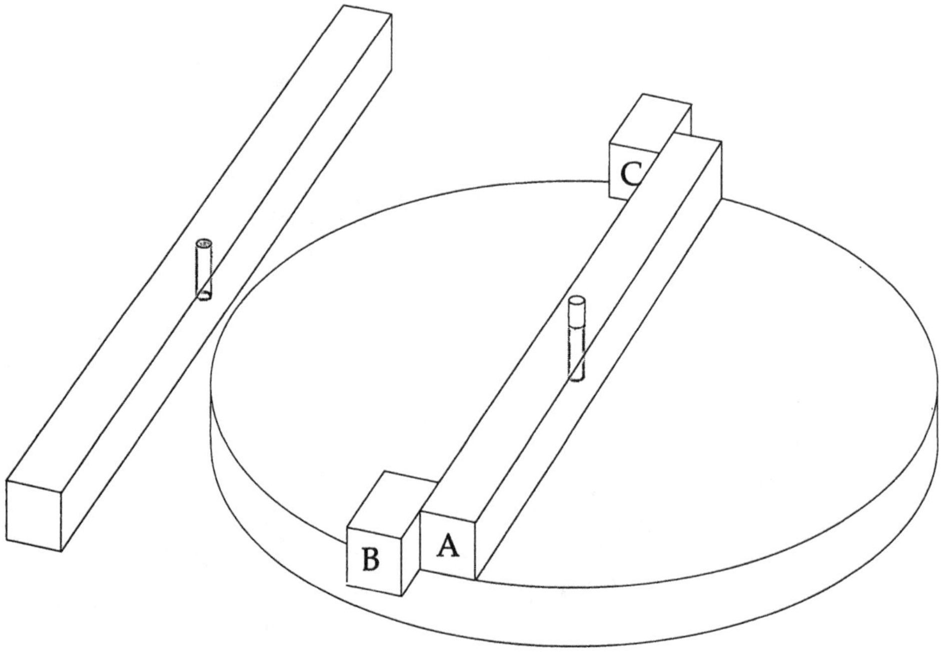

figure 7.2.7

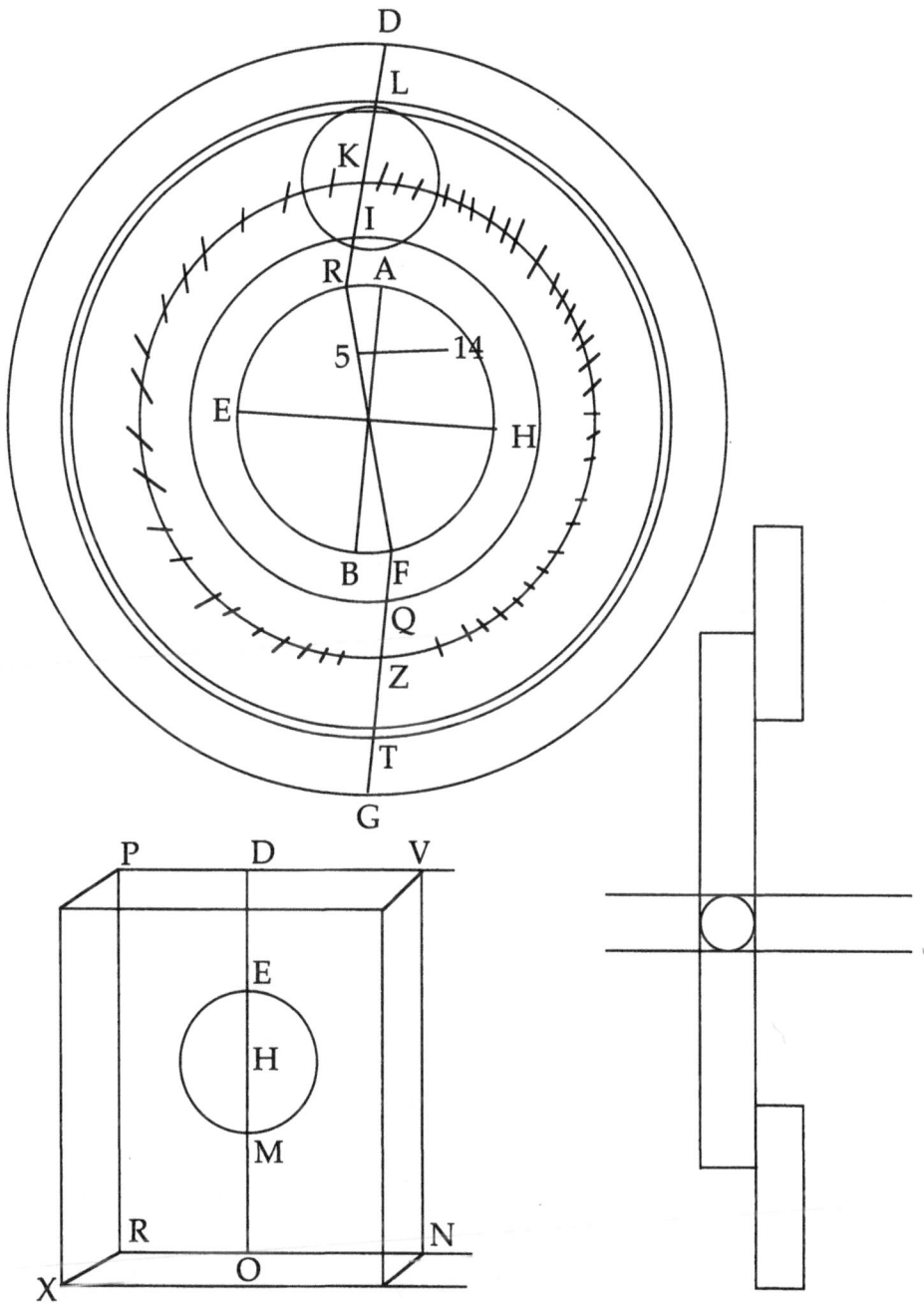

figure 7.2.7a

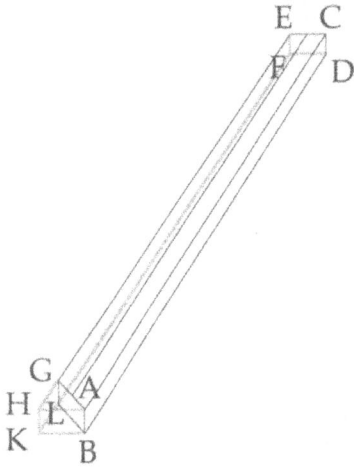

figure 7.2.8

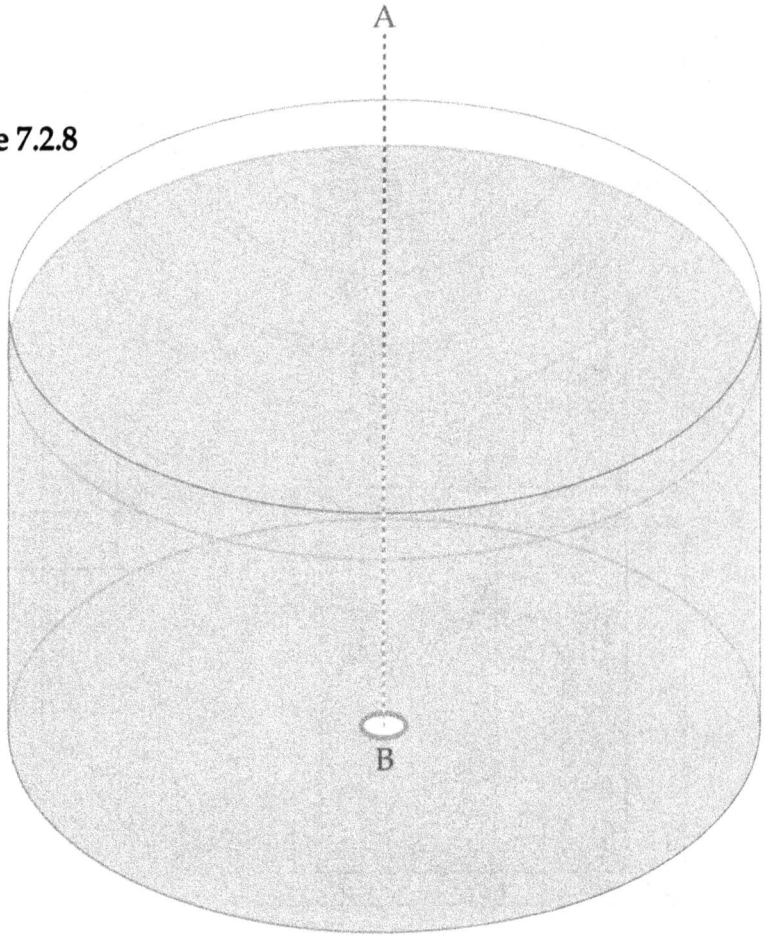

figure 7.2.9

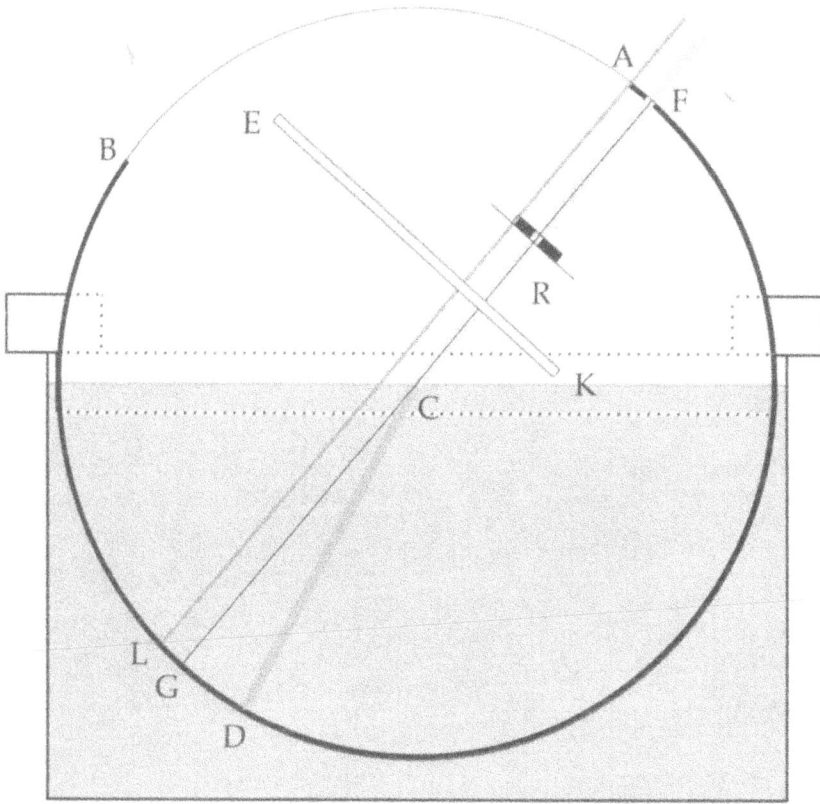

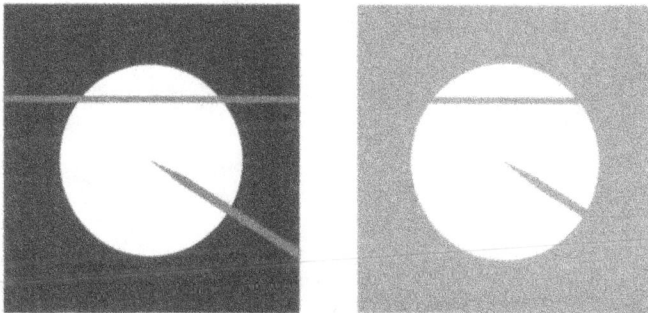

figure 7.2.10

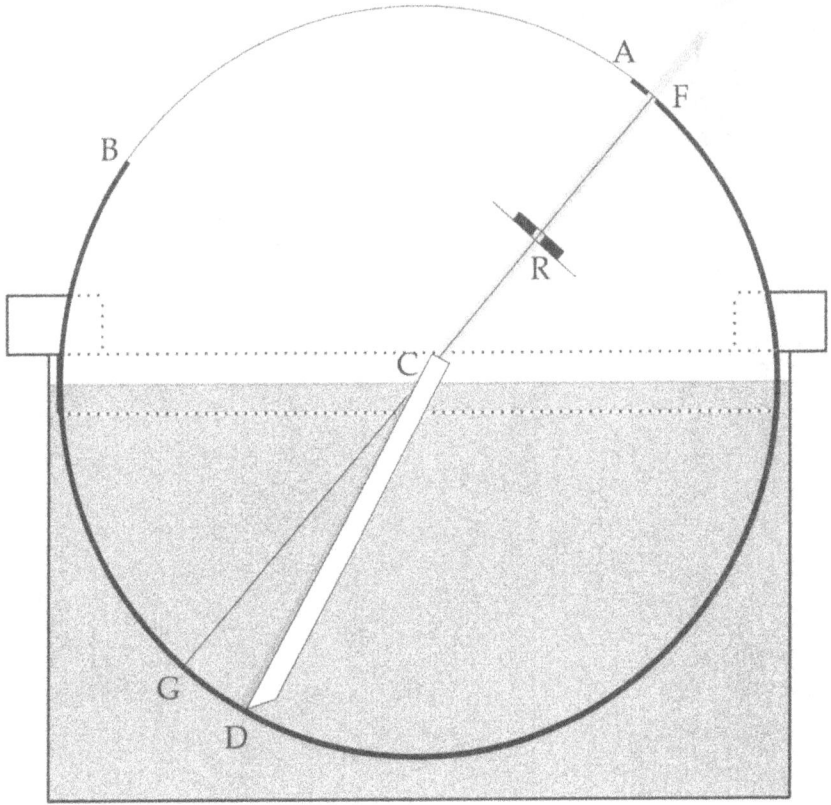

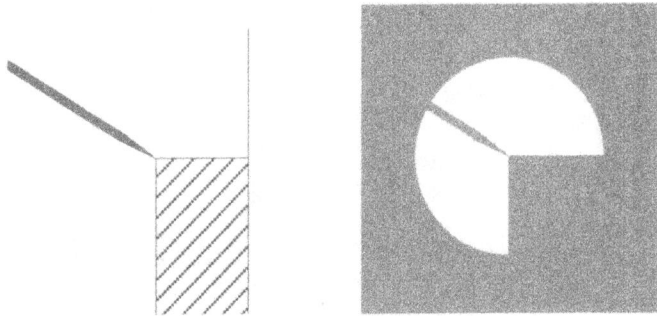

figure 7.2.11

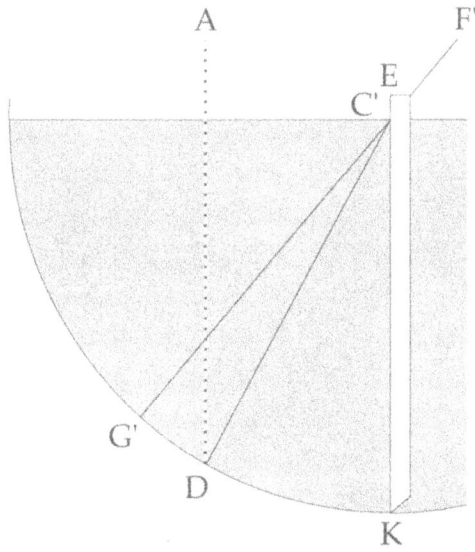

figure 7.2.12

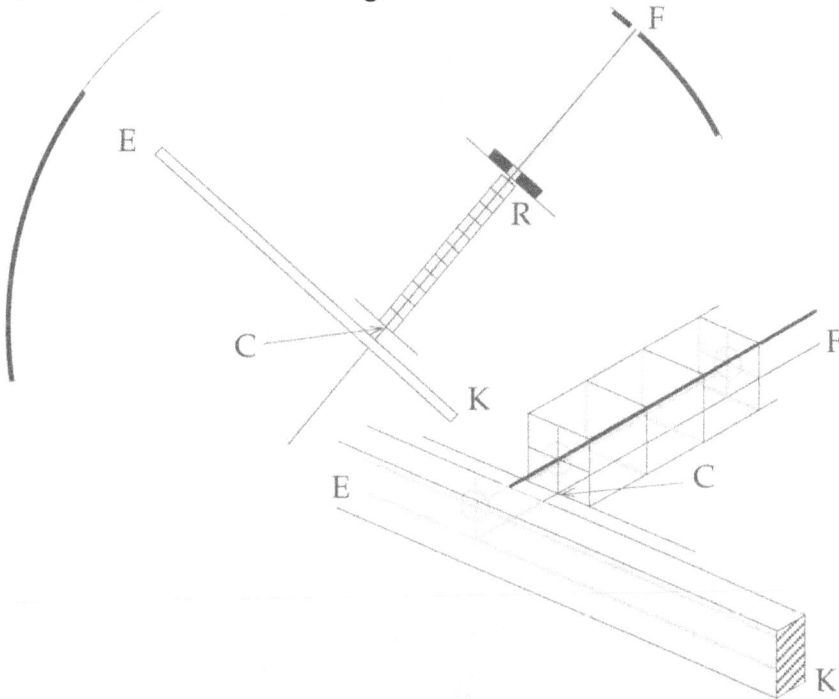

figure 7.2.13

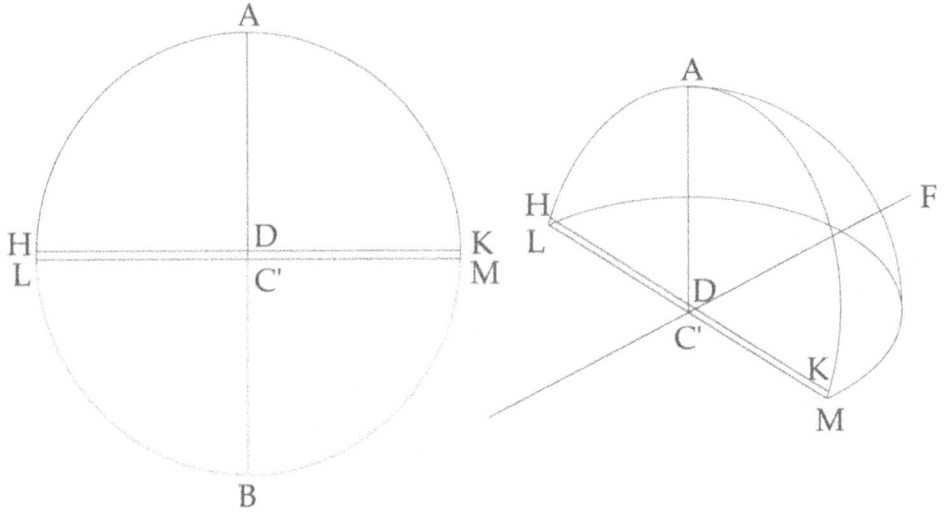

figure 7.2.14

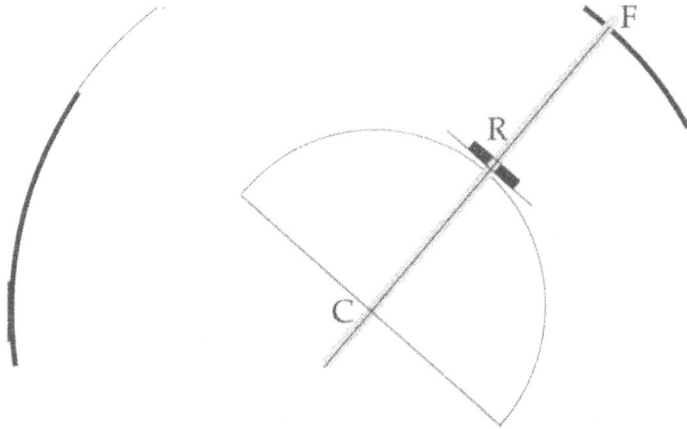

figure 7.2.15

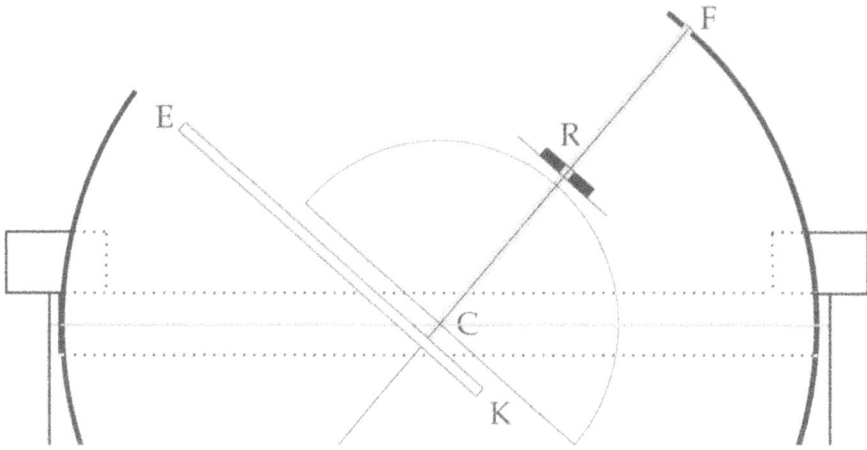

figure 7.2.16

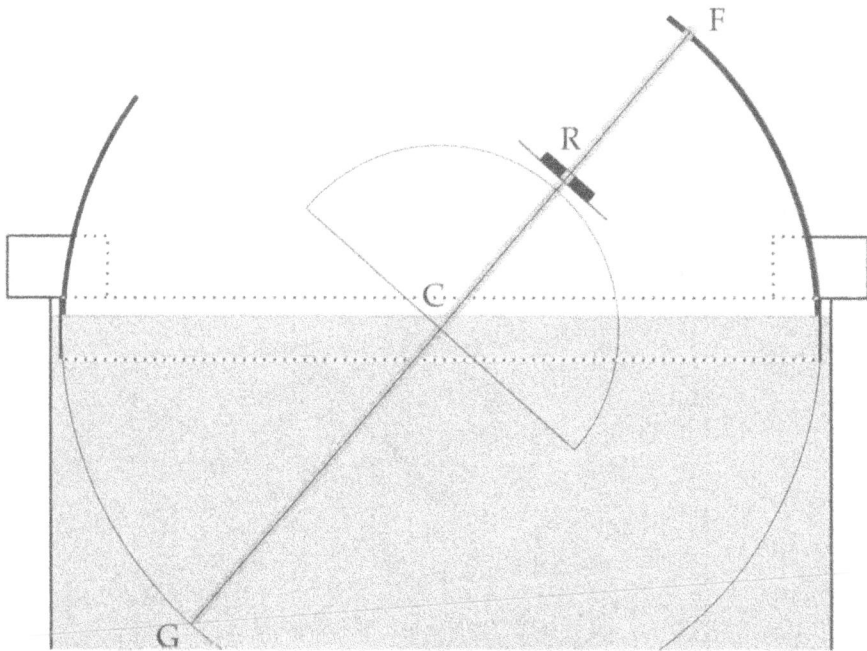

figure 7.2.17

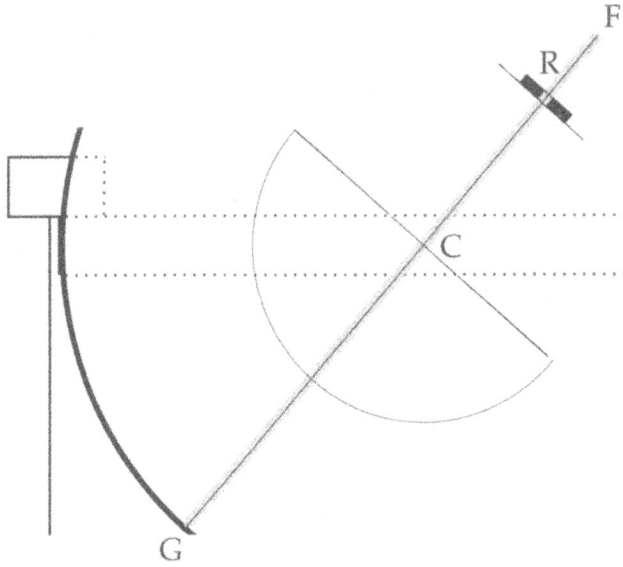

figure 7.2.18

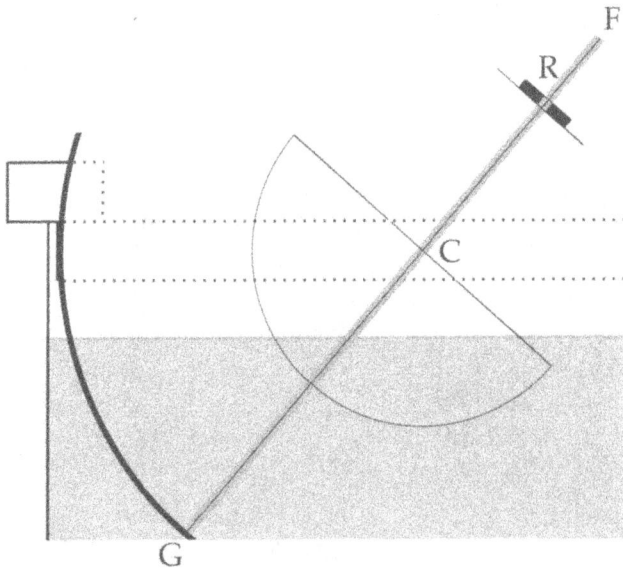

figure 7.2.19

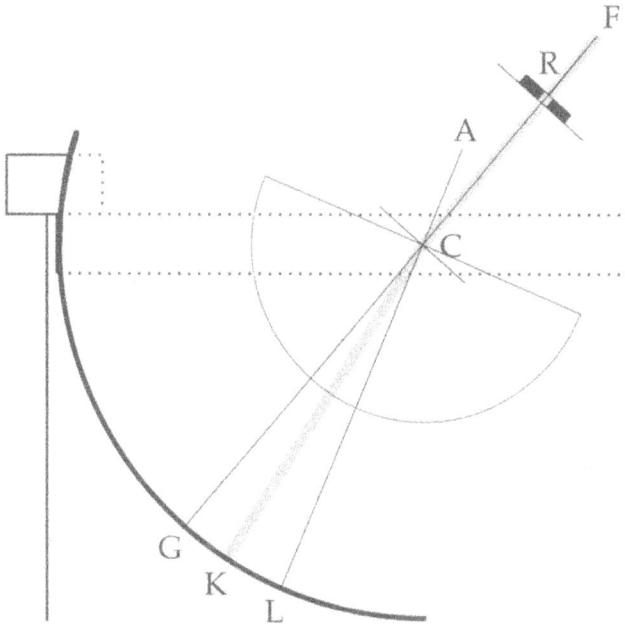

figure 7.2.20

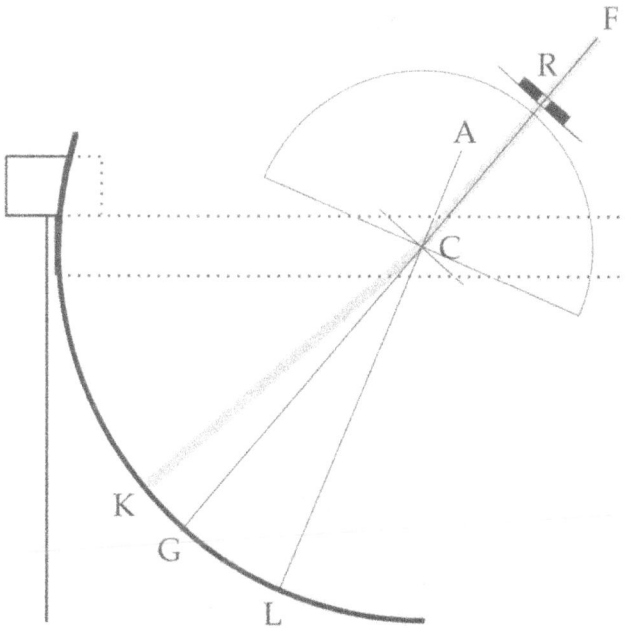

figure 7.2.21

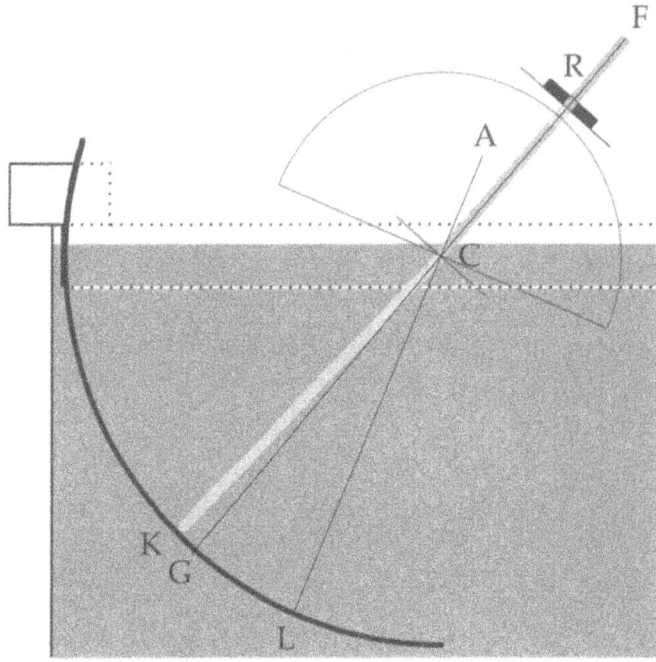

figure 7.2.22

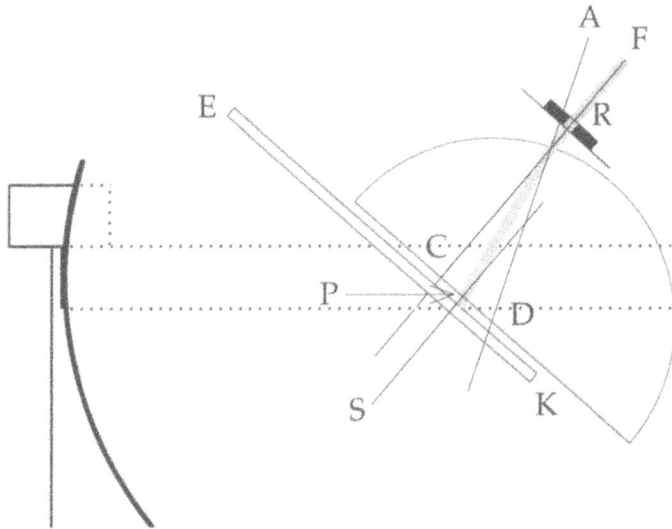

figure 7.2.23

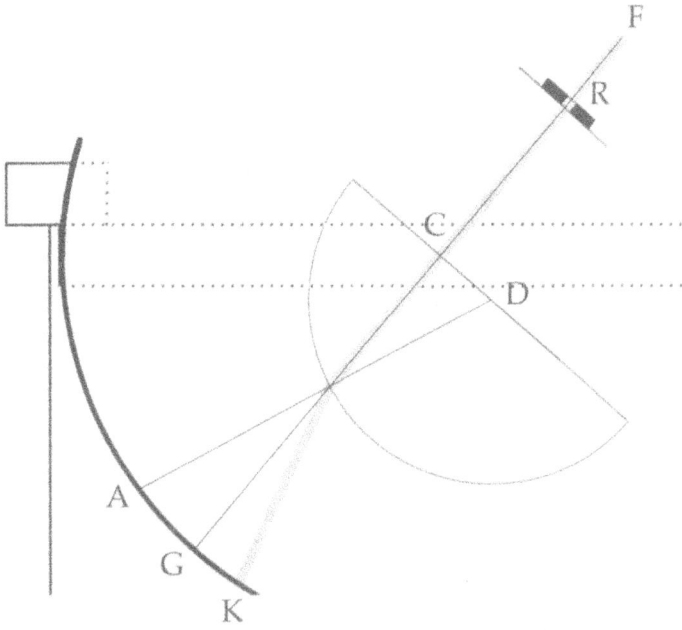

figure 7.2.24

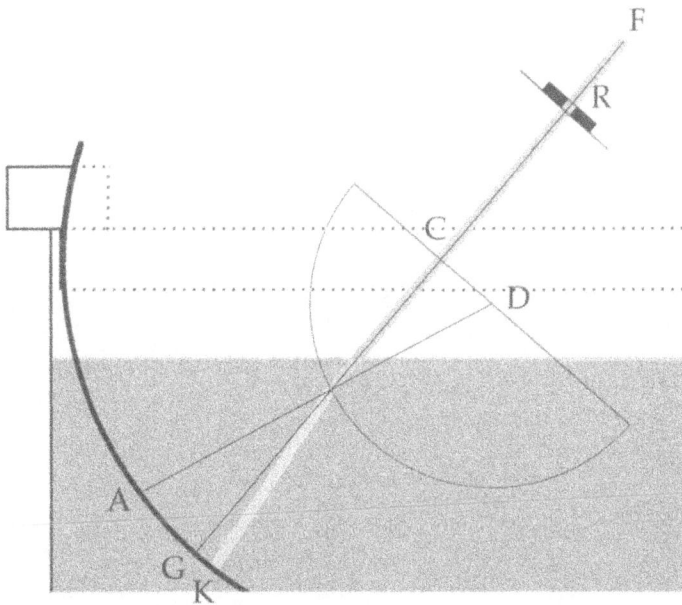

figure 7.2.25

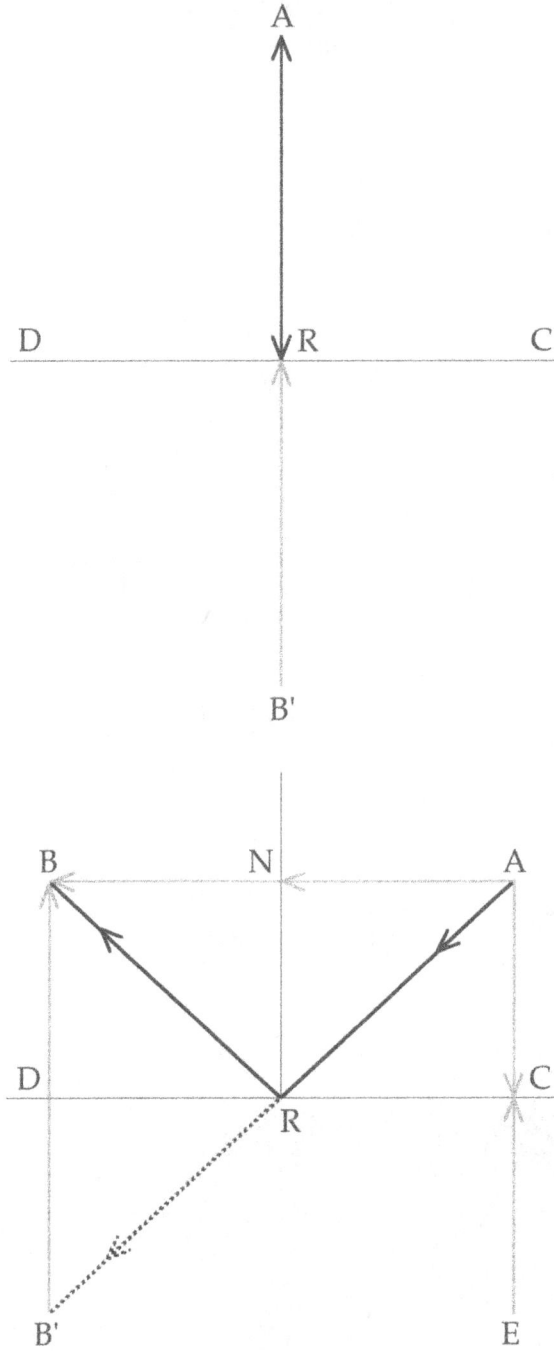

figure 7.2.26

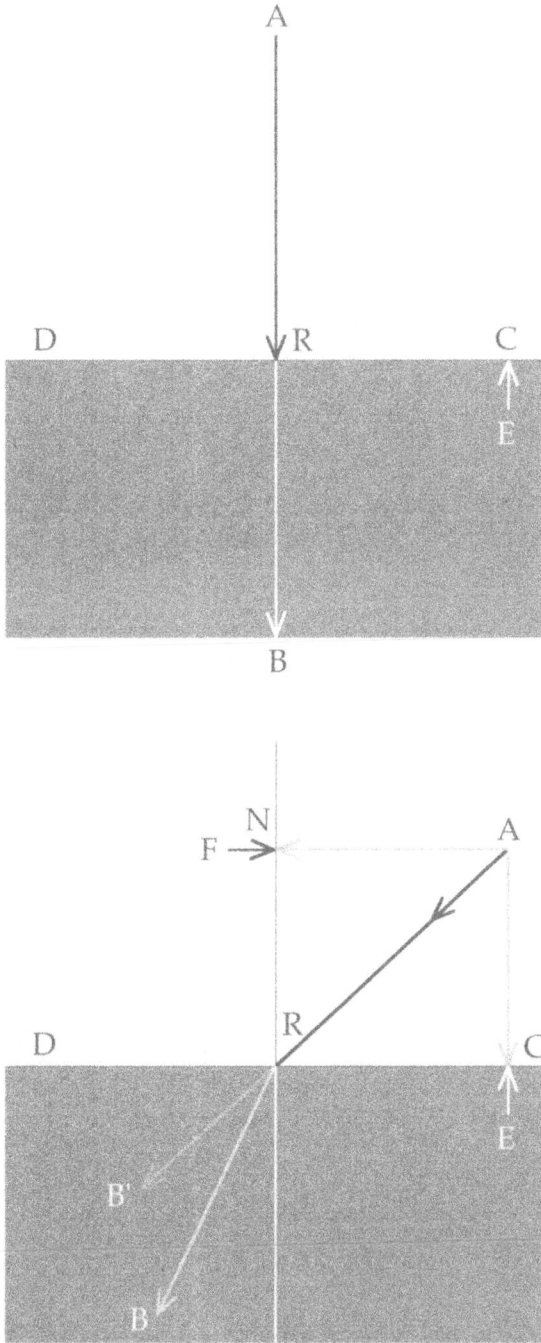

figure 7.2.27

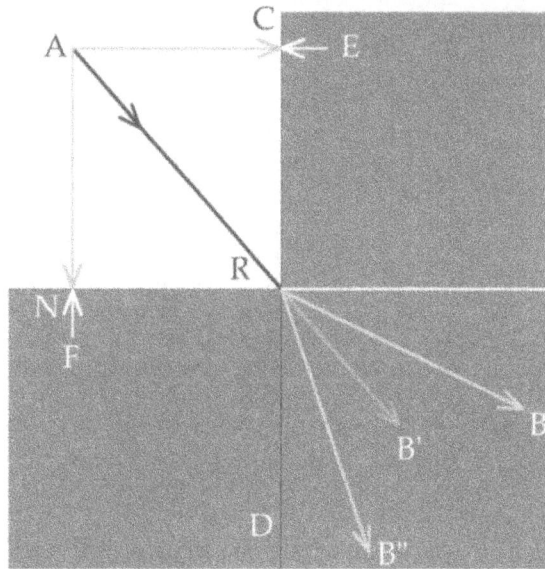

figure 7.2.28

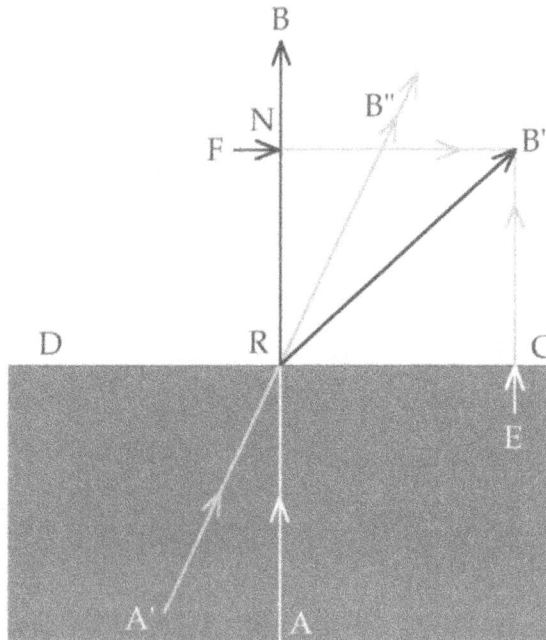

figure 7.2.29

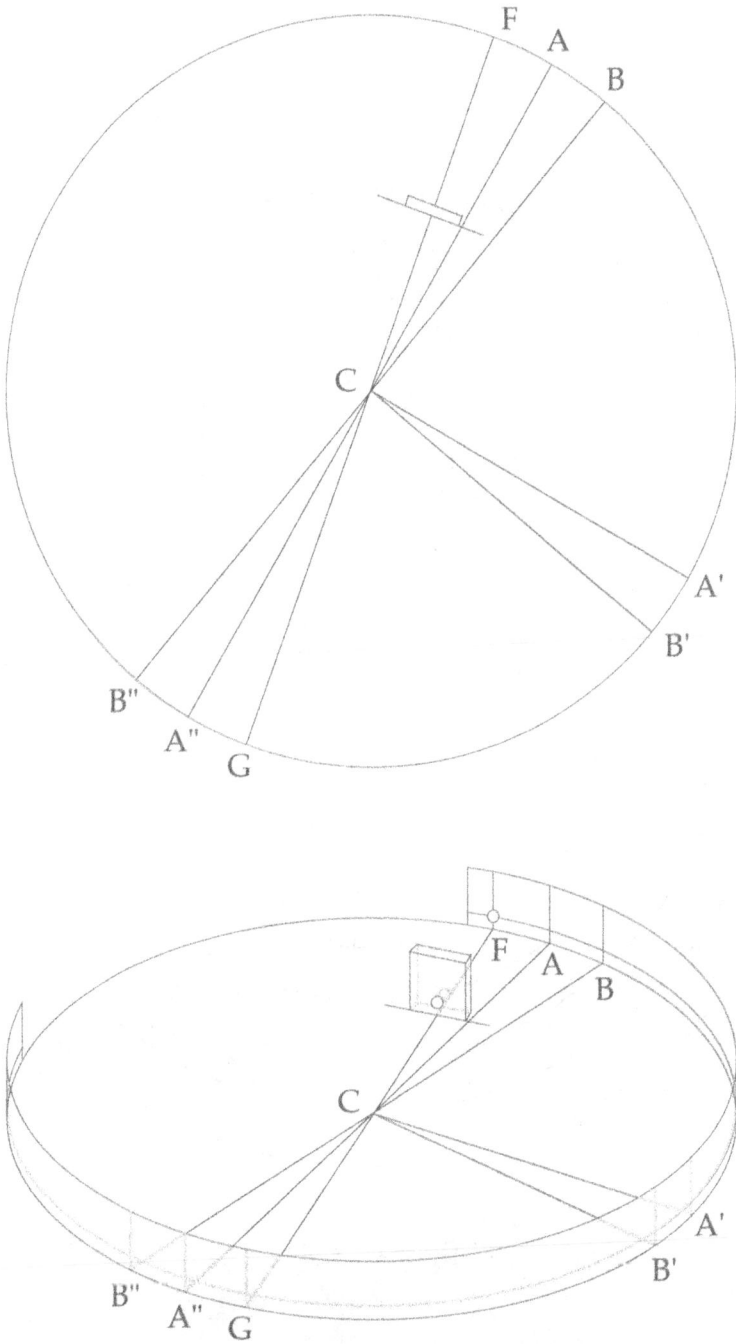

figure 7.3.30

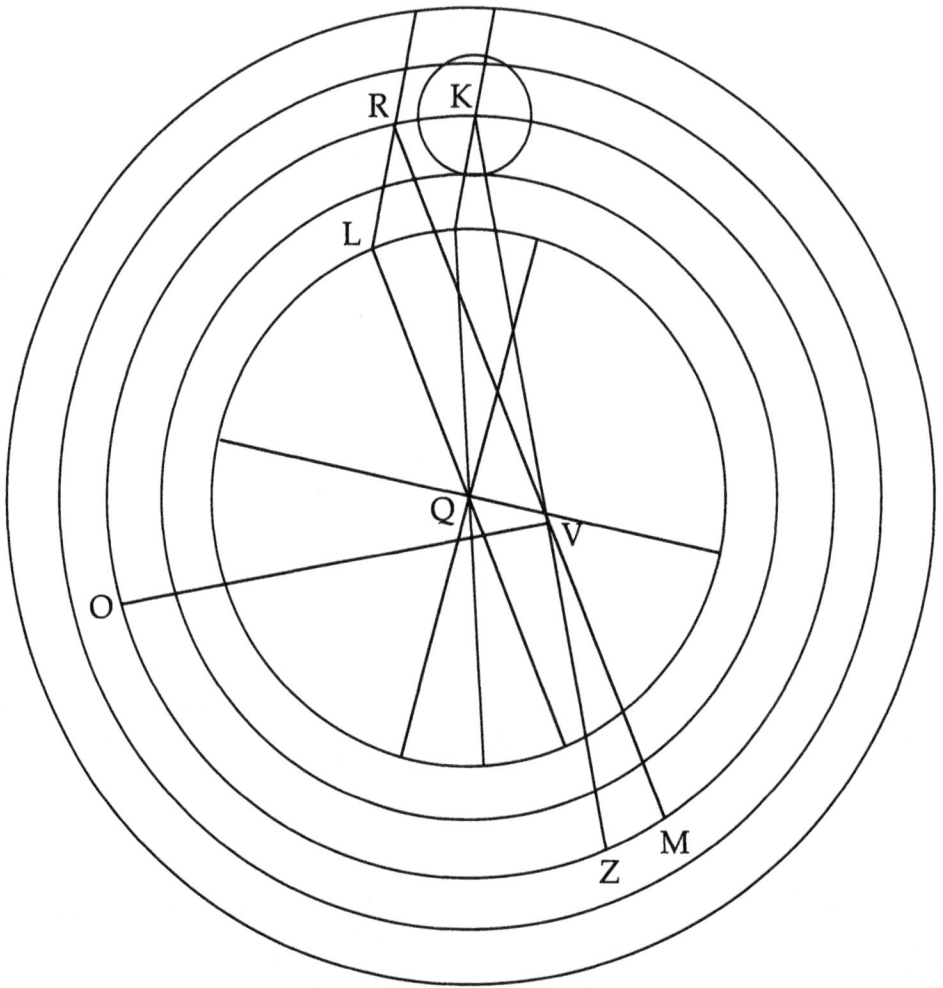

figure 7.3.30a

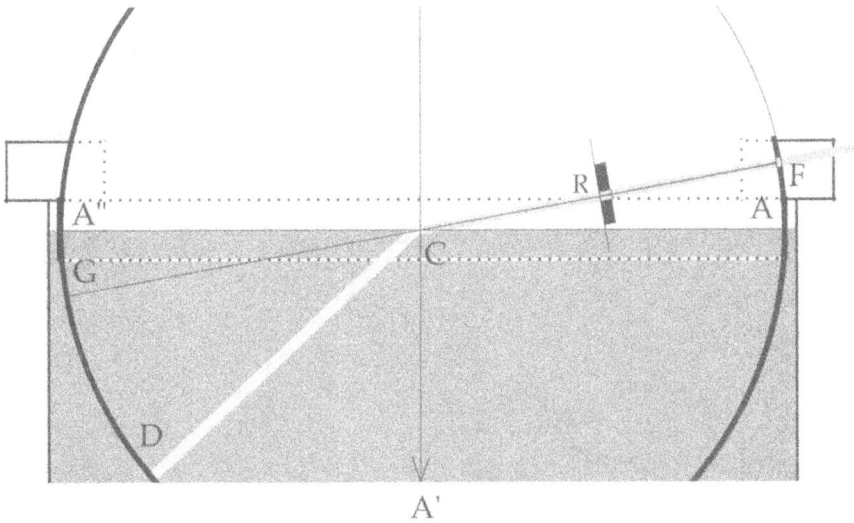

figure 7.3.31

AIR-WATER

Incid.	Inclin.	R (Mod.)	R (Alh.)
80	10	47.8	32.2
70	20	45.0	25.0
60	30	40.6	19.4
50	40	35.2	14.8
40	50	28.9	11.1
30	60	22.1	7.9
20	70	14.9	5.1
10	80	7.5	2.5

figure 7.3.32

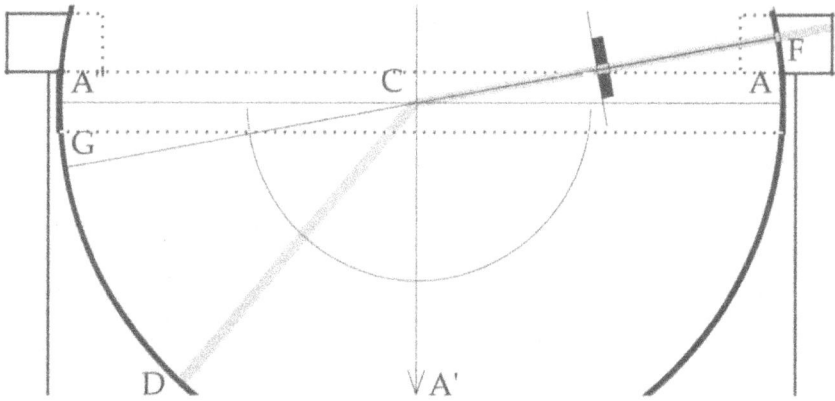

figure 7.3.33

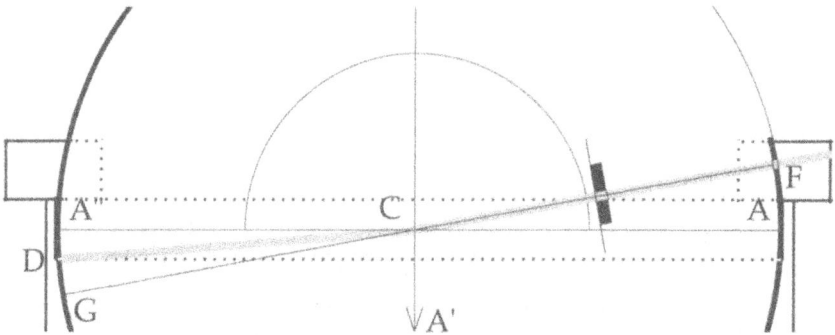

figure 7.3.34

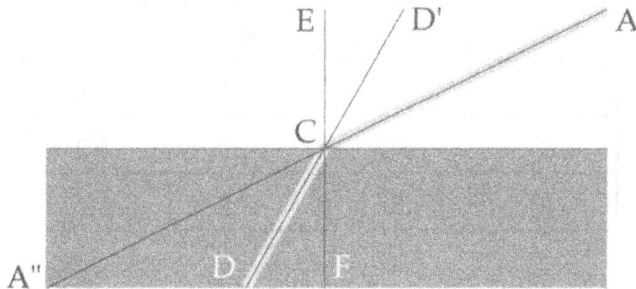

figure 7.3.35

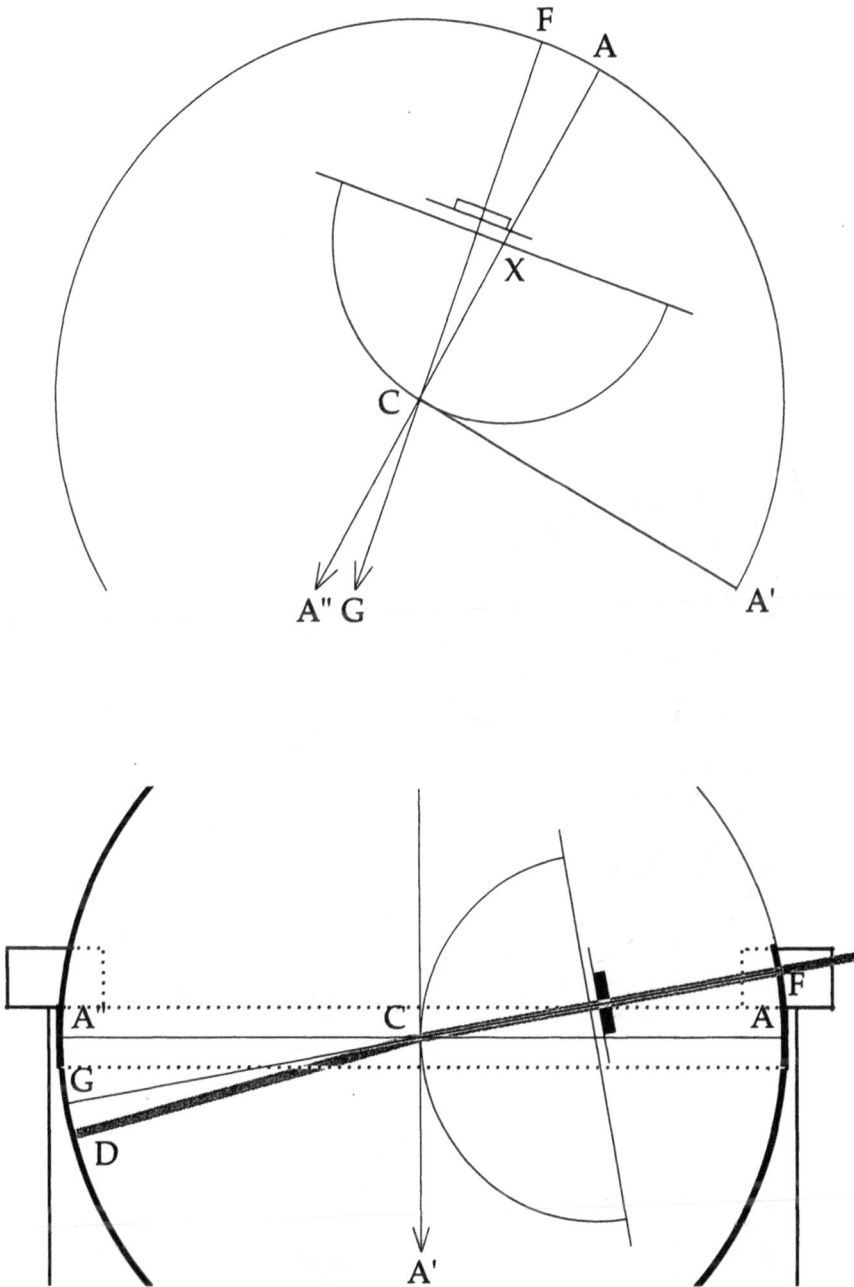

figure 7.3.36

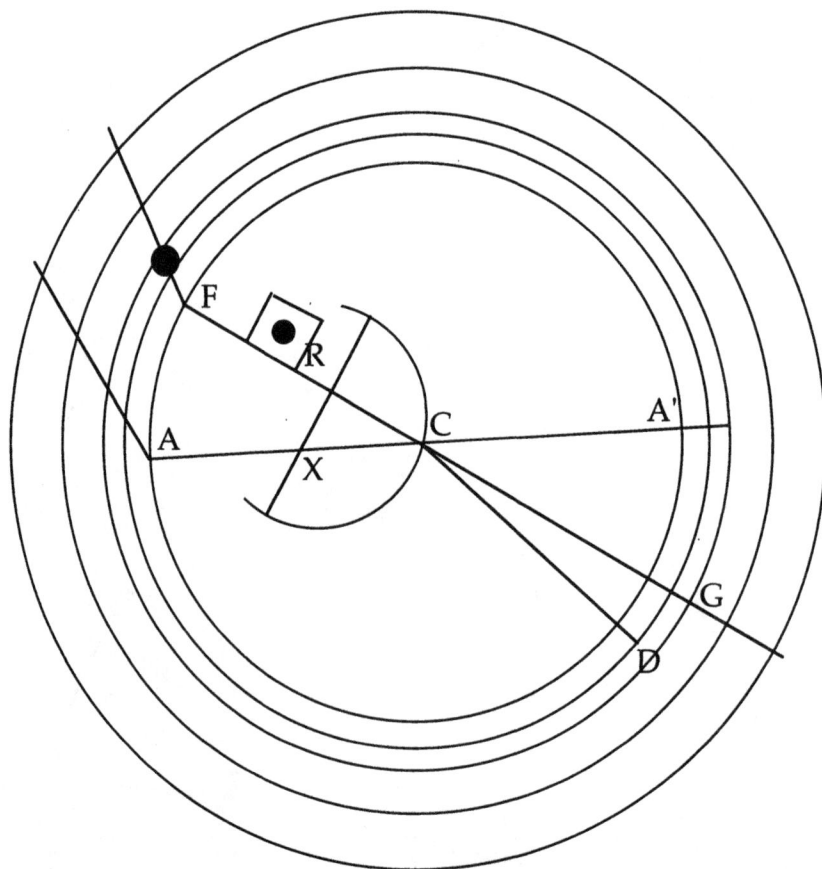

figure 7.3.36a

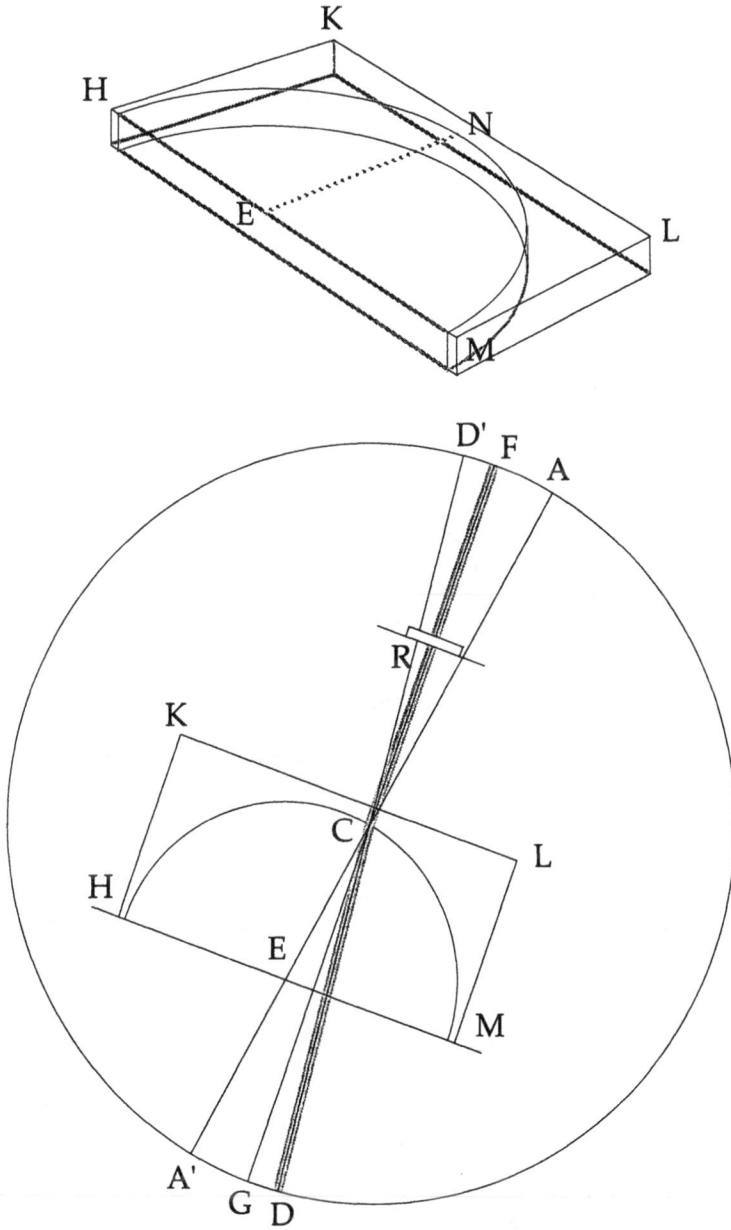

figure 7.3.37

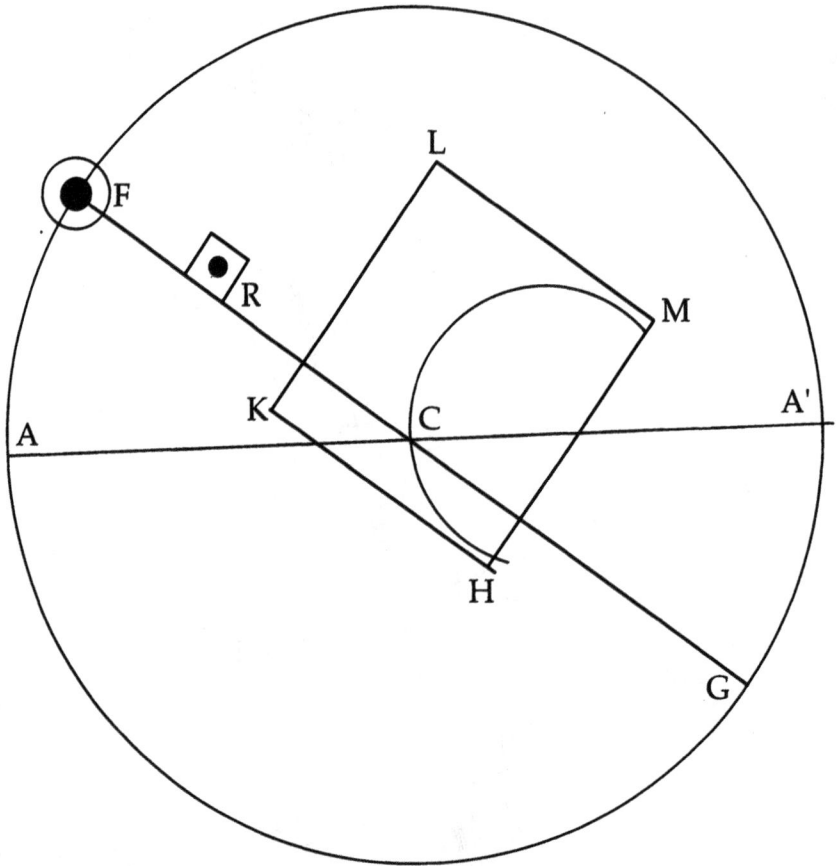

figure 7.3.37a

AIR-GLASS

Incid.	Inclin.	R (Mod.)	R (Alh.)
80	10	41.0	39.0
70	20	38.8	31.2
60	30	35.3	24.7
50	40	30.7	19.3
40	50	25.4	14.6
30	60	19.5	10.5
20	70	13.2	6.8
10	80	6.6	3.3

GLASS-AIR

Incid.	Inclin.	R (Mod.)	R (Alh.)
40	50	74.6	34.6
30	60	48.6	18.6
20	70	30.9	10.9
10	80	15.1	5.1

GLASS-WATER

Incid.	Inclin.	R (Mod.)	R (Alh.)
60	30	77.6	17.6
50	40	59.8	9.8
40	50	46.5	6.5
30	60	34.3	4.3
20	70	22.7	2.7
10	80	11.3	1.3

figure 7.3.38

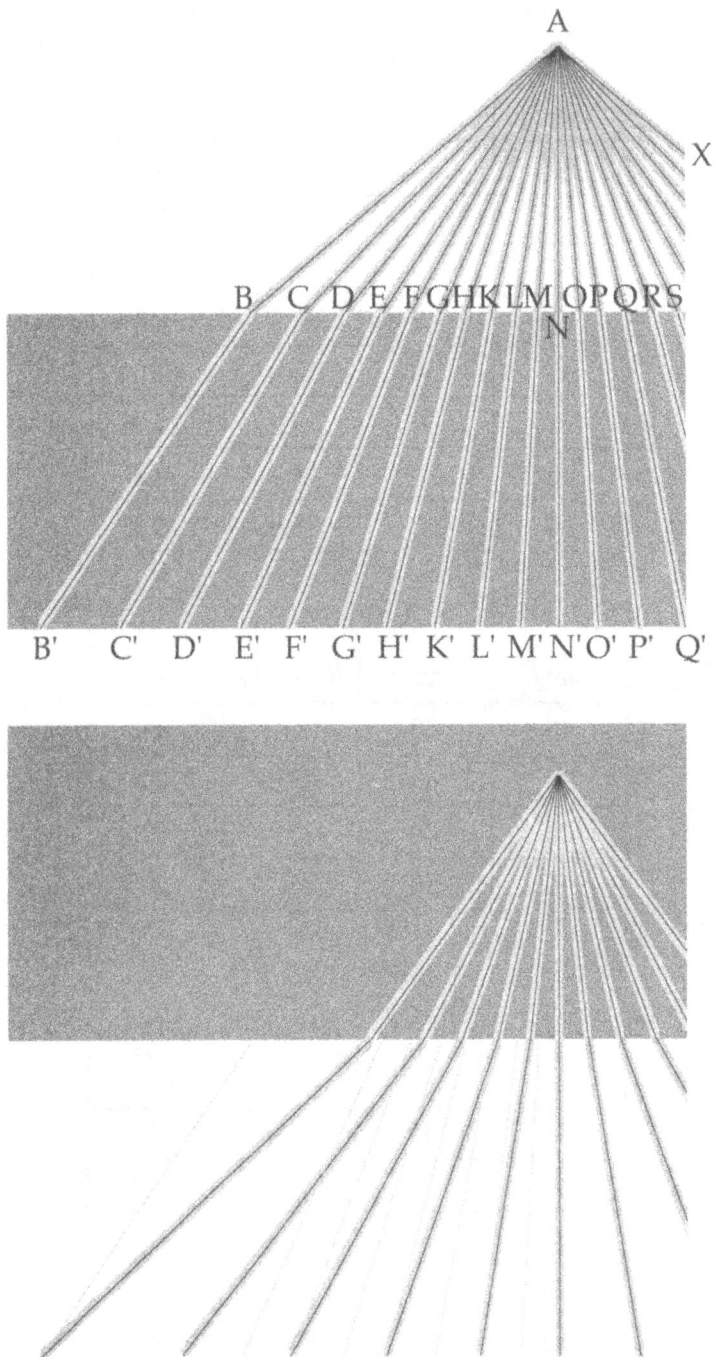

figure 7.4.39

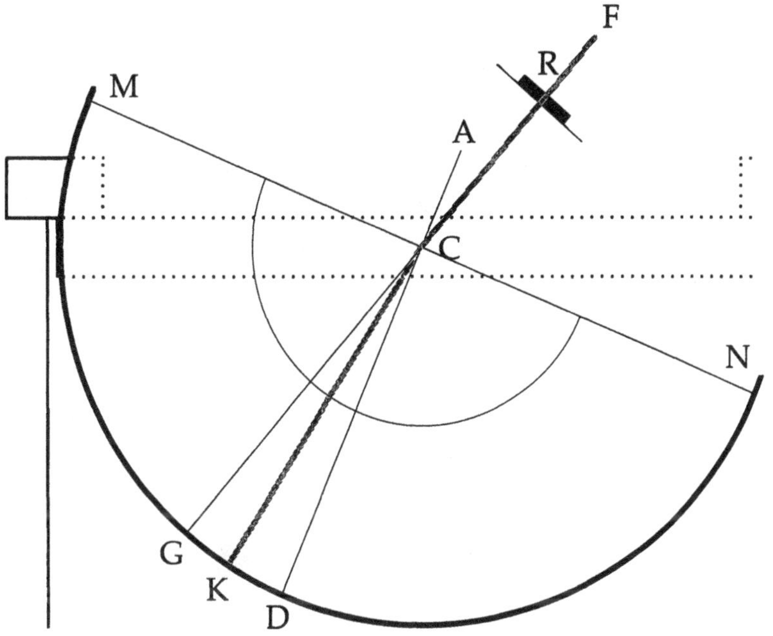

figure 7.4.40

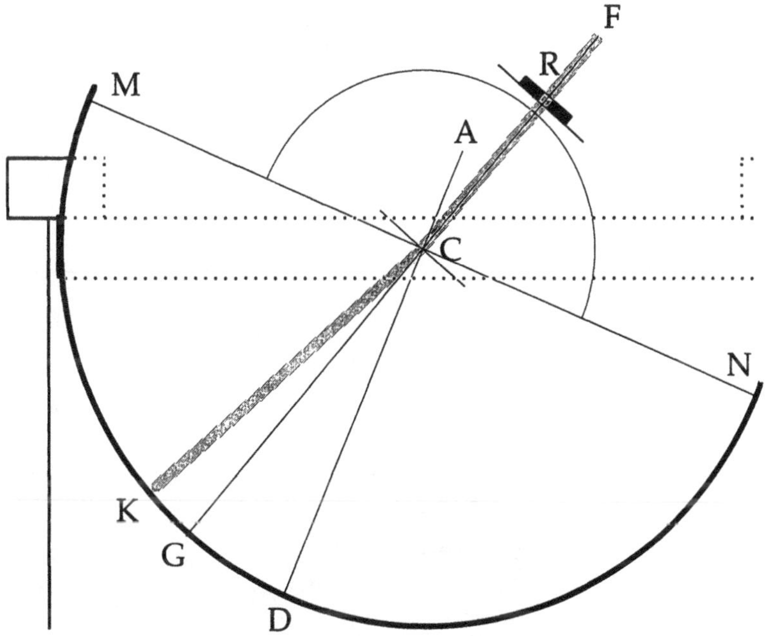

figure 7.4.41

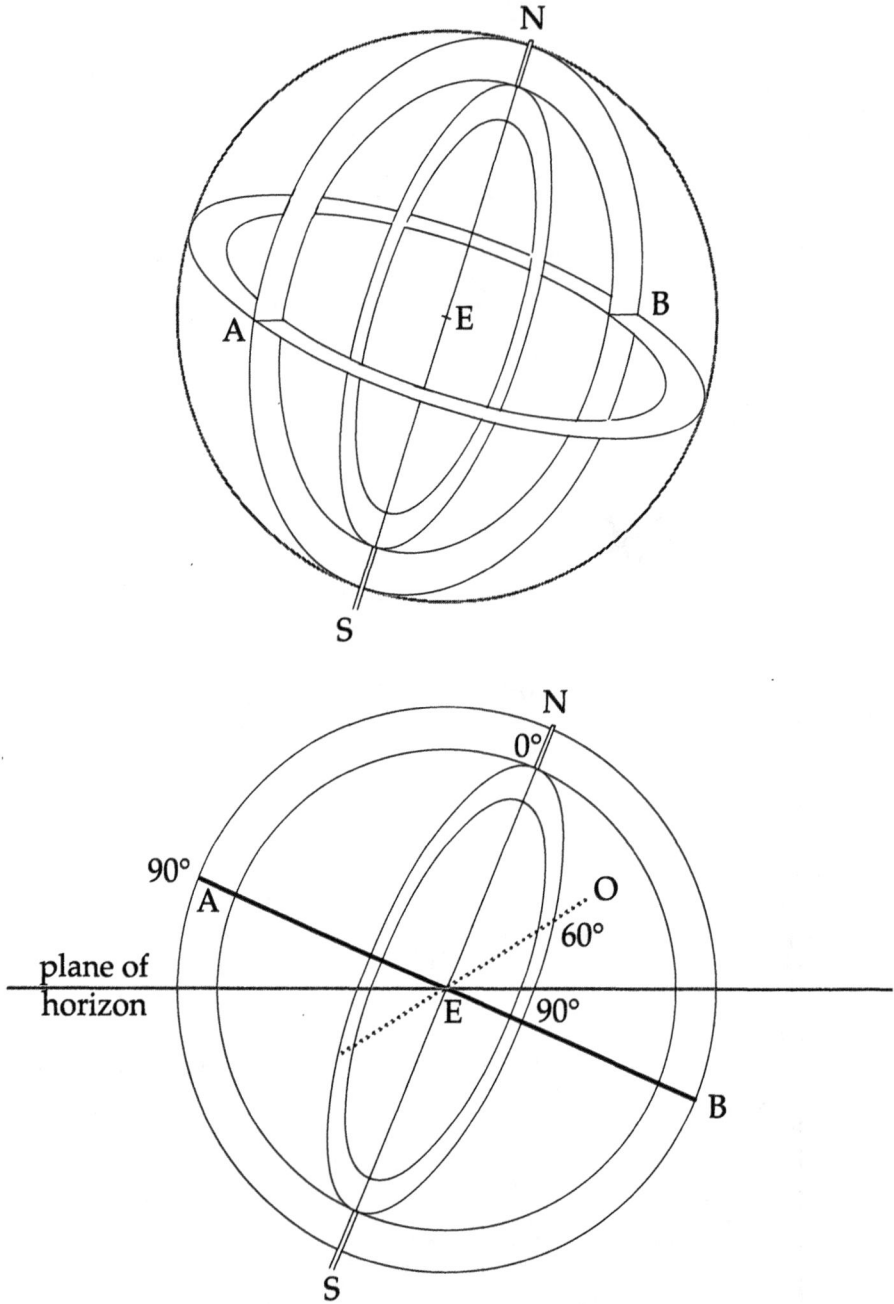

figure 7.4.42

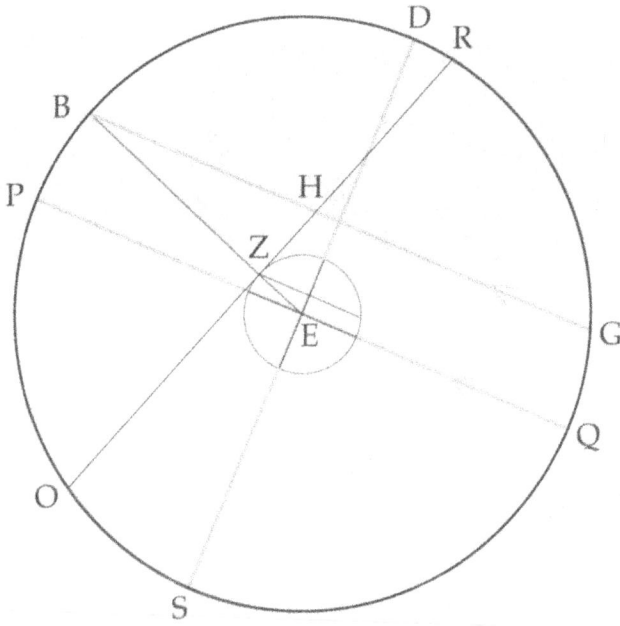

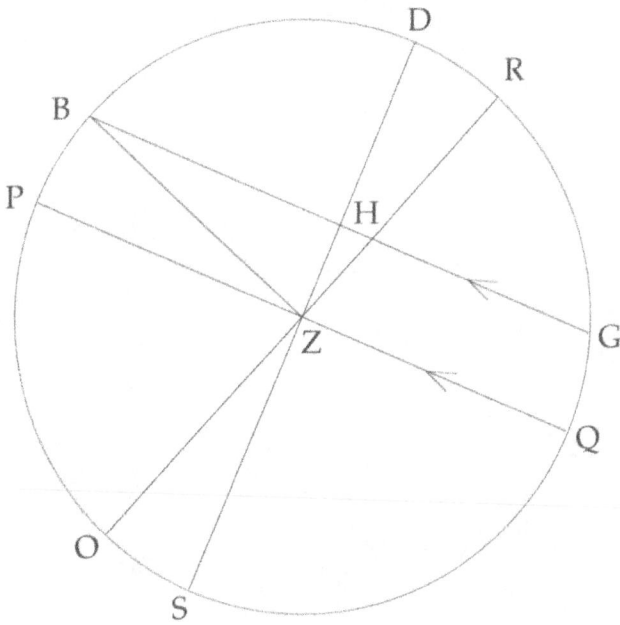

figure 7.4.43

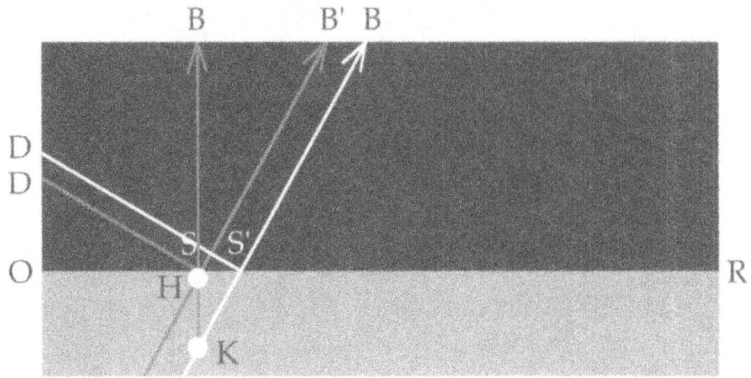

figure 7.4.44

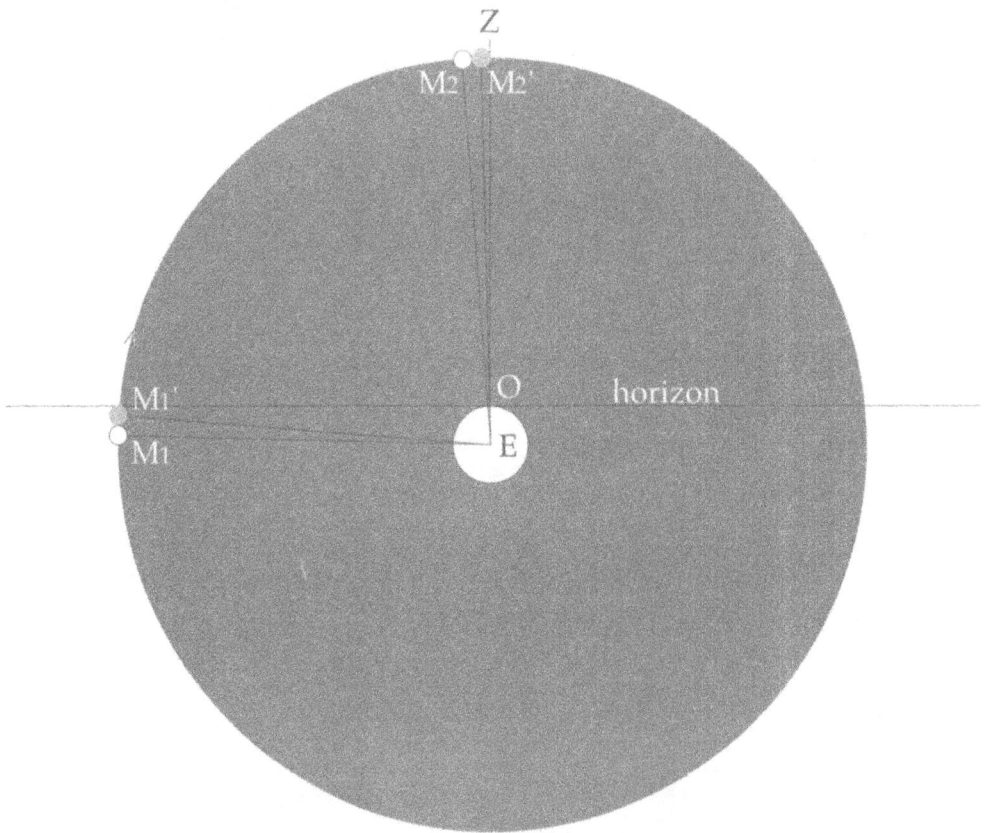

figure 7.4.44a

figure 7.4.44b

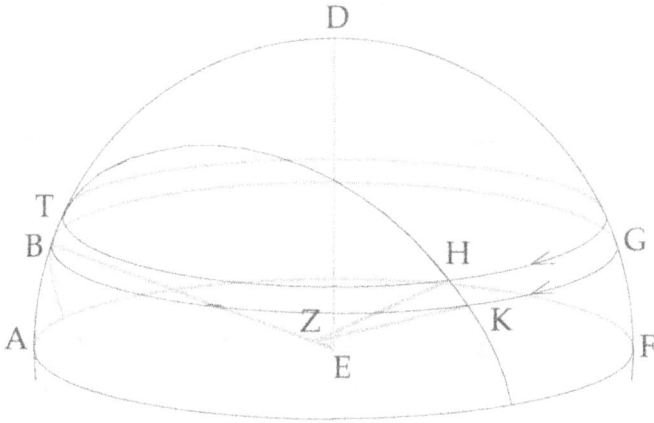

figure 7.4.45

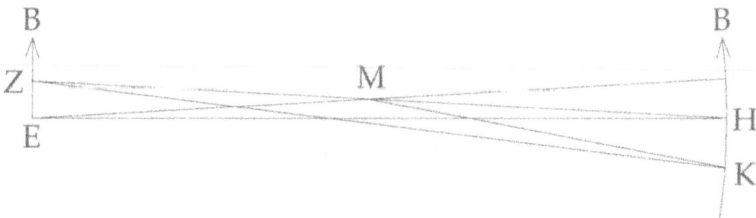

figure 7.4.45a

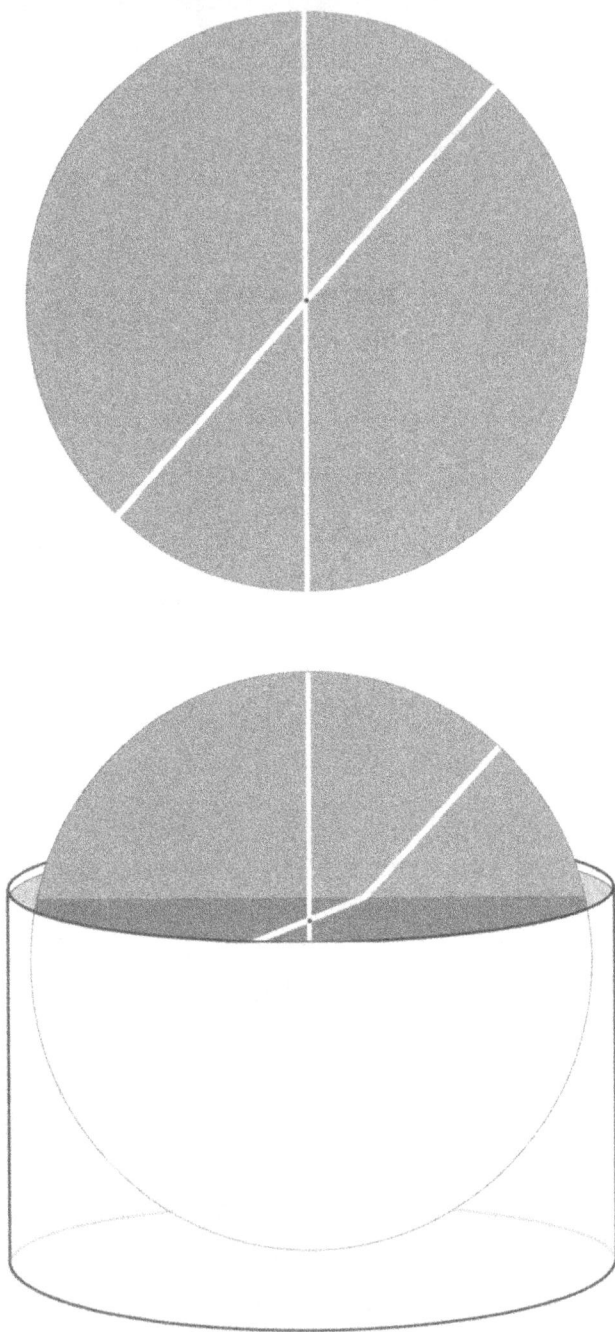

figure 7.5.46

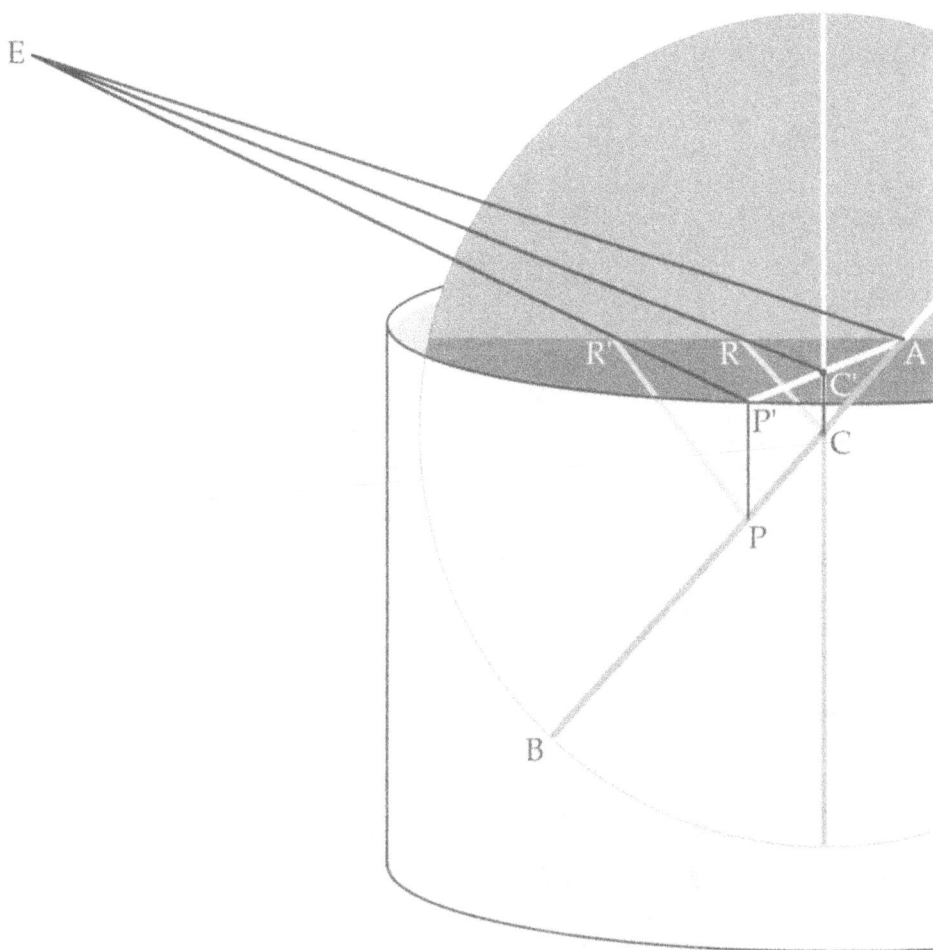

figure 7.5.46a

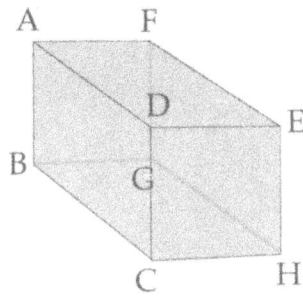

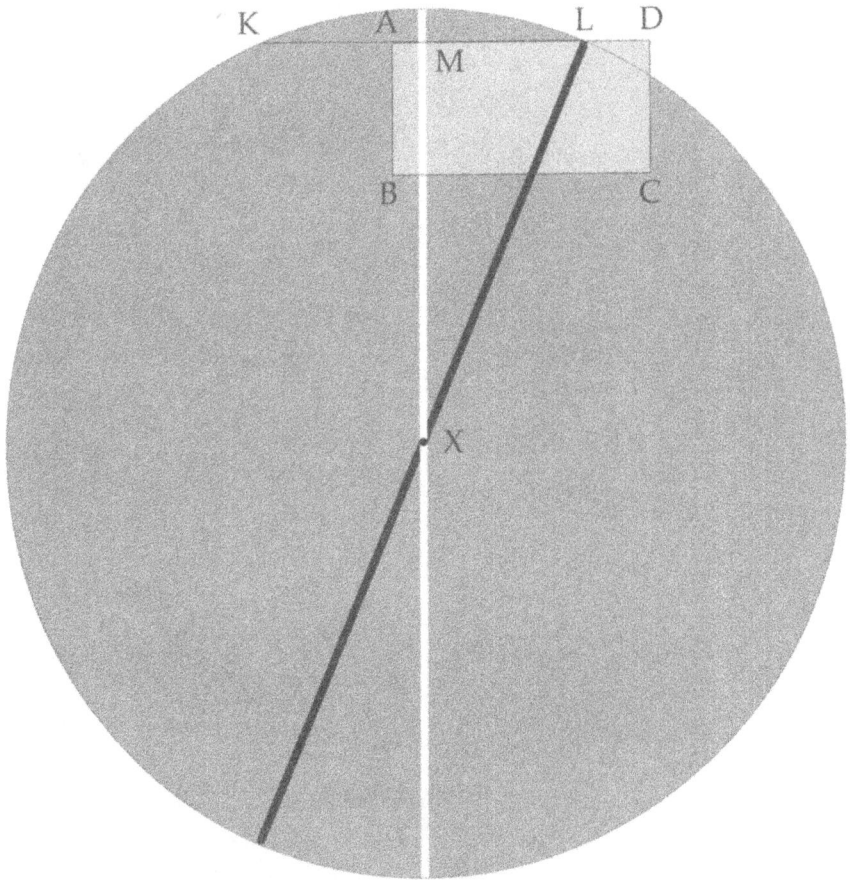

figure 7.5.47

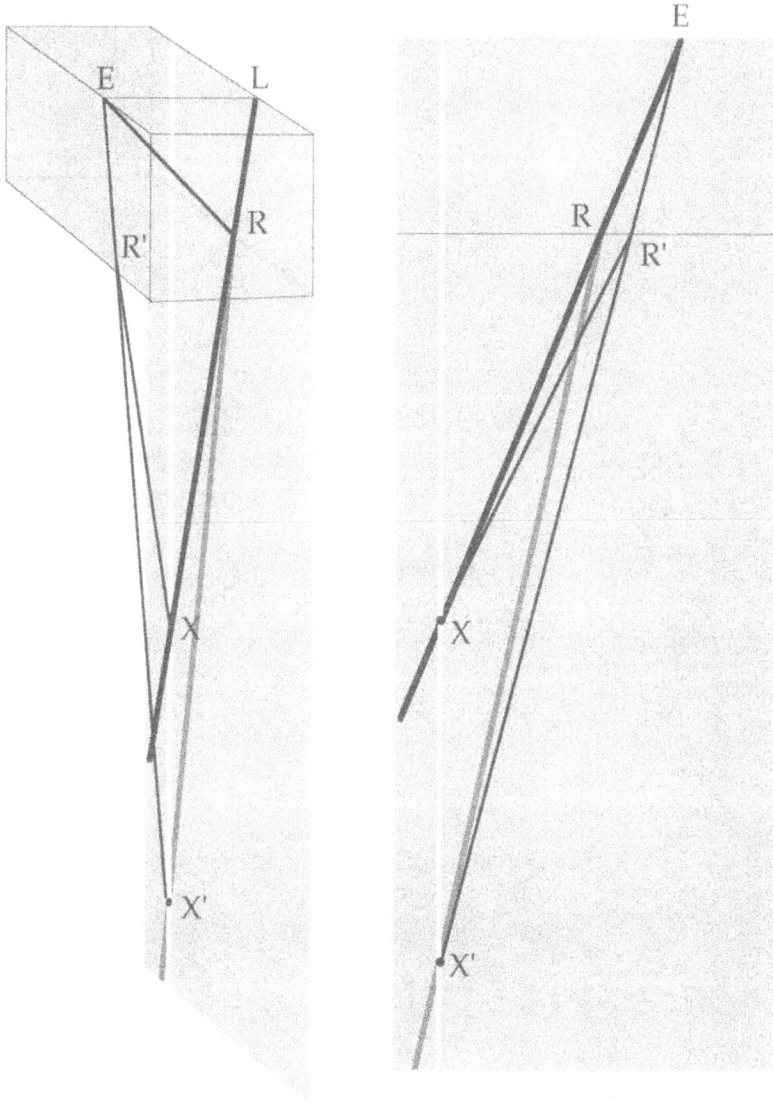

figure 7.5.47a

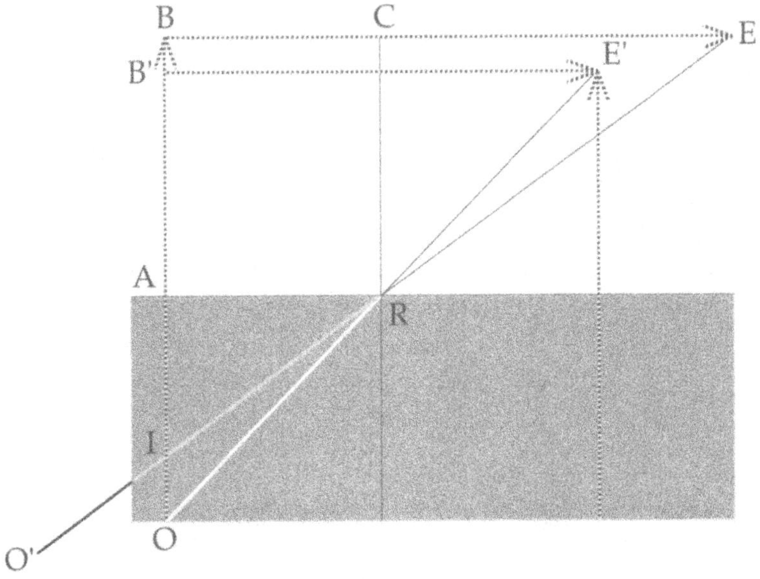

figure 7.5.48

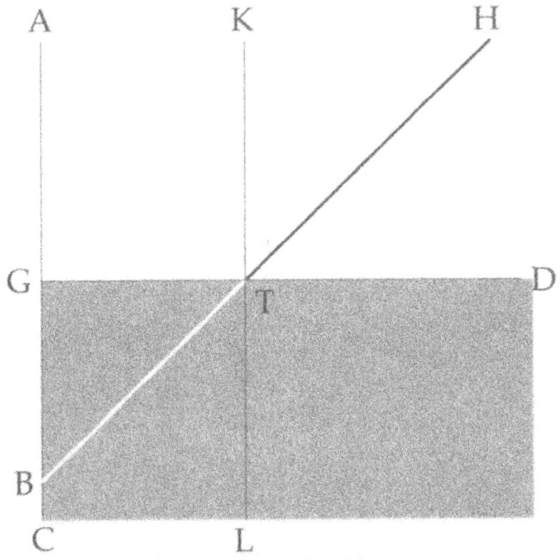

figure 7.5.49

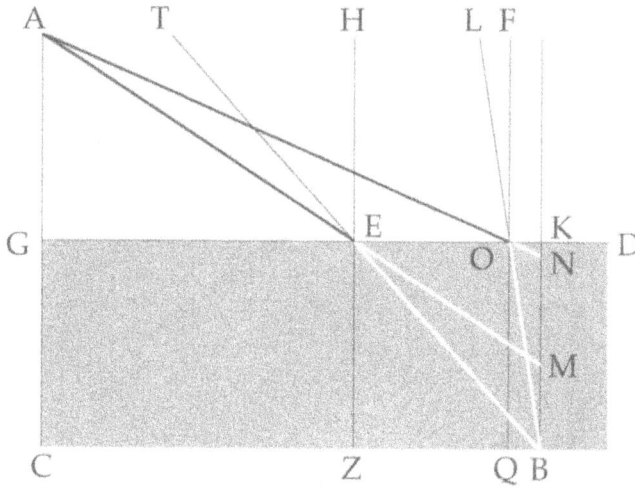

figure 7.5.50

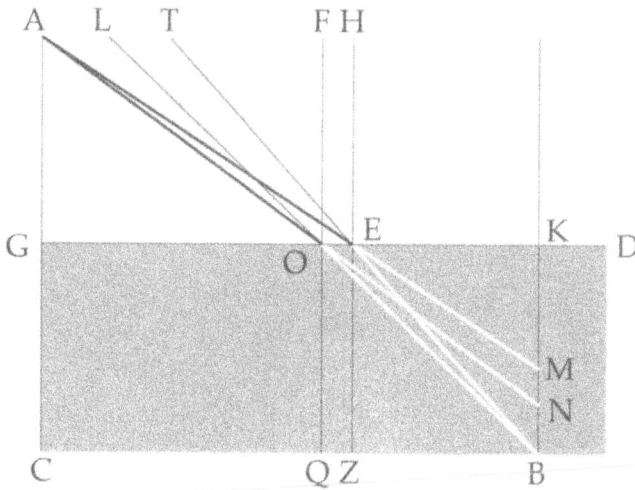

figure 7.5.50a

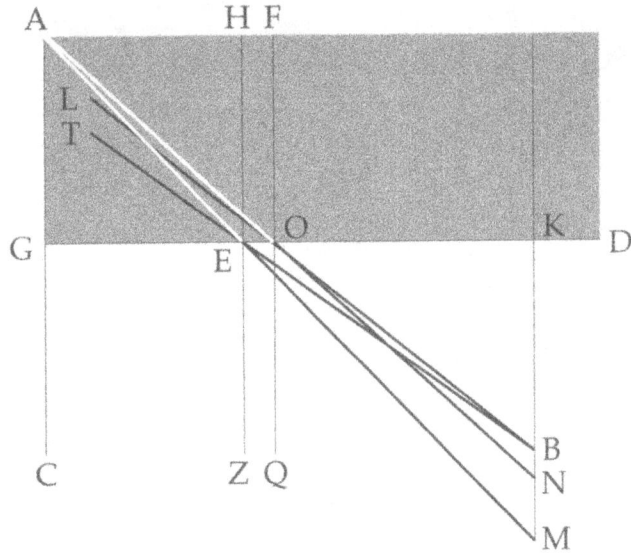

figure 7.5.51

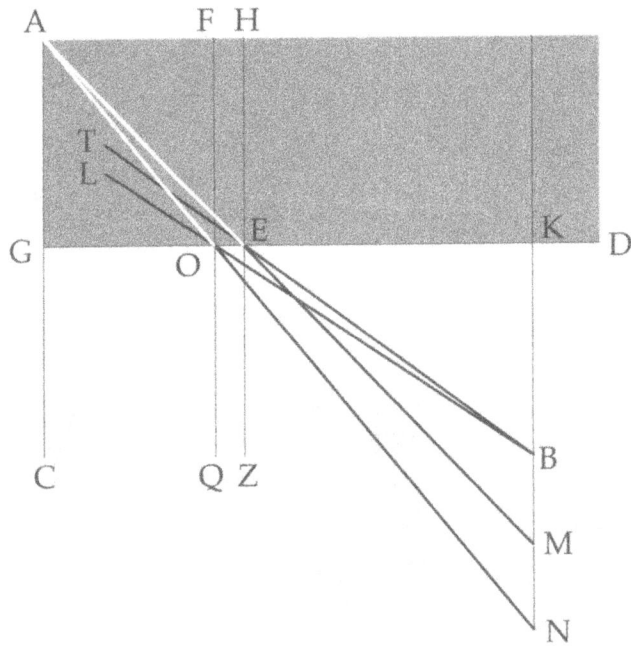

figure 7.5.51a

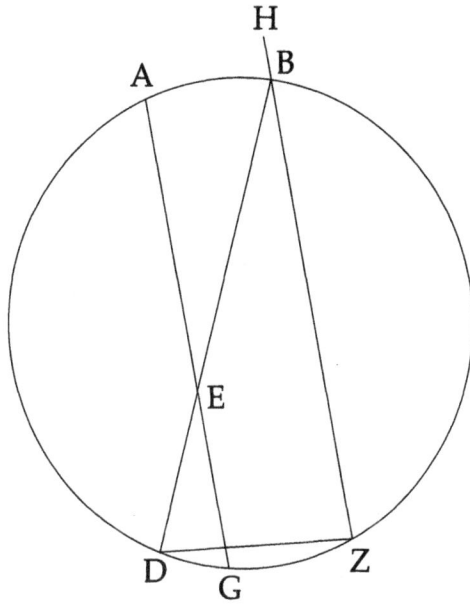

figure 7.5.52

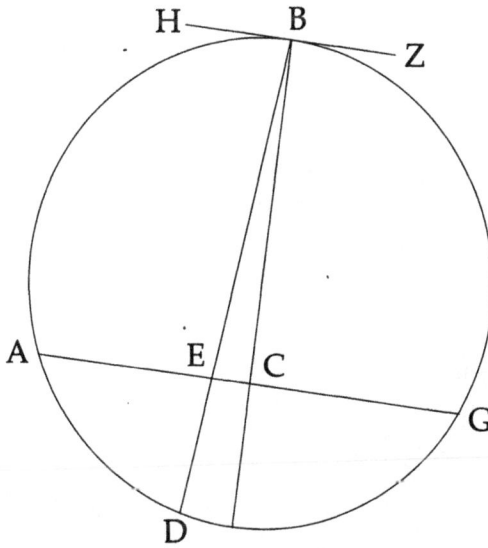

figure 7.5.52a

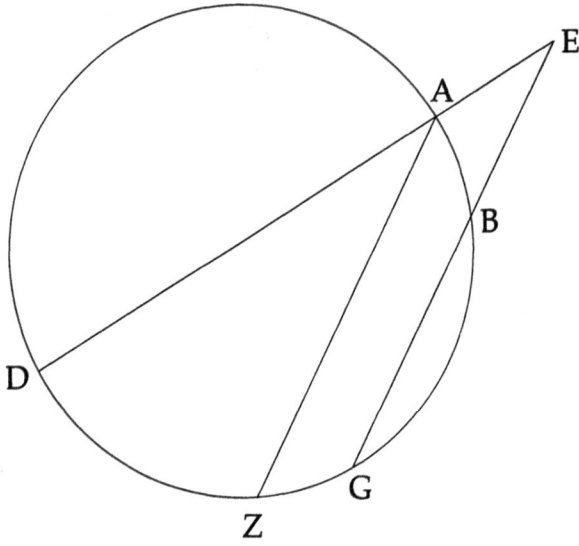

figure 7.5.52b

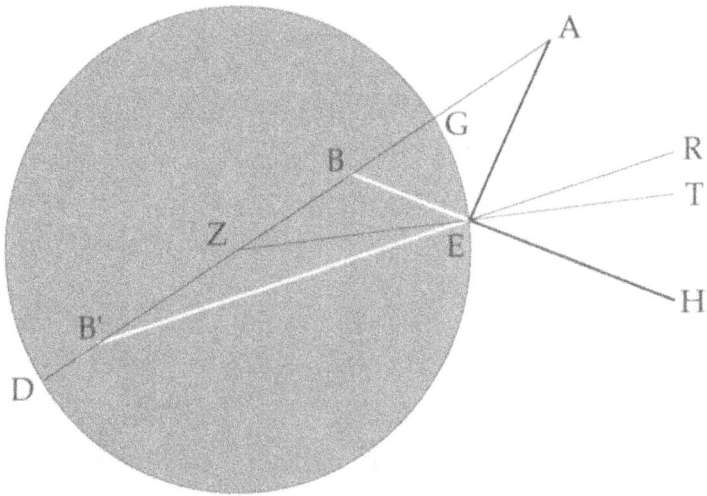

figure 7.5.53

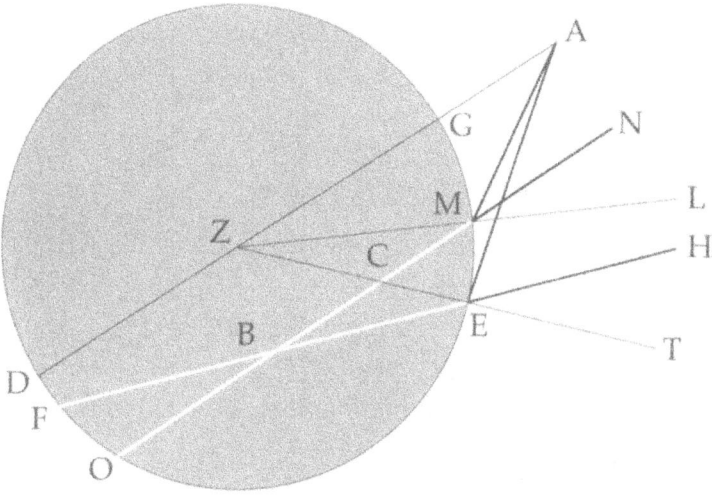

figure 7.5.53a

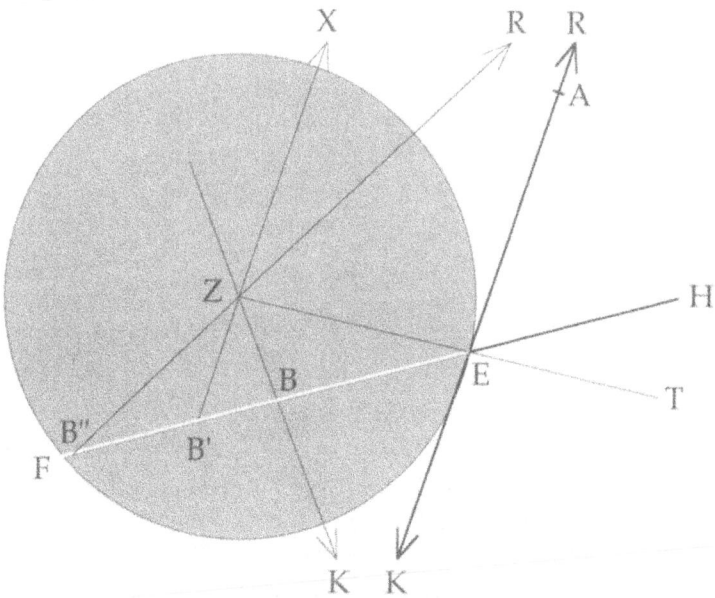

figure 7.5.53b

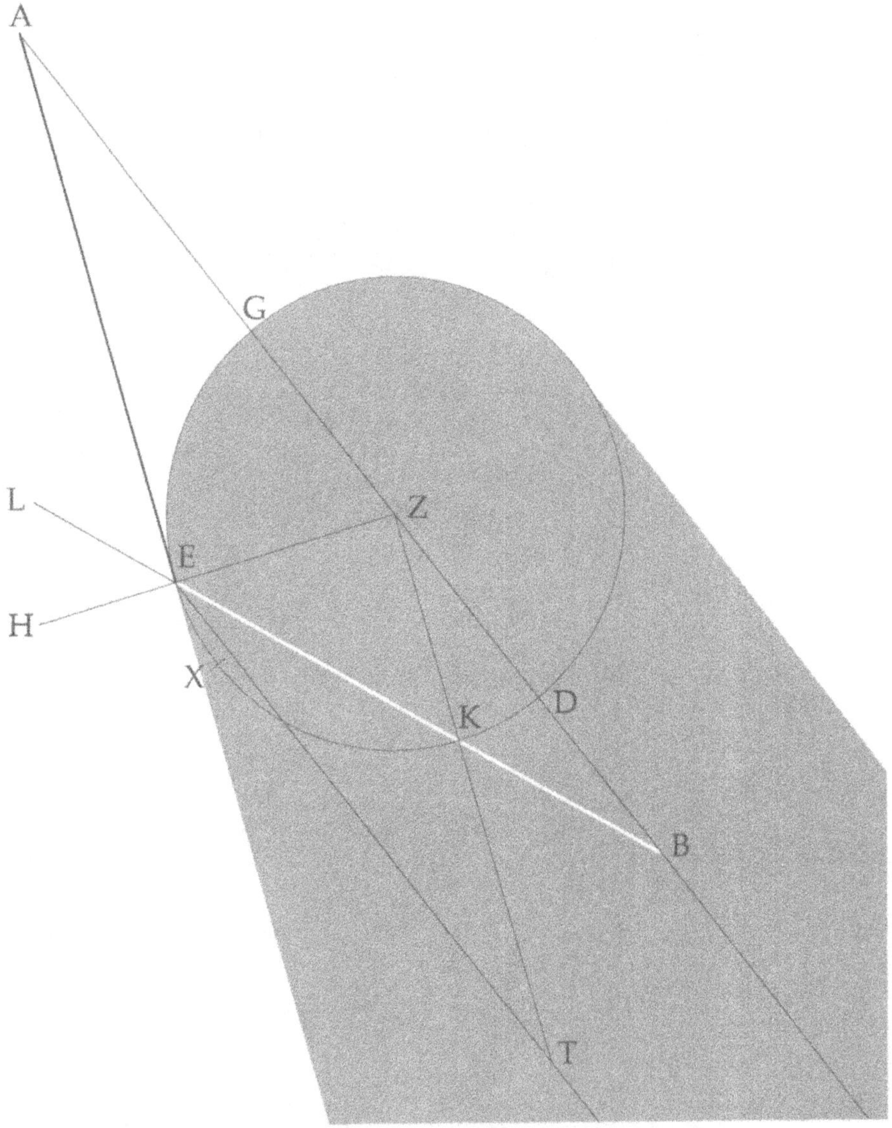

figure 7.5.54

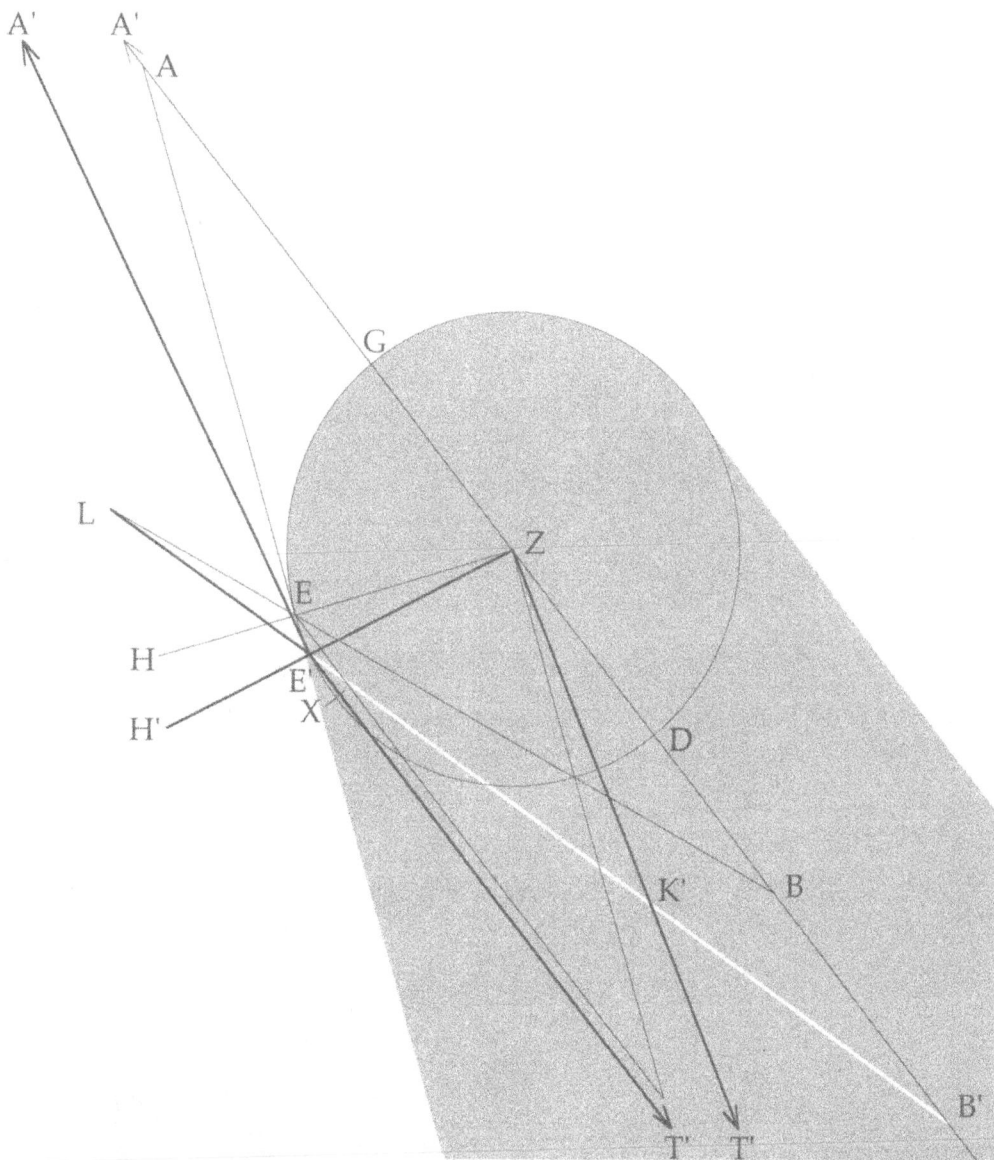

figure 7.5.54a

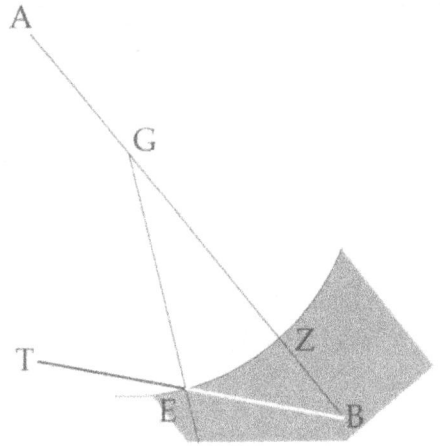

figure 7.5.55

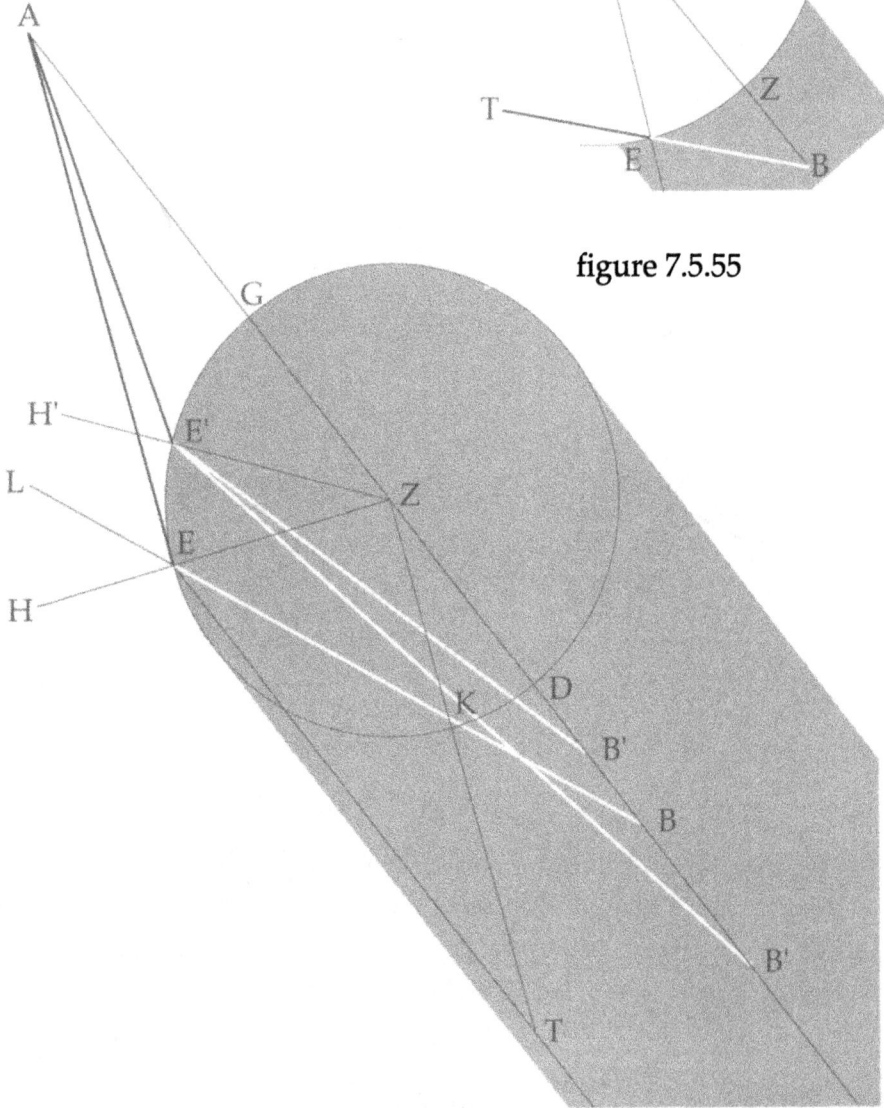

figure 7.5.54b

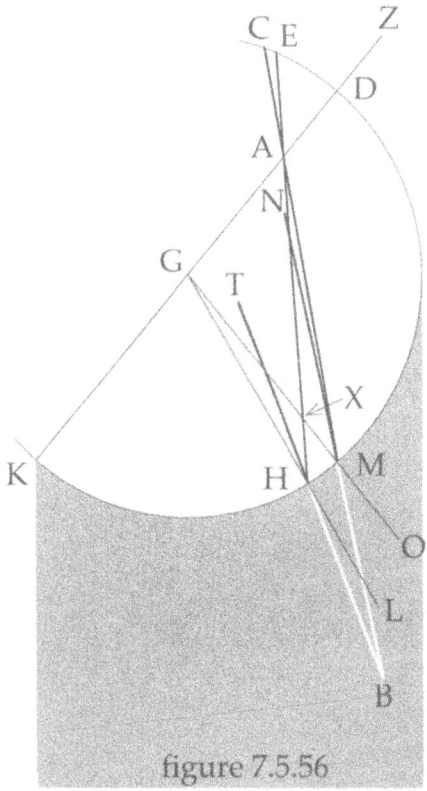

figure 7.5.56

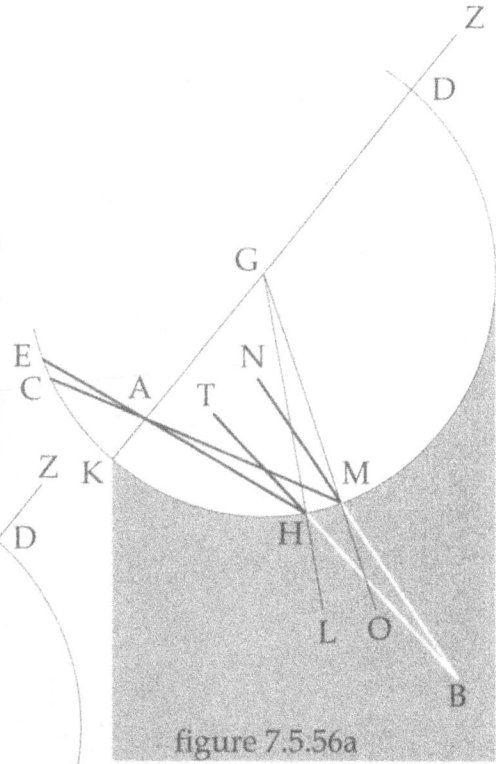

figure 7.5.56a

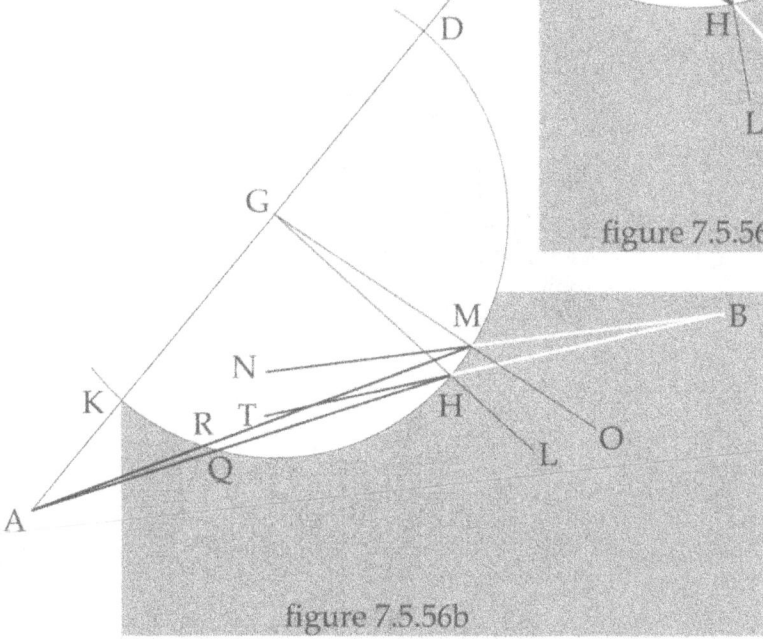

figure 7.5.56b

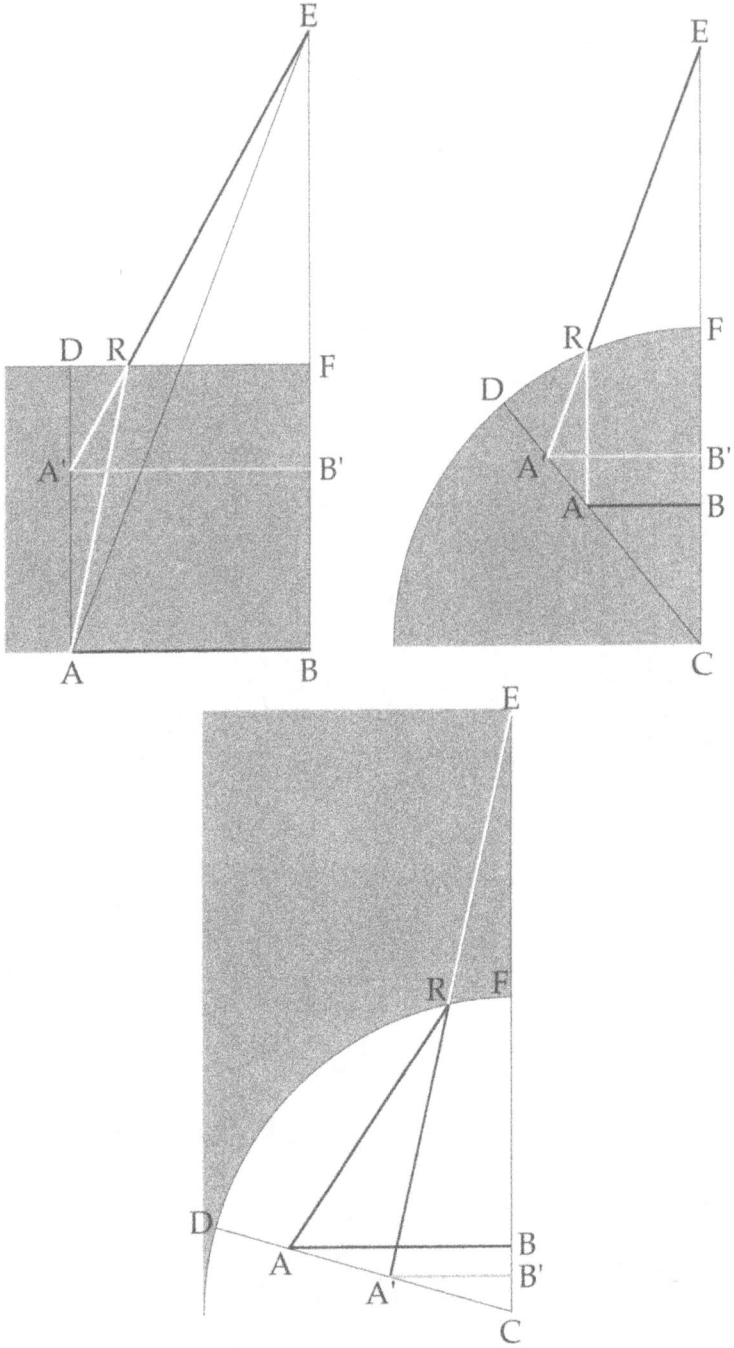

figure 7.6.57

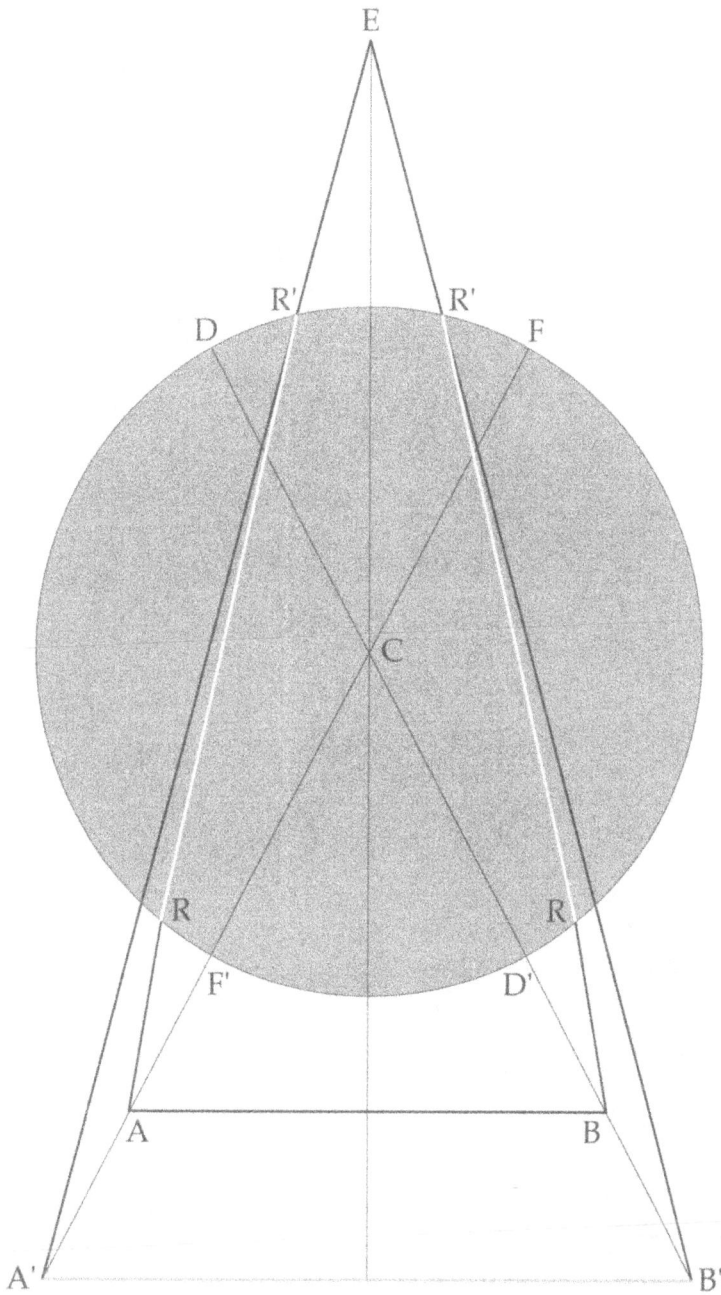

figure 7.6.57a

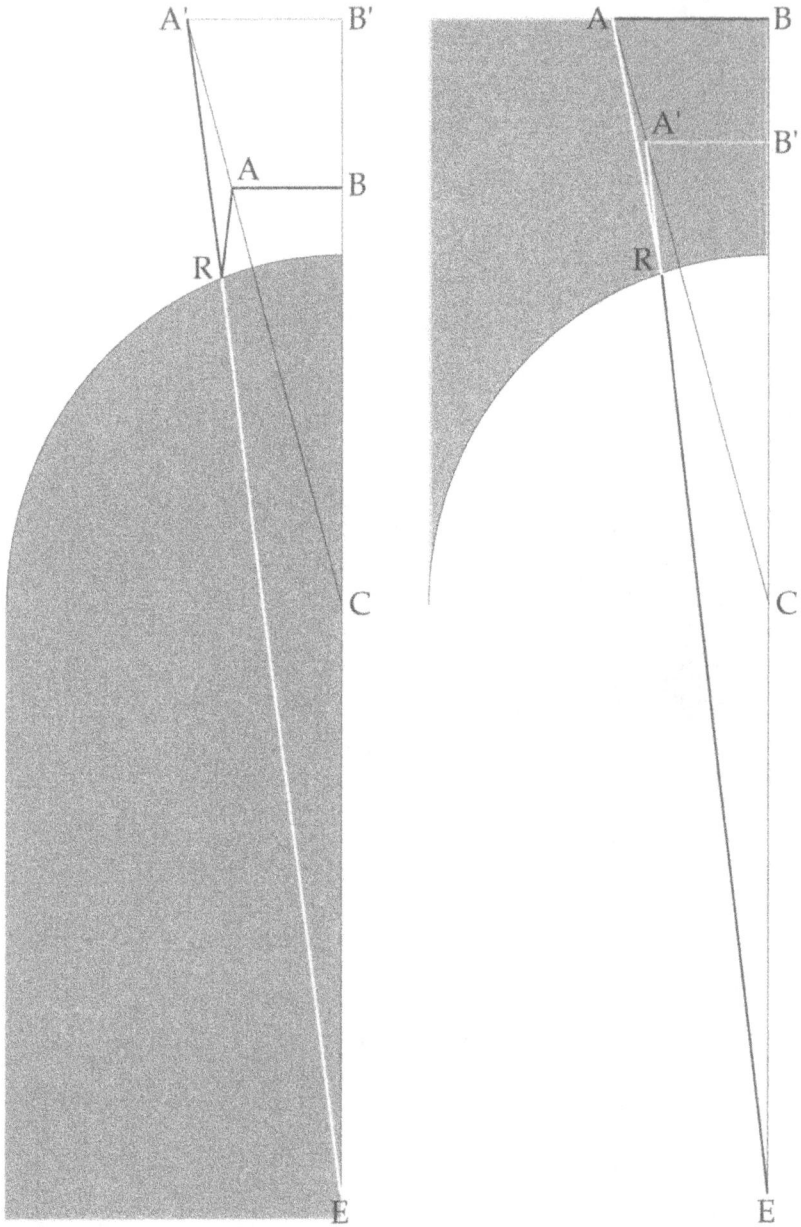

figure 7.6.57b

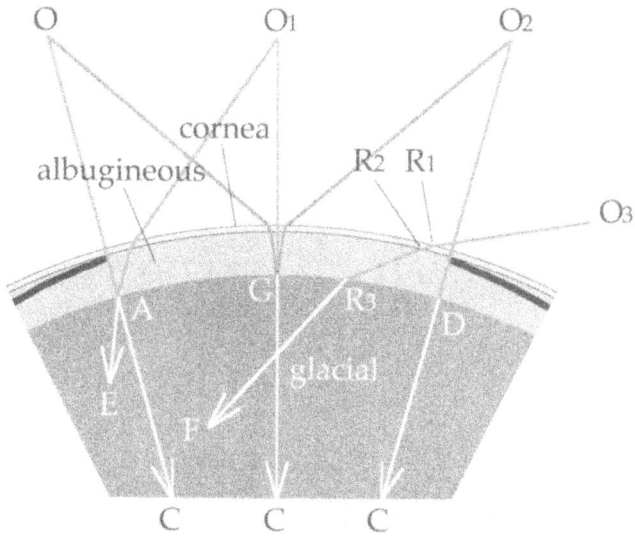

figure 7.6.58

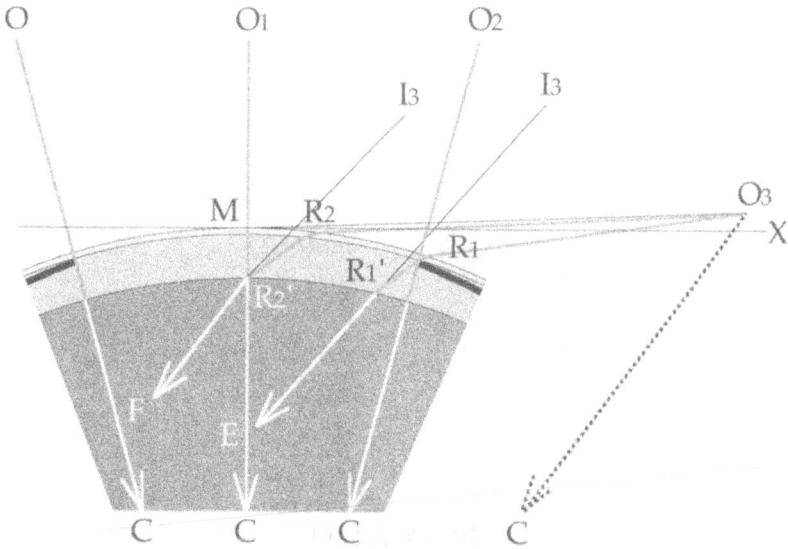

figure 7.6.59

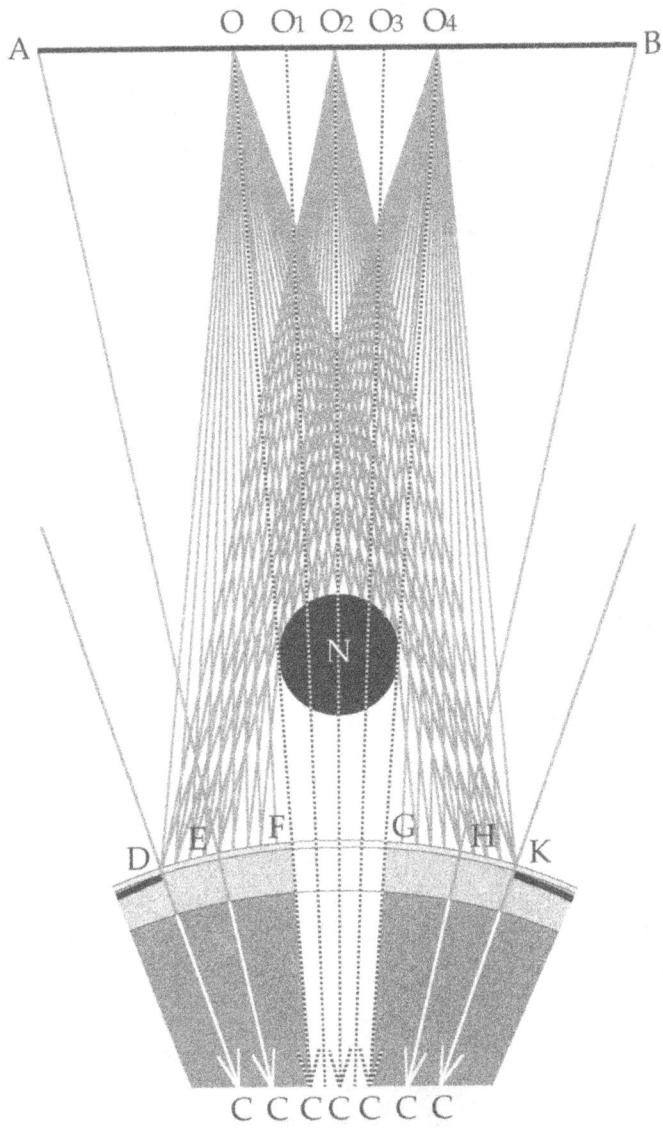

figure 7.6.60

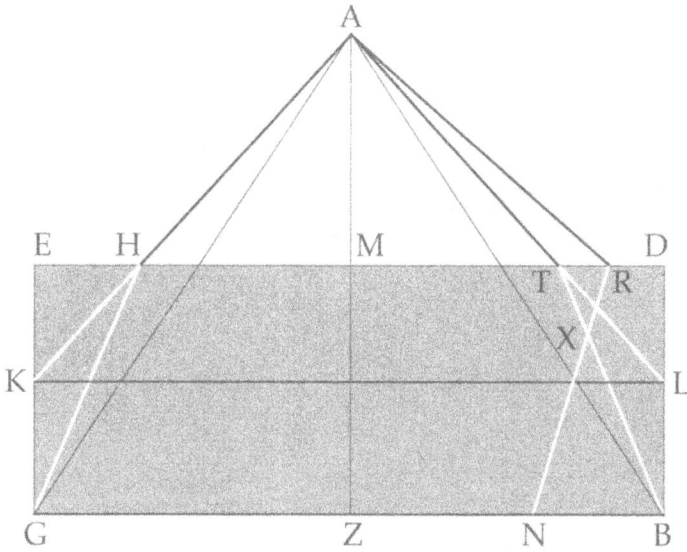

figure 7.7.61

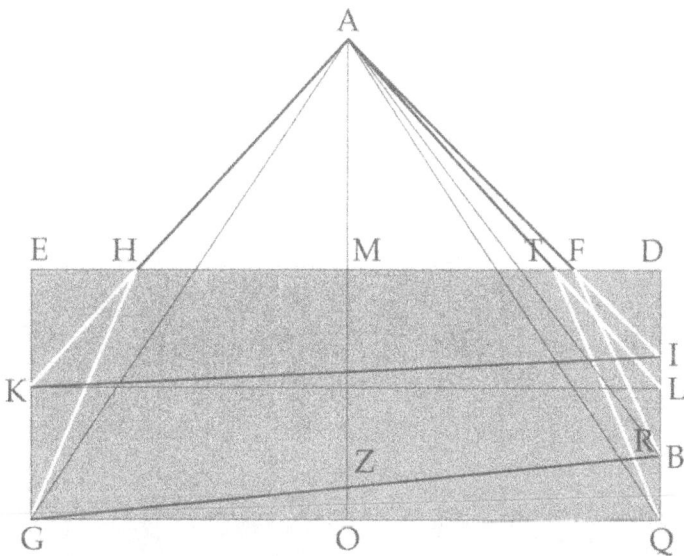

figure 7.7.61a

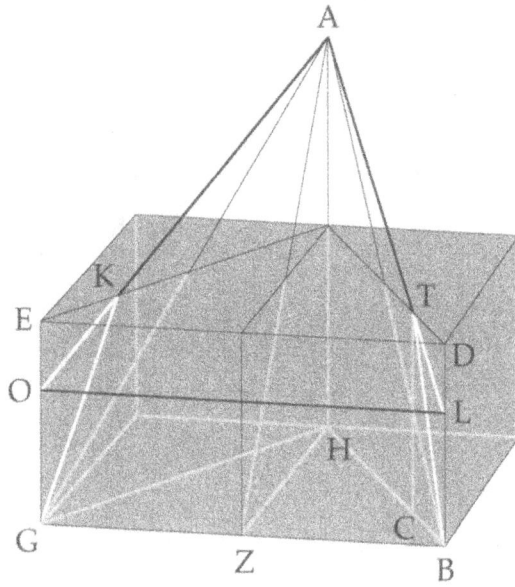

figure 7.7.62

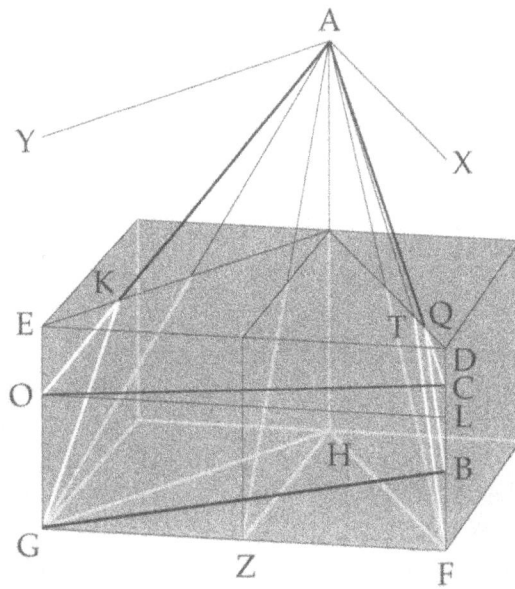

figure 7.7.62a

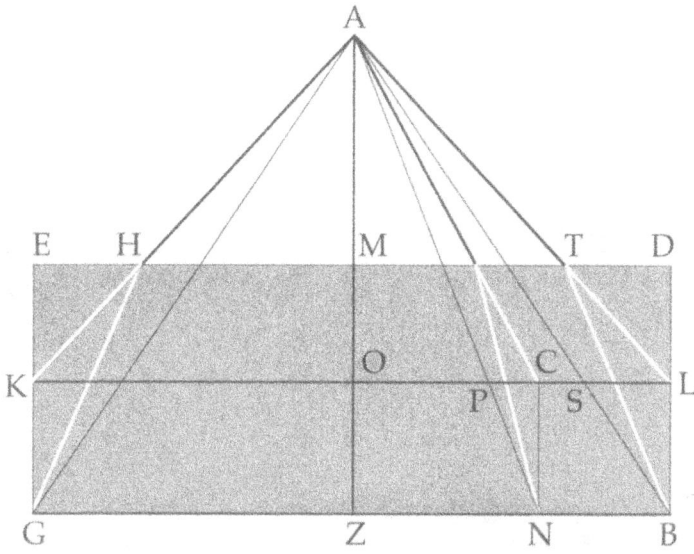

figure 7.7.63

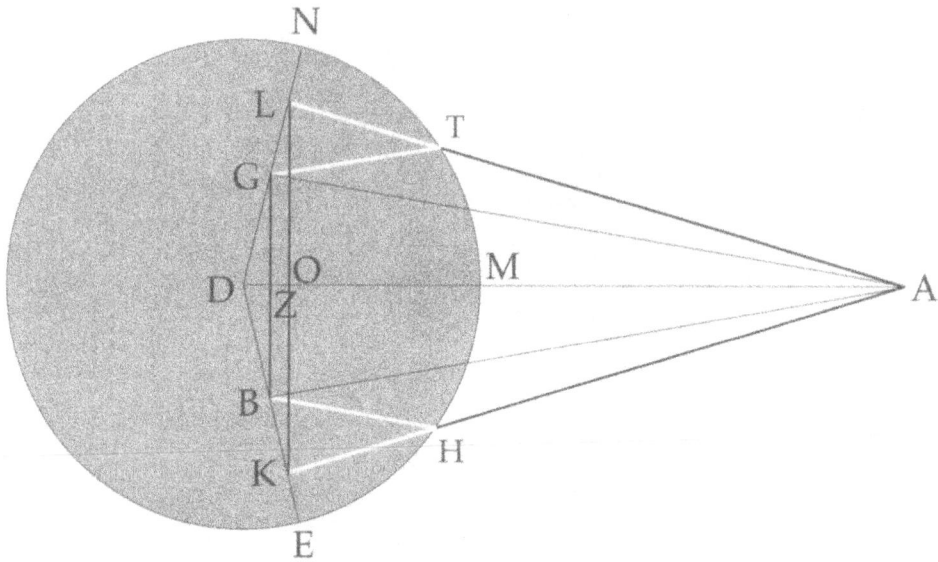

figure 7.7.64

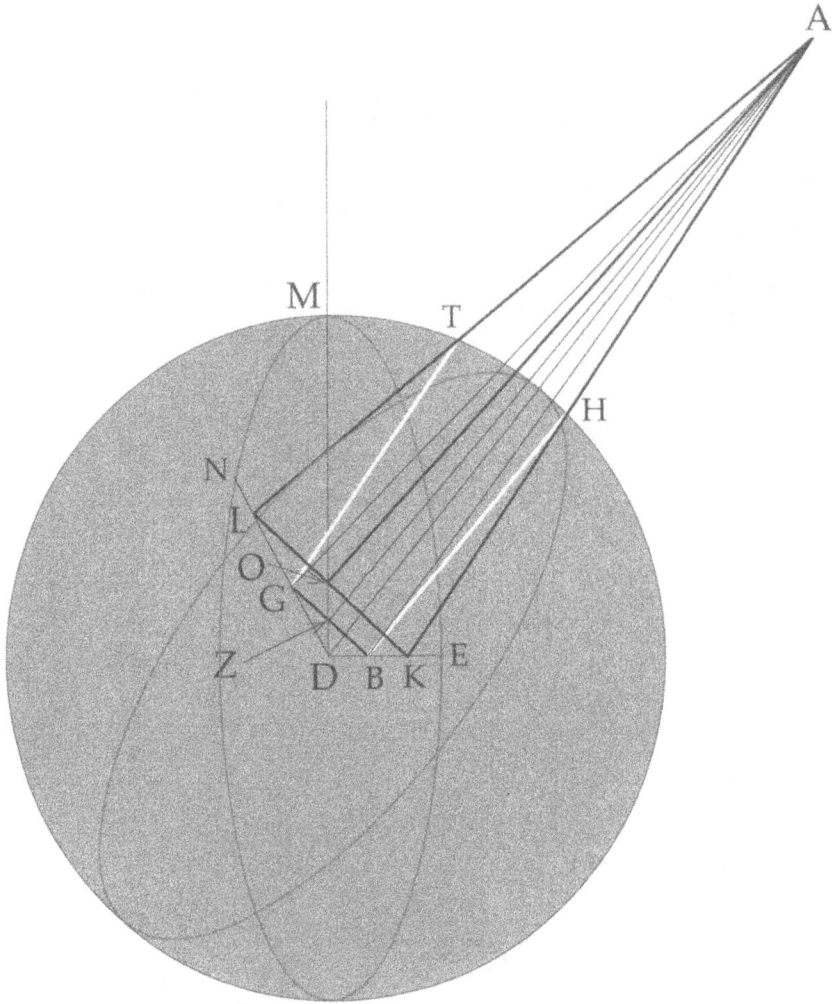

figure 7.7.65

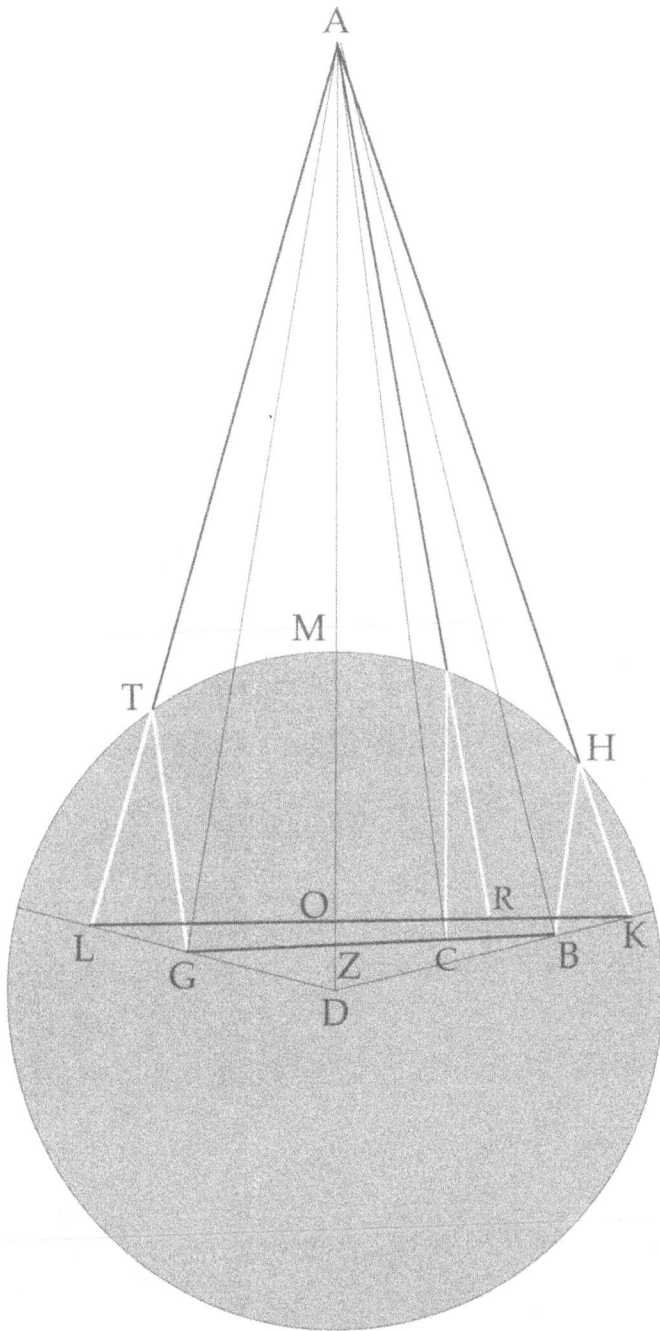

Figure 7.7.65a

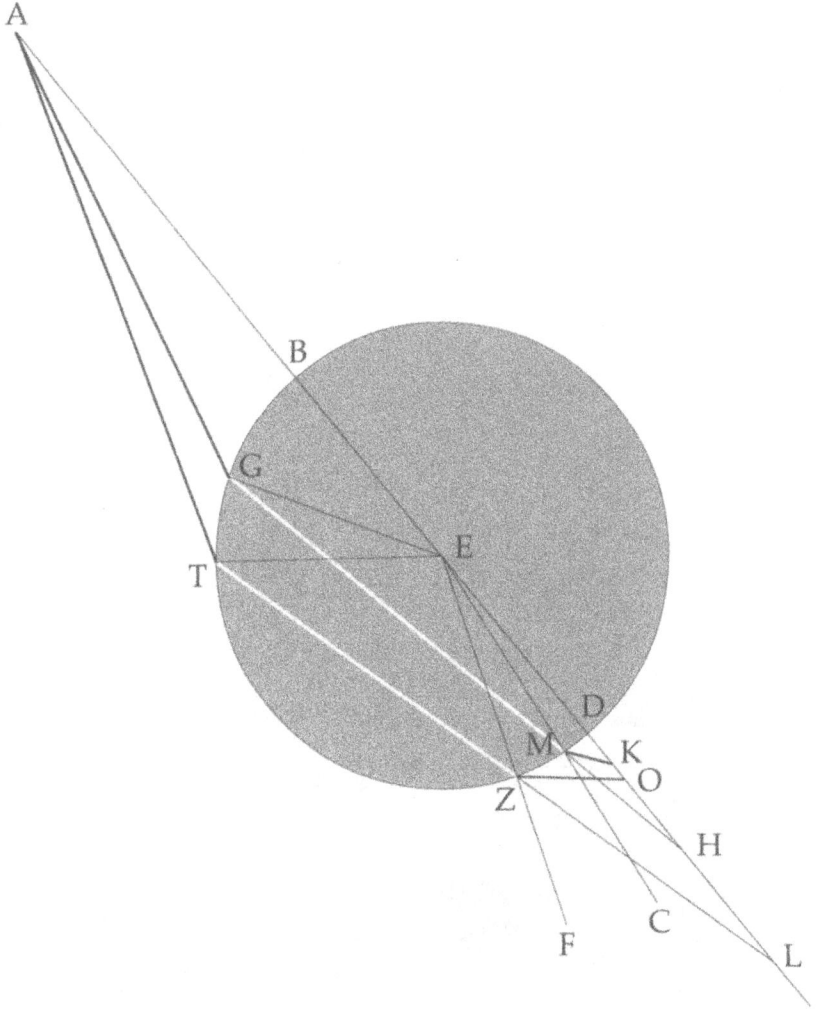

figure 7.7.66

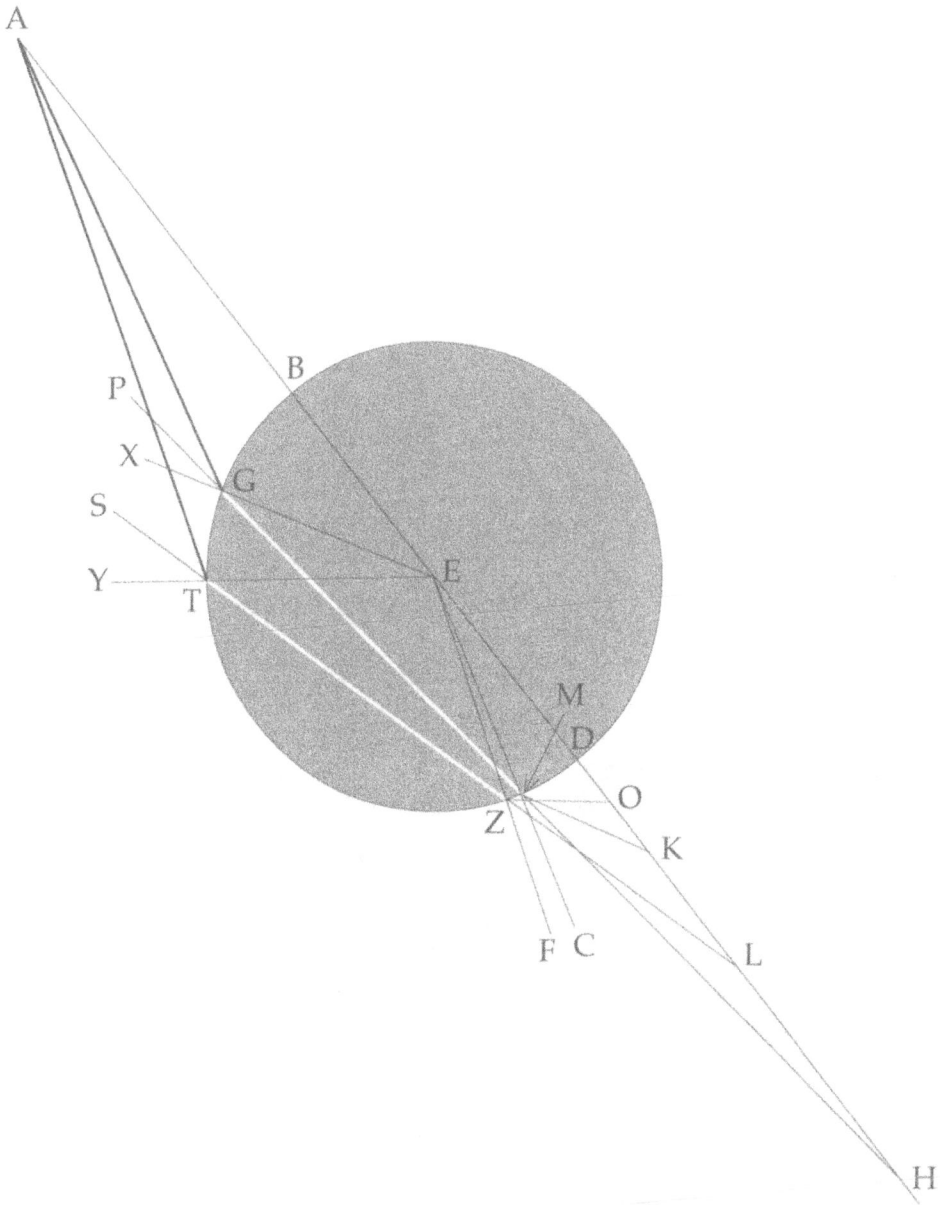

figure 7.7.66a

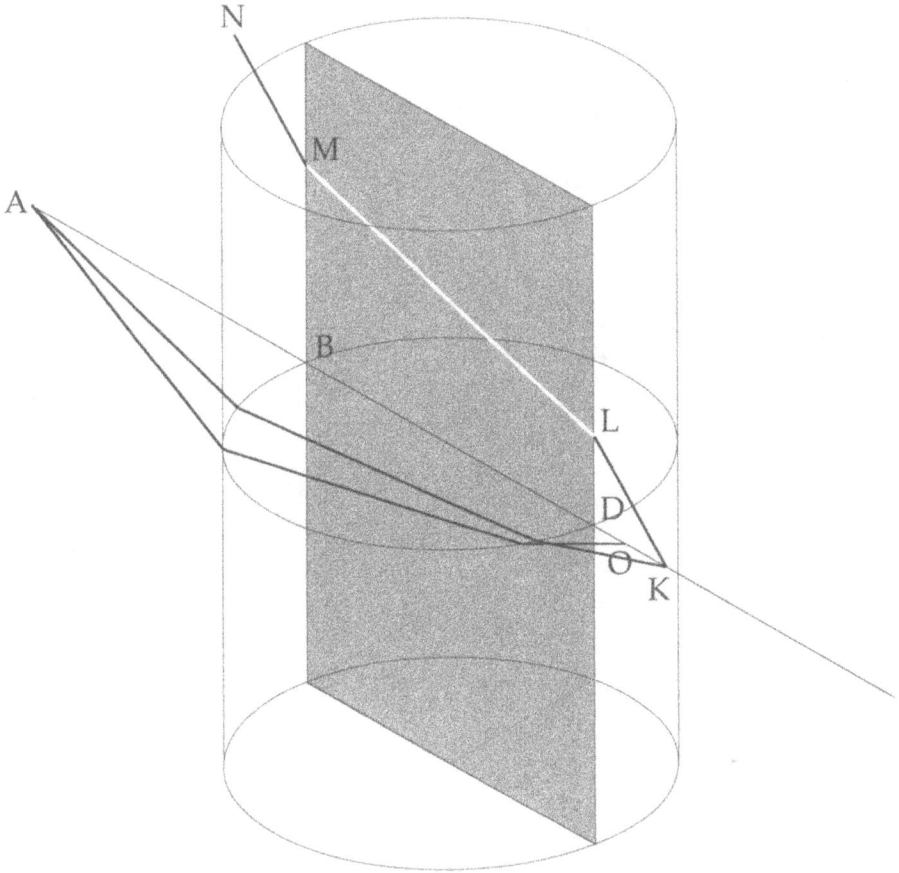

figure 7.7.66b

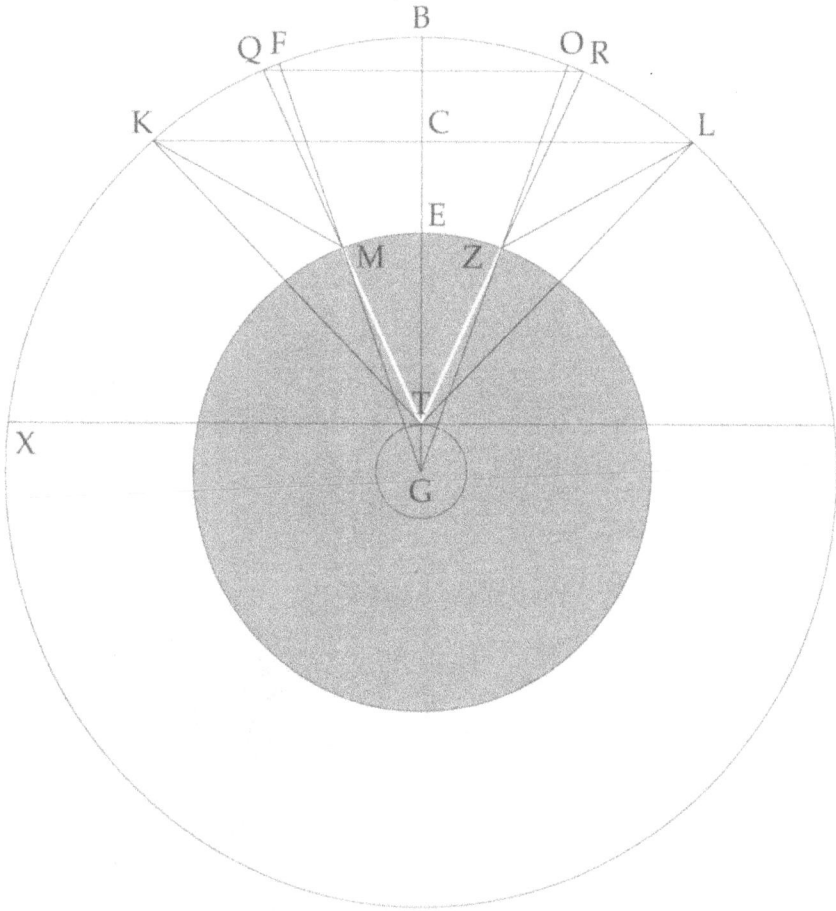

figure 7.7.67

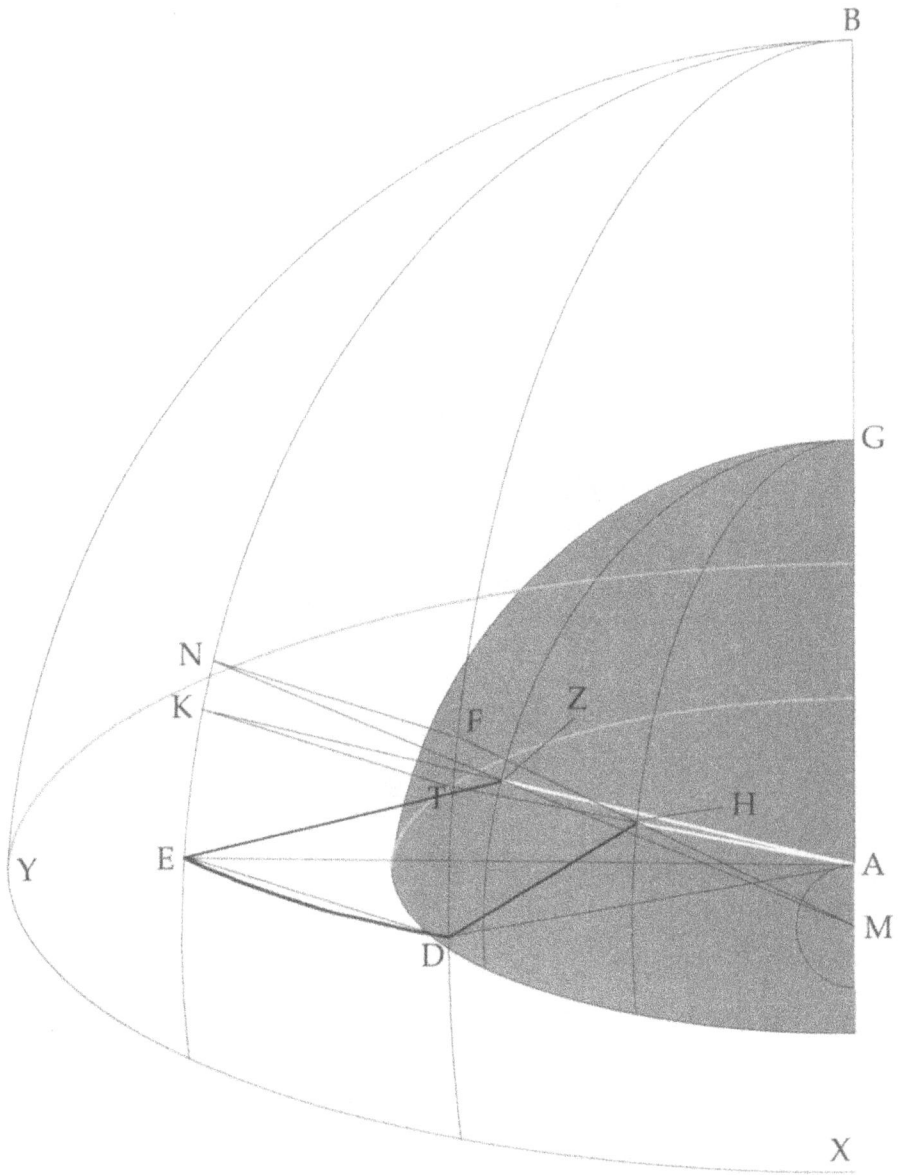

figure 7.7.68

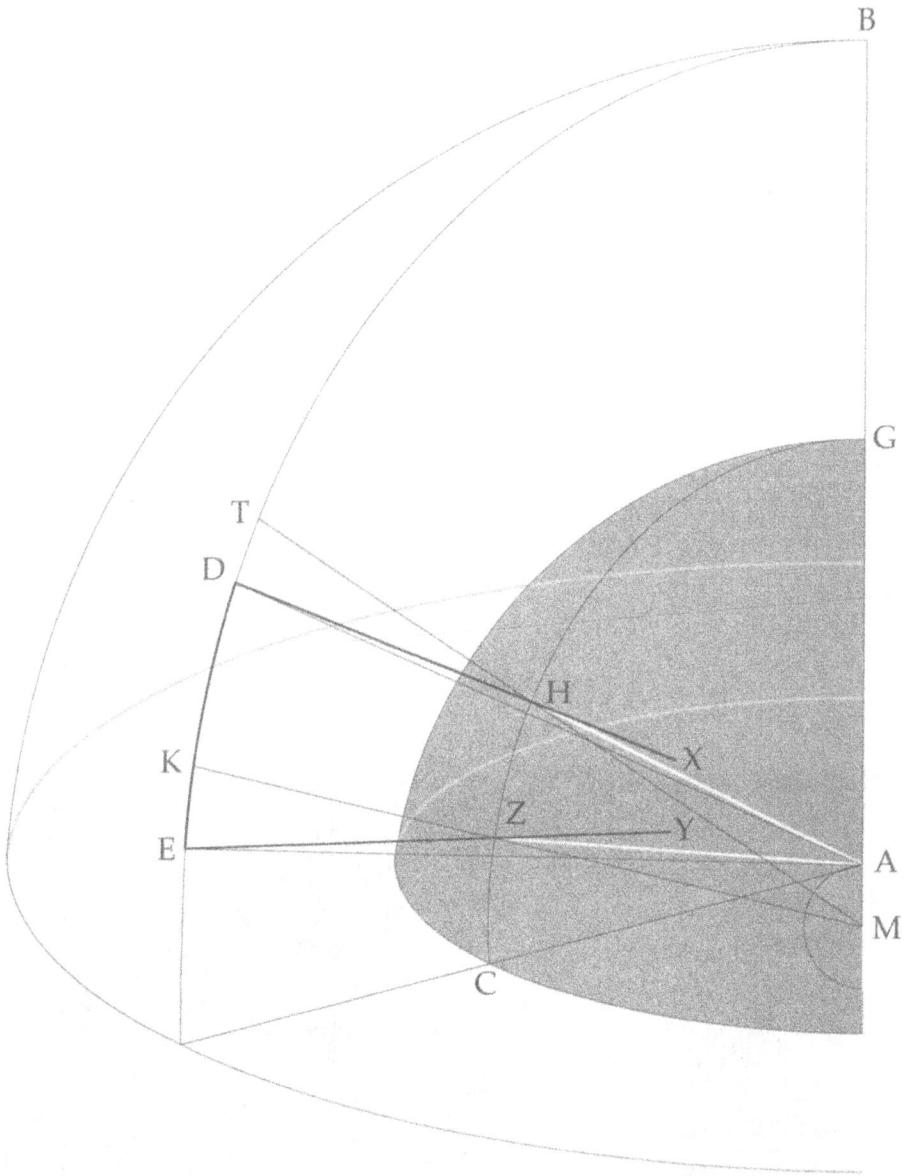

figure 7.7.69

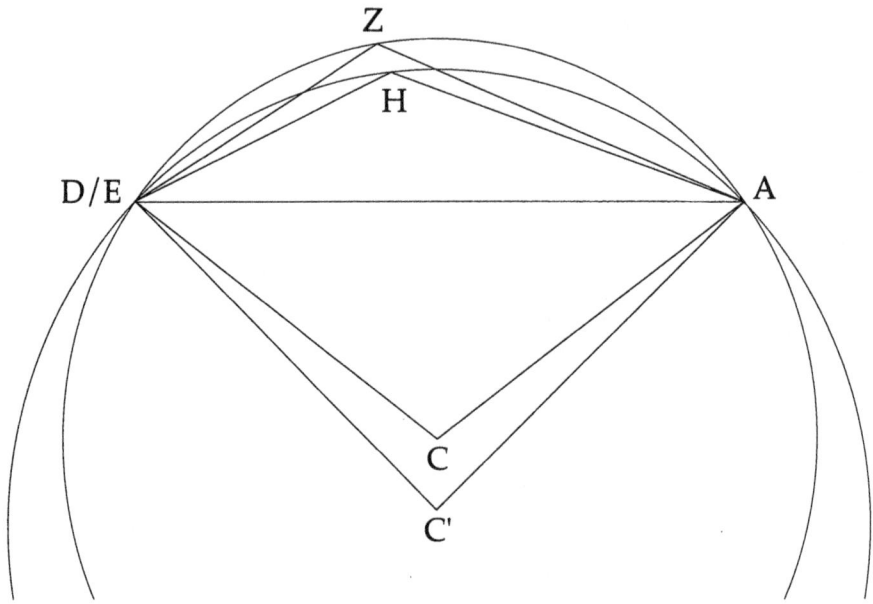

figure 7.7.69a

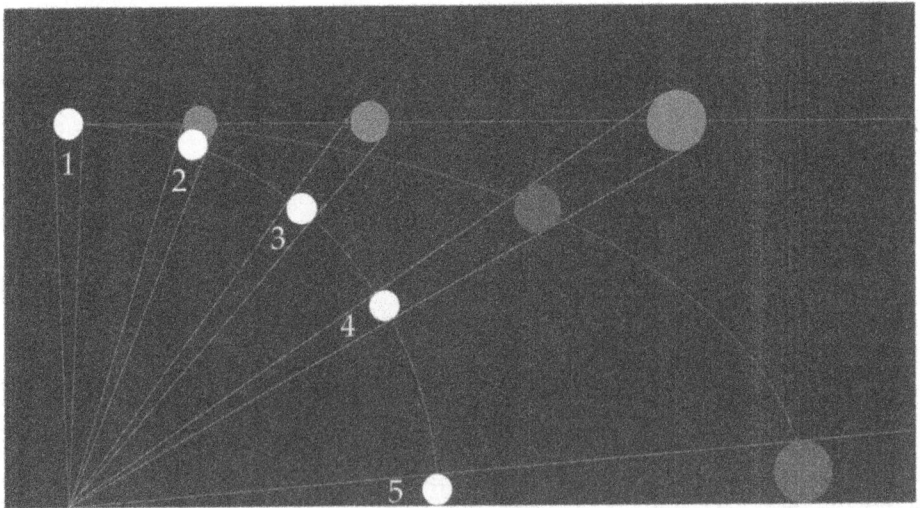

figure 7.7.70

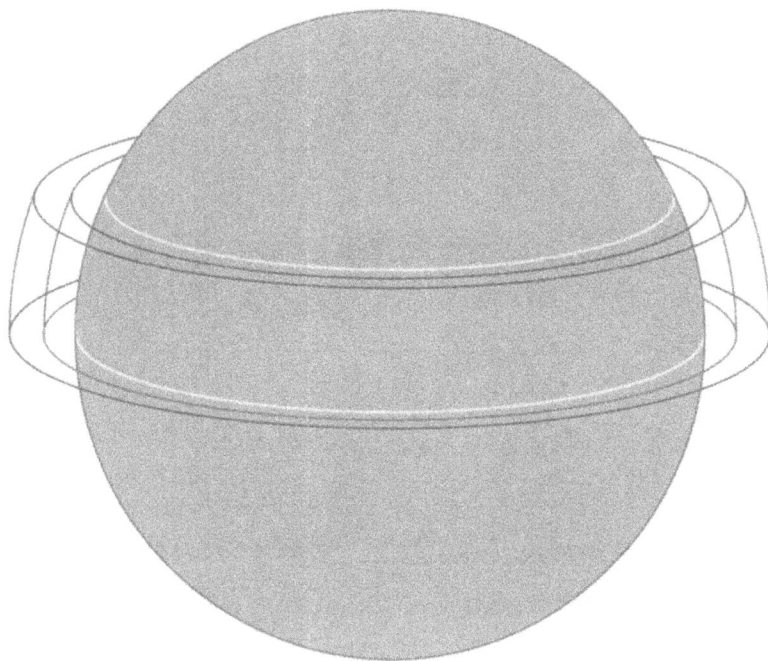

figure 7.7.71

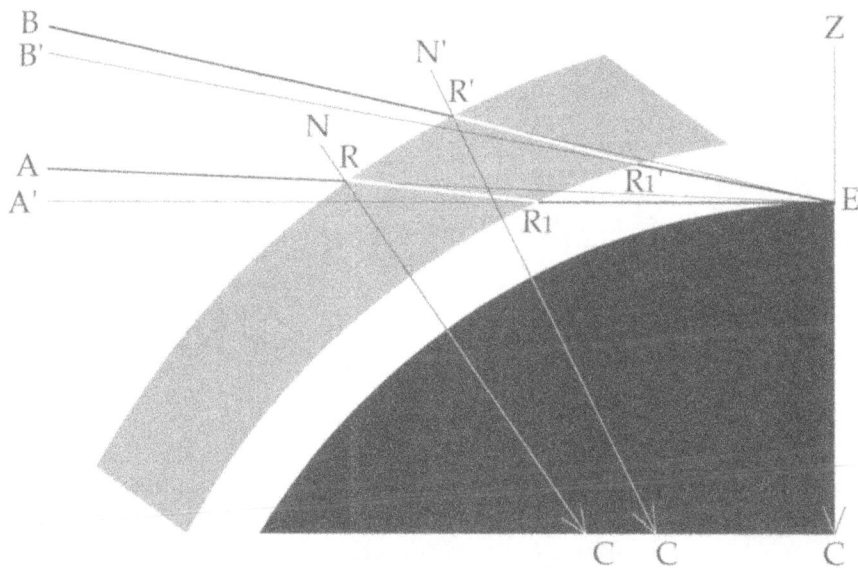

figure 7.7.72

A. Mark Smith

ON

Alhacen—2000 to 2010

- ◆ *Alhacen's Theory of Visual Perception*
- ◆ *Alhacen on the Principles of Reflection*
- ◆ *Alhacen on Image-Formation and Distortion in Mirrors*
- ◆ *Alhacen on Refraction*

A Critical Edition, with English Translation and Commentary, of Alhacen's *De Aspectibus*, the Medieval Latin Translation of Ibn-al-Haytham's *Kitāb al-Manāzir*

ALHACEN'S THEORY OF VISUAL PERCEPTION

A Critical Edition, with English Translation and Commentary, of the First Three Books of Alhacen's *De Aspectibus*, the Medieval Latin Version of Ibn al-Haytham's *Kitāb al-Manāzir*

Volume One

Introduction and Latin Text

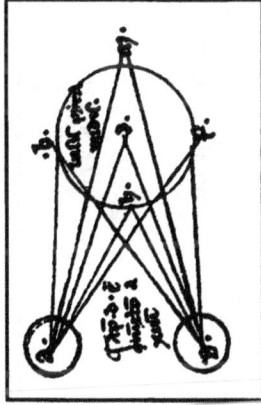

A. Mark Smith

ALHACEN'S THEORY OF VISUAL PERCEPTION

A Critical Edition, with English Translation and Commentary, of the First Three Books of Alhacen's *De Aspectibus*, the Medieval Latin Version of Ibn al-Haytham's *Kitāb al-Manāzir*

Volume Two

English Translation

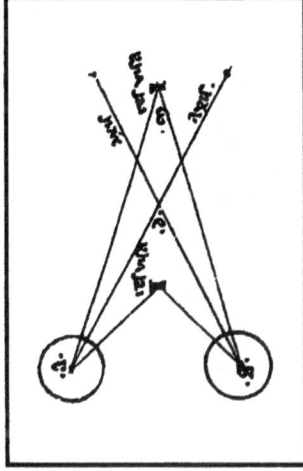

A. Mark Smith

ALHACEN ON THE PRINCIPLES OF REFLECTION

A Critical Edition with English Translation and Commentary, of Books 4 and 5 of Alhacen's *De Aspectibus*, the Medieval Latin Version of Ibn al-Haytham's *Kitāb al-Manāzir*

Volume One
Introduction and Latin Text

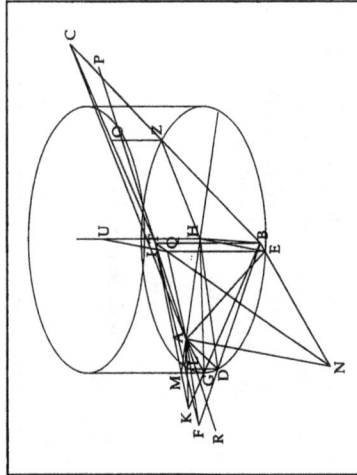

A. Mark Smith

ALHACEN ON THE PRINCIPLES OF REFLECTION

A Critical Edition with English Translation and Commentary, of Books 4 and 5 of Alhacen's *De Aspectibus*, the Medieval Latin Version of Ibn al-Haytham's *Kitāb al-Manāzir*

Volume Two
English Translation

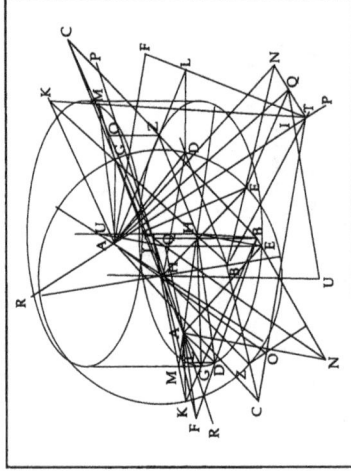

A. Mark Smith

ALHACEN ON IMAGE-FORMATION AND DISTORTION IN MIRRORS

A Critical Edition, with English Translation
and Commentary, of Book 6 of Alhacen's *De Aspectibus*,
the Medieval Latin Version of Ibn al-Haytham's
Kitāb al-Manāẓir

Volume Two
English Translation

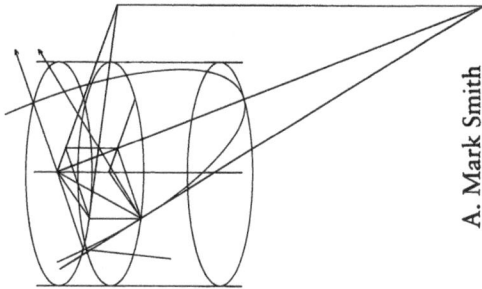

A. Mark Smith

ALHACEN ON IMAGE-FORMATION AND DISTORTION IN MIRRORS

A Critical Edition, with English Translation
and Commentary, of Book 6 of Alhacen's *De Aspectibus*,
the Medieval Latin Version of Ibn al-Haytham's
Kitāb al-Manāẓir

Volume One
Introduction and Latin Text

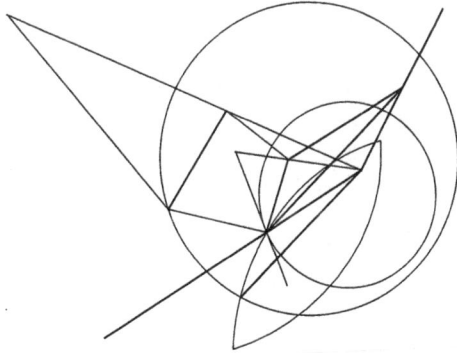

A. Mark Smith

ALHACEN ON REFRACTION

A Critical Edition, with English Translation
and Commentary, of Book 7 of Alhacen's *De Aspectibus*,
the Medieval Latin Version of Ibn al-Haytham's
Kitāb al-Manāzir

Volume Two
English Translation

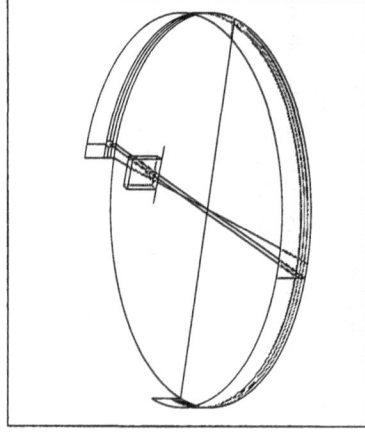

A. Mark Smith

American Philosophical Society

ALHACEN ON REFRACTION

A Critical Edition, with English Translation
and Commentary, of Book 7 of Alhacen's *De Aspectibus*,
the Medieval Latin Version of Ibn al-Haytham's
Kitāb al-Manāzir

Volume One
Introduction and Latin Text

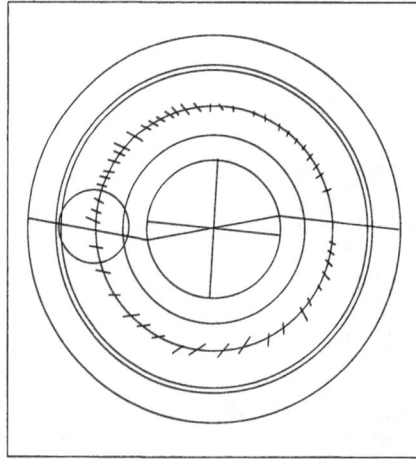

A. Mark Smith

American Philosophical Society

www.ingramcontent.com/pod-product-compliance
Lightning Source LLC
Chambersburg PA
CBHW081338190326
41458CB00018B/6044